TIC

Technology
Innovation
Centre

FOR REFERENCE ONLY

NOT TO BE TAKEN AWAY

Dictionary of Materials and Testing

Second Edition

Joan L. Tomsic
Editor

with contributions by
Robert S. Hodder
Manager, Metallurgical services,
retired, Latrobe Steel

INTERNATIONAL
Society of Automotive Engineers, Inc.
Warrendale, Pa.

Library of Congress Cataloging-in-Publication Data

Dictionary of materials and testing / Joan L. Tomsic, editor ;
with contributions by Robert S. Hodder.--2nd ed.
  p. cm.
 ISBN 0-7680-0531-0
 1. Materials--Dictionaries. 2. Materials--Testing--Dictionaries.
I. Tomsic, Joan L. II. Hodder, Robert S.

TA402 .C83 2000
620.1'1'03--dc21
                                                99-089906

Copyright © 2000   Society of Automotive Engineers, Inc.
                    400 Commonwealth Drive
                    Warrendale, PA 15096-0001 U.S.A.
                    Phone: (724)776-4841
                    Fax: (724)776-5760
                    E-mail: publications@sae.org
                    http://www.sae.org

ISBN 0-7680-0531-0

SAE Order No. R-257

Preface to the Second Edition

Rapid advances in the sciences, and especially in the fields of materials and testing, have introduced hundreds of new concepts and the words that gave them currency. It was with a focus on these current concepts and technology that this extensive and scholarly revision was initiated.

As part of the expanded coverage provided by this second edition, technical terms from potentially related fields have been included. These, of course, include computer-related language, computer hardware and software technology related to testing instrumentation, and others. The dictionary should prove useful to students, teachers, scientists, and engineers in research and industry.

The editor gratefully thanks Robert S. Hodder, Sr. for his technical review and contributions to this edition of the *Dictionary of Materials and Testing*.

Joan Tomsic
Editor

Preface to the First Edition

The *Dictionary of Materials and Testing* is the first such volume that was specifically developed for engineers. The emphasis is on "engineered" materials that can withstand stress or unusual environments for an extended period of time.

Testing terms and definitions cover the evaluation of engineering materials properties and characteristics. Also included are terms and definitions for PC hardware and software technology related to testing instrumentation.

In total, this entirely new volume has nearly 11,000 engineering terms from the following sources:

> 1,543 terms and definitions from SAE Standards.

> 6,411 terms and definitions developed by Engineering Resources, Inc., a leading company in engineering terminology and usage.

> 2,912 terms and definitions related to PC hardware and software.

This volume is designed for the practicing engineer, and those that support them. Students and professors will also appreciate this fully comprehensive dictionary on engineered materials.

William H. Cubberly
Editor

How to Use This Dictionary

Basic Format

The format for a defined term provides the term in boldface and the definition in regular typeface. A term may have more than one definition, in which case the definition is preceded by a number. These multiple definitions are presented in order of their relevance to materials and testing.

Alphabetization

The terms are alphabetized on a letter-by-letter basis. Hyphens, commas, word spacing, etc., in a term are ignored in the sequencing of terms. To aid the user in finding a term, the first and last terms on each page are shown in bold type at the top of the page.

A

a *See* ampere.

A *See* angstrom.

ABA copolymers Copolymers with three sequences and only two domains.

A-basis The mechanical property value "A" is the value above which 99% of the test values is expected to fail.

aberration *1.* In astronomy, the apparent angular displacement of the position of a celestial body in the direction of motion of the observer, caused by the combination of the velocity of the observer and the velocity of light. *2.* In optics, a specific deviation from perfect imagery, for example, spherical aberration, coma, astigmatism, curvature of field, or distortion.

abhesive A film or coating that resists adhesion or sticking.

ablation *1.* The breaking down of material caused by high temperature, pressure, or oxidation. *2.* A design loss of surface material to protect an underlying material.

ablative materials Materials, especially coating materials, designed to provide thermal protection to a body in a fluid stream through the loss of mass.

ablative plastic A material that is designed to absorb heat on the surface that is exposed to high temperature.

ABL bottle A pressure test vessel used to determine the quality and properties of filament-wound materials.

abnormal steel Steel that does not produce test results consistent with what you would expect from its composition and method of manufacture.

abort To cut short or break off an action, operation, or procedure with an aircraft, space vehicle, or the like, especially because of equipment failure, for example, to abort a takeoff, abort a mission, or abort a launch.

abrade To prepare a surface by roughening it by sanding or any other means.

abrasion *1.* Removal of surface material by sliding or rolling contact with hard particles of the same substance or another substance; the particles may be loose or may be part of another surface in contact with the first. *2.* A surface blemish caused by roughening or scratching.

abrasive *1.* Particulate matter, usually having sharp edges or points, that can be used to shape and finish workpieces in grinding, honing, lapping, polishing, blasting, or tumbling processes. *2.* A material formed into a solid mass, usually fired or sintered, and used to grind or polish workpieces.

ABS *See* acrylonitrile butadiene styrene.

absolute accuracy error The deviation of the analog value at any code from its theoretical value after the full-scale range has been calibrated. Expressed in percent, ppm, or fraction of 1 LSB.

absolute alarm An alarm caused by the detection of a variable that has exceeded a set of prescribed high- or low-limit conditions.

absolute encoder An electronic or electromechanical device that produces a unique digital output (in coded form) for each value of an analog or digital input.

absolute feedback In numerical control, assignment of a unique value to each possible position of machine slide or actuating member.

absolute filter The filter downstream of the unit under test, the purpose of which is to retain the contaminant passed by the unit under test.

absolute humidity The weight of water vapor in a gas/water vapor mixture per unit volume of space occupied, expressed, for example, in grains or pounds per cubic foot.

absolute particle retention rating *See* largest particle passed test.

absolute pressure *1.* The pressure measured relative to zero pressure (vacuum). *2.* Gage pressure plus barometric pressure in the same units.

absolute programming In numerical control, using a single point of reference for determining all positions and dimensions.

absolute rating A theoretical size designation which is an estimation of the largest particle, by length, that can pass through a filter with a specific rating.

absolute sealing A level of sealing that requires all seams, slots, holes, and fasteners passing through the seal plane to be sealed. (All integral fuel tanks require absolute sealing.)

absolute stability Condition of a linear system in which there exists a limiting value of the open-loop gain such that the system is stable for all lower values of that gain, and unstable for all higher values.

absolute value error The magnitude of the error disregarding the algebraic sign, or, for a vectorial error, disregarding its function.

absolute viscosity A measure of the internal shear properties of fluids, expressed as the tangential force per unit area at either of two horizontal planes separated by one unit thickness of a given fluid, one of the planes being fixed and the other moving with unit velocity.

absolute zero Temperature of $-273.16°C$ or $-459.69°F$ or 0K at which molecular motion vanishes and a body has no heat energy.

absorb To take in or assimilate (for example, sound) with little or none being transmitted or reflected.

absorbance An optical property expressed as $\log (1/T)$, where T is the transmittance.

absorbate A material that is absorbed by another.

absorbed horsepower Total horsepower absorbed by the absorption unit of the dynamometer and by the frictional components of the dynamometer.

absorbed horsepower at 50 mph (80.5 km/h) road lane The dynamometer setting values for various inertia weight vehicles published in the Federal Register.

absorption *1.* The penetration of one substance into another. *2.* The process whereby energy is expended within a material in a field of radiant energy. *3*. The attraction of a liquid adhesive film into a substrate.

absorption band A region of the electromagnetic spectrum where a given substance exhibits a high absorption coefficient compared to adjacent regions of the spectrum. *See also* absorption spectra.

absorption coefficient An inherent material property expressed as the fractional loss in radiation intensity per unit mass or per unit thickness determined over an infinitesimal thickness of the given material at a fixed wavelength and band width.

absorption curve A graph of the variation of transmitted radiation through a fixed sample of material of a given thickness while the wavelength is changed at a uniform rate.

absorption dynamometer A device for measuring mechanical force or power by converting the mechanical energy to heat in a friction mechanism or bank of electrical resistors.

absorption-emission pyrometer An instrument for determining gas temperature by measuring the radiation emitted by a calibrated reference source both before and after the radiation passes through the gas, where it is partly absorbed.

absorption factor The ratio of the light absorbed to the incident light.

absorption hygrometer An instrument for determining water vapor content of the atmosphere by measuring the amount absorbed by a hygroscopic chemical.

absorption meter An instrument for measuring the quantity of light transmitted through a transparent

medium by means of a photocell or other light-detecting device.

absorption spectra The arrays of absorption lines and absorption bands that result from the passage of radiant energy from a continuous source through a selectively absorbing medium cooler than the source.

absorption spectroscopy The study of the wavelengths of light absorbed by materials and the relative intensities at which different wavelengths are absorbed. This technique can be used to identify materials and measure their optical densities.

absorption tower A vertical tube in which a gas rising through a falling stream of liquid droplets is partially absorbed by the liquid.

absorptive index *See* absorptivity.

absorptivity The capacity of a material to absorb incident radiant energy, measured as the absorptance of a specimen of material thick enough to be completely opaque, and having an optically smooth surface.

AC or a-c *See* alternating current.

Ac$_1$ *See* transformation temperature.

Ac$_3$ *See* transformation temperature.

Ac$_4$ *See* transformation temperature.

accelerate To make or become faster; to cause to happen sooner.

accelerated life test *1*. A life test under test conditions that are more severe than usual operating conditions. In an accelerated life test, it is necessary that a relationship between test severity and the probability distribution of life be ascertainable. *2*. Test conditions that are increased in magnitude to reduce the time necessary to attain results. *3*. To reproduce in a short time damaging effects that could be attained under normal service conditions.

accelerating agent *1.* A substance that increases a chemical reaction rate. *2.* A chemical that hastens the curing of rubber, plastic, cement, or adhesives, and may also improve their properties. Also known as an accelerator.

accelerated aging *See* artificial aging.

accelerating electrode An auxiliary electrode in an electron tube that is maintained at an applied potential to accelerate electrons in a beam.

acceleration *1.* The time rate of change of velocity; the second derivative of a distance function with respect to time. *2.* The rate of change of velocity. The act or process of accelerating or the state of being accelerated.

acceleration error The maximum difference, at any measurand value within the specified range, between output readings taken with and without the application of specified constant acceleration along specified axes.

acceleration factor *1.* The factor by which the failure rate can be increased by an increased environmental stress. *2.* The ratio between the times necessary to obtain the same portion of failure in two equal samples under two different sets of stress conditions, involving the same failure modes and mechanisms.

acceleration time *1.* The span of time it takes a mechanical component of a computer to go from rest to running speed. *2.* The measurement of time for any object to reach a predetermined speed.

accelerator A chemical additive used to hasten a chemical reaction under specific conditions.

accelerometer An instrument for measuring acceleration or an accelerating force such as gravity.

acceptable quality level (AQL) The maximum percent defective that can be considered satisfactory as a process average, or the percent defective whose probability of rejections is designated by a.

acceptance number The largest number of defects that can occur in an acceptance sampling plan with the lot still being accepted.

acceptance sampling plan An accept/reject test whose purpose is to accept or reject a lot of items or material.

acceptance test A test to demonstrate the degree of compliance of a device or material with purchaser's requirements.

acceptance test properties Test properties that are required to be included on each certification accompanying the delivered product, for example, tensile, stress rupture, hardness, and metallographic test results.

access Pertaining to the ability to place information into, or retrieve information from, a storage device.

access control The procedures for providing systematic, unambiguous, orderly, reliable, and generally automatic use of communication lines, channels, and networks for information transfer.

accessibility A measure of the relative ease of admission to the various areas of an item for the purpose of operation or maintenance.

access method Any of the data-management techniques available to the user for transferring data between main storage and an input/output device.

access procedures The procedure by which the devices attached to the network gain access to the medium.

access, random Pertaining to the process of obtaining data from, or placing data into, storage, where the time required for such access is independent of the location of the data most recently obtained or placed in storage.

access time The interval between a request for stored information and the delivery of the information; often used as a reference to the speed of memory.

Ac$_{cm}$ *See* transformation temperature.

accommodation coefficient The ratio of the average energy actually transferred between a surface and impinging gas molecules scattered by the surface to the average energy that would theoretically be transferred if the impinging molecules reached complete thermal equilibrium with the surface before leaving the surface.

accumulation test A leak test method in which pressurized gas is used to determine the leak integrity of a component.

accumulator *1.* In computing, a device where one of the operands for arithmetic and logic operations is commonly held, with the result of the operation becoming the new stored data. *2.* A device or apparatus that accumulates or stores up, for example, a contrivance in a hydraulic system that stores fluid under pressure (energy).

accumulator, compensating An accumulator that, in addition to its high-pressure volume, incorporates low-pressure volumetric capacity which will accommodate a like volume of fluid to that discharged from the high-pressure chamber. The sum of the volumes of the high- and low-pressure chambers remains constant.

accumulator metal Lead alloys used in the manufacture of lead-acid storage battery plates.

accuracy *1.* The degree of freedom from error; that is, the degree of conformity to truth or to a rule. Accuracy is contrasted with precision. For example, four-place numerals are less precise than six-place numerals; nevertheless, a properly computed four-place numeral might be more accurate than an improperly computed six-place numeral. *2.* The ratio of the error to the full-scale output or the ratio of the error to the output, as specified, expressed in percent. *Note 1*: Accuracy may be expressed in terms of units of measurand, or as within percent of full scale output. *Note 2*: Use of the term accuracy should be limited to generalized descriptions of characteristics. It should not be used in specifications. The term error is preferred in specifications and other specific descriptions of transducer performance. *3.* Quantitatively, the difference between the measured value and the most probable value for the same quantity, when the latter is determined from all available data, critically adjusted for sources of error. *4.* The degree of conformity of a measured or calculated value to some recognized standard or specified value. This concept involves the systemic error of an operation or process, which is seldom negligible. Contrast with precision.

accuracy, measured The degree to which an indicated value matches the actual value of a measured variable.

accuracy, total The deviation, or error, by which an actual output varies from an expected ideal or absolute output. Quantitatively, the total accuracy is the difference between the measured value and the most probable value for the same quantity, when the latter is determined from all available data.

acenaphthylene A thermoplastic derived from an acenaphthene monomer, which resembles polystyrene.

acetal A thermoplastic material derived from formaldehyde. Acetals are strong, rigid materials with good dimensional stability and resilience.

acetal copolymers Highly crystalline thermoplastics that are heat and alkali resistant.

acetal resins Polyformaldehyde and polyxymethylene resins with outstanding fatigue life, resilience, solvent and chemical resistance, and good electrical attributes.

acetate A fiber derived from cellulose acetate.

acetone Commonly used wipe solvent. Used for cleaning composite surfaces prior to bonding and also metal surfaces prior to other treatments. Acetone can also be coupled with dry ice to create a medium for a cryogenic heat-treatment cycle.

AC generator *See* alternator.

achieved reliability The reliability demonstrated at a given point in time under specified conditions of use and environment.

achromatic Optical elements that are designed to refract light of different wavelengths at the same angle. Typically, achromatic lenses are made of two or more components of different refractive index, and are designed for use at visible wavelengths only.

acicular ferrite (in steel) A highly substructed nonequiaxed ferrite that forms upon continuous cooling by a mixed diffusion and shear mode of transformation that begins at a temperature slightly higher than the temperature transformation range for upper bainite. Acicular ferrite is distinguished from bainite in that it has a limited amount of carbon available; thus, there is only a small amount of carbide present.

acicular structure A structure containing needle-shaped microconstituents, such as martensite in steels.

acid A substance that yields hydrogen ions when dissolved in water. Opposite of base.

acid-acceptor A compound that chemically combines with and neutralizes any acid that may be present in a plastic, or an acid that may be formed later by decomposition.

acid copper Electrodeposited copper derived from copper sulfate.

acid embrittlement Acid treatment of certain metals resulting in a type of hydrogen embrittlement.

acidic Describes any solution having a pH less than 7.

acidity Represents the amount of free carbon dioxide mineral acids and salts (especially sulfates of iron and aluminum) that hydrolyze to give hydrogen ions in water. Reported as milliequivalents per liter of acid, or ppm acidity as calcium carbonate, or pH, a measure of hydrogen ions concentration.

acid wash A chemical solution containing phosphoric acid which is used to neutralize residues from alkaline cleaners and to simultaneously produce a phosphate coating that protects the metal surface from rusting and prepares it for painting.

a-c input module I /O module that converts process switched a-c to logic levels for use in the PC.

acoustic Pertaining to sound, the sense of hearing, or the science of sound.

acoustical ohm The unit of measure for acoustic resistance, reactance, or impedance. One acoustical ohm is equivalent to a sound pressure of one microbar producing a volume velocity of one cubic centimeter per second.

acoustic coupler A type of communications device that converts digital signals into audio tones which can be transmitted by telephone.

acoustic dispersion Separation of a complex sound wave into its various frequency components, usually due to variation of wave velocity in the medium with sound frequency; usually expressed in terms of the rate of change of velocity with frequency.

acoustic emission *1.* A transient elastic wave generated by the rapid release of energy from a localized source or sources within a material. The emission may be the result of any of several changes taking place in the material. A crack may be growing, the material may be undergoing permanent deformation, the internal structure may be changing due to heat treatment, or, in the case of composite materials, the fibers that strengthen the material may be breaking. *2.* The stress and pressure waves generated during dynamic processes in materials and used in assessing structural integrity in machined parts.

acoustic excitation The process of inducing vibration in a structure by exposure to sound waves.

acoustic generator A transducer for converting electrical, mechanical, or some other form of energy into sound waves.

acoustic holography A technique for detecting flaws or regions of inhomogeneity in a part by subjecting it to ultrasonic energy, producing an interference pattern on the free surface of water in an immersion tank, and reading the interference pattern by laser holography to produce an image of the test object.

acoustic impedance The complex quotient obtained by dividing sound pressure on a surface by the flux through the surface.

acoustic inertance A property related to the kinetic energy of a sound medium which equals $Z_a/2\pi f$, where Z_a is the acoustic reactance and f is sound frequency.

acoustic interferometer An instrument for measuring either the velocity or frequency of sound pressure in a standing wave established in a liquid or gas medium between a sound source and reflector as the reflector is moved or the frequency is varied.

acoustic microscopes Instruments that use acoustic radiation at microwave frequencies to allow visualization of microscopic detail exhibited in elastic properties of objects.

acoustic reactance The imaginary component of acoustic impedance.

acoustic resistance The real component of acoustic impedance.

acoustics *1.* The technology associated with the production, transmission, and utilization of sound, and the science associated with sound and its effects. *2.* The architectural quality of a room—especially a concert hall, theater, or auditorium—which influences the ability of a listener to hear sound clearly at any location.

acoustic spectrometer An instrument for analyzing a complex sound wave by determining the volume (intensity) of sound-wave components having different frequencies.

acoustic stiffness A property related to the potential energy of a medium or its boundaries which equals $2\pi f/Z_a$, where Z_a is the acoustic reactance and f is sound frequency; the usual units of measure are dyne/cm^5.

acousto-optic An interaction between an acoustic wave and a lightwave passing through the same material. Acousto-optic devices can serve for beam deflection, modulation, signal processing, and Q switching.

acousto-optic glass Glass with a composition designed to maximize the acousto-optic effect.

a-c output module I/O module that converts PC logic levels to output switch action for a-c load control.

acrylate resins *See* acrylic resins.

acrylate styrene acrylonitrile (ASA) A thermoplastic that has good resistance to ultraviolet light and has intermediate-range impact strength.

acrylic Resin polymerized from acrylic acid, methacrylic acid, esters of these acids, or acrylonitrile.

acrylic plastic A thermoplastic polymer created from acrylic acid and its derivatives.

acrylic resins Polymers of acrylic or methacrylic esters with good clarity and optical attributes.

acrylonitrile A monomer used to create copolymers such as synthetic rubber and strong fibers.

acrylonitrile-butadiene-styrene (ABS) A group of strong, heat- and chemical-resistant thermoplastics made up of acrylonitrile, butadiene, and styrene.

actinicity The ability of radiation to induce chemical change.

actinium A radioactive metal with atomic number 89.

activation A process that makes a surface more receptive to a coating material.

activation analysis A method of determining composition, especially the concentration of trace elements, by bombarding the composite substance with neutrons and measuring the wavelengths and intensities of characteristic gamma rays emitted from activated nuclides.

activator A chemical additive used to initiate the chemical reaction in a specific chemical mixture.

active A state in which a metal tends to corrode. Opposite of passive.

active potential The potential of a corroding metal.

actual horsepower The load horsepower, which includes the friction in the dynamometer bearings and inertia simulation mechanism.

actuate To put into action or motion.

actuation signal The setpoint minus the controlled variable at a given instant. Same as error.

actuator(s) *1.* A device responsible for actuating a mechanical device such as a control valve. *2.* A device that actuates.

A/D *See* analog-to-digital.

adaptation The adjustment, alteration, or modification of an organism to fit it more perfectly for existence in its environment.

adapter(s) *1.* Devices or contrivances used or designed primarily to fit or adjust one thing to another. *2.* Devices, appliances, or the like used to alter something so as to make it suitable for a use for which it was not originally designed.

adaptive control A control system that adjusts its response to its inputs based on its previous experience. Automatic means are used to change the type or influence (or both) of control parameters in such a way as to improve the performance of the control system.

ADC *See* analog-to-digital converter.

A-D converter (ADC) A hardware device that converts analog data into digital form; also called an encoder.

adder *1.* Switching circuit that combines binary bits to generate the sum and carry of these bits. *2.* A device that forms, as output, the sum of two or more numbers presented as inputs. Often no data-retention feature is included, i.e., the output signal remains only as long as the input signals are present. Related to accumulator.

adder-subtractor A device whose output is a representation of either the arithmetic sum or difference, or both, of the quantities represented by its operand inputs.

addition polymerization A chemical reaction in which simple monomers are added together to form long-chain polymers without resulting byproducts.

additive Generally, any substance added to another substance to enhance its properties, such as flame retardants or plasticizers in plastics, or brighteners and antipitting agents in metals.

additive removal test A test conducted by recirculating a specified additive type of oil through a test element for a specified period of time. The amount of ash type additive removed by the element is determined by oil analysis.

adduct *1.* A phase (often a compound) formed by direct combination, generally in simple proportions, or two or more different compounds or elements. *2.* Chemical compound with weak bonds, for example occlusive or van der Waals bonds.

adhere To cause two surfaces to be held together by adhesion.

adherend *1.* Any part held to another part by an adhesive. *2.* A part ready for bonding.

adhesion Surfaces that are held together by chemical or mechanical means.

adhesion promoter A material applied to a surface for the purpose of chemically enhancing adhesion of a sealant to the surface. Present adhesion promoters for polysulfide integral fuel tank sealant also contains an organic solvent cleaner.

adhesive Any substance capable of bonding two materials. Can be a liquid, film, or paste.

adhesive bonding A commercial process for fastening parts together in an assembly using only glue, cement, resin, or other adhesive.

adhesive failure Rupture of an adhesive bond such that the separation appears to be at the adhesive-adherent interface.

adhesive film A synthetic resin adhesive, with or without a carrier fabric, usually of the thermosetting type, in the form of a thin film of resin, used under heat and pressure in the production of bonded structures.

adhesive joint The point or area in which two structures are held together by an adhesive.

adhesive strength The strength of an adhesively bonded joint, usually measured in tension (perpendicular to the plane of the bonded joint) or in shear (parallel to the plane of the joint).

adhesive wear Wear that develops when two surfaces slide across one another under pressure.

adiabatic Refers to a process that takes place without any exchange of heat between the process system and another system or its surroundings.

adiabatic curing Curing concrete or mortar under conditions whereby heat is neither gained nor lost.

adiabatic demagnetization cooling Process in which paramagnetic salts are cooled to the boiling point of helium in a strong magnetic field, then thermally isolated and removed from the field to demagnetize them and attain temperatures of 10^{-3} K.

adiabatic rate The rate that results when there is no heat transfer to or from the gas during spring deflection. The adiabatic rate is usually approached during rapid spring deflection when there is insufficient time for heat transfer.

adiabatic temperature The theoretical temperature that would be attained by the products of combustion provided the entire chemical energy of the fuel, the sensible heat content of the fuel, and combustion air above the datum temperature were transferred to the products of combustion.

adjustment, span In instrumentation, the change in slope of the input-output curve.

adjustment, zero In instrumentation, the reading at which the value of the measured quantity is zero.

admixture The addition and dispersion of individual material ingredients before curing.

adsorbents Materials that take up gases by adsorption.

adsorption The concentration of molecules of one or more specific elements or compounds at a phase boundary, usually at a solid surface bounding a liquid or gaseous medium containing the specific element or compound.

adsorption chromatography A test method based on different degrees of adsorption of sample compounds.

advanced composites Composite materials that are reinforced with very strong, continuous fibers, for example, metal matrix or ceramic matrix composites.

advanced filaments Continuous filaments made from high-strength, high-modulus materials for use as constituents of advanced composites.

advection The process of transport of an atmospheric property solely by the mass motion of the atmosphere; also, the rate of change of the value of the advected property at a given point.

Ae$_1$ *See* transformation temperature.

Ae$_3$ *See* transformation temperature.

Ae$_4$ *See* transformation temperature.

Ae$_{cm}$ *See* transformation temperature.

AEM *See* analytical electron microscopy.

aerodynamic angle of attack (α_a) The angle between the vehicle x-axis and the trace of the resultant air velocity vector on a vertical plane containing the vehicle x-axis.

aerodynamic drag force A force opposite to the direction of travel due to aerodynamic resistance, measured in the nominal plane of symmetry of the vehicle. For some engineering purposes, aerodynamic drag force can be defined, alternatively, parallel to the relative wind vector.

aerodynamics The science that deals with the motion of air and other gaseous fluids, and the forces acting on bodies when they move through such fluids, or when such fluids move against or around the bodies.

aerodynamic sideslip angle (β_a) The angle between the traces on the vehicle x-y plane of the vehicle x-axis and the resultant air velocity vector at some specified point in the vehicle.

aerograph Any self-recording instrument carried aloft to take meteorological data.

aerology The study of the free atmosphere throughout its vertical extent, as distinguished from studies confined to the layer of the atmosphere adjacent to the earth's surface.

aerometer An instrument for determining the density of air or other gases.

aerosol A finely divided, non-condensable liquid and/or solid dispersed in a gas.

aerothermoelasticity The study of the response of elastic structures to the combined effects of aerodynamic heating and loading.

aerozine A rocket fuel consisting of a mixture of hydrazine and unsymmetrical dimethylhydrazine (UDMH).

aged beta A beta matrix in which alpha, typically fine, has precipitated as a result of aging.

age hardening Raising the strength and hardness of an alloy by heating a supersaturated solid solution at a relatively low temperature to induce precipitation of a finely dispersed second phase. Also known as aging or precipitation hardening.

age sensitive Describes an elastomer that is subject to deterioration by oxygen, ozone, sunlight, heat, rain, and similar factors experienced in the normal environmental exposure subsequent to vulcanization.

age softening aluminum alloys Spontaneous decrease of strength and hardness that take place at room temperature in certain strain-hardened alloys.

age, threshold The time before which inspection of the condition of an item is required, or beyond which inspection is considered to provide useful condition information.

agglomerate Two or more particles that are in intimate contact and cannot be separated by gentle stirring and from the small shear forces thus generated.

agglomeration Any process for converting a mass of relatively fine solid material into a mass of larger lumps.

aggregate *1.* A relatively stable assembly of dry particles formed under the influence of physical forces. *2.* A hard, coarse mineral usually combined with an epoxy binder and used in plastic tools or flooring. *3.* Natural sand, gravel, or crushed stone mixed with cement to make mortar or concrete.

aging *1.* A generic term denoting a time-temperature-dependent change in the properties of certain alloys. Except for strain aging and age softening, aging is the result of precipitation from a solid solution of one or more compounds whose solubility decreases with decreasing temperature. For each alloy system susceptible to aging, there is a unique range of time-temperature combinations to which the system will respond. *See* age hardening, artificial aging, age softening, natural aging, overaging, peak aging, precipitation hardening, precipitation heat treatment, quench aging, step aging, strain aging, and underaging. *2.* A term applied to changes in physical and mechanical properties of low carbon steel that occur with the passage of time. These changes adversely affect formability. Aging accelerates as the temperature is raised. *3.* Alteration of the characteristics of a device due to use. *4.* Operating a product before shipping it to stabilize component functions or detect early failures. *5.* Curing or stabilizing parts or materials by long-term storage outdoors or under closely controlled storage conditions.

aging test(s) The long-term environmental exposure or accelerated aging at high temperature to determine material and performance changes with time.

agitation Movement of parts or material, or circulation of liquid media around parts or material.

agitator A device for mixing, stirring, or shaking liquids or liquid-solid mixtures to keep them in motion.

AI *See* artificial intelligence.

air The mixture of oxygen, nitrogen, and other gases, which with varying amounts of water vapor, forms the atmosphere of the earth.

air bind An air pocket in a pump, conduit, or piping system that prevents liquid from flowing past it. Also called a liquid trap.

airborne Carried in the atmosphere—either by being transported in an aircraft or by being dispersed in the atmosphere.

air-bubbler liquid-level detector A device for indirectly measuring the level of liquid in a vessel—especially a corrosive liquid, viscous liquid, or liquid containing suspended solids.

air-bubbler specific-gravity meter Any of several devices that measure specific gravity by determining differential pressure between two air-purged bubbler columns.

air-bubble void Entrapment of air between plies of a bonded structure.

air compressor A machine that raises the pressure of air above atmospheric pressure and normally delivers it to an accumulator or distribution system.

air condenser *1.* A heat exchanger for converting steam to water where the heat-transfer fluid is air. Also known as air-cooled condenser. *2.* A device for removing oil or water vapors from a compressed-air line.

air density (ρ) $\rho = 1.2250 \text{ kg}^3$ at standard day conditions.

air, dry *1.* Air with which no water vapor is mixed. This term is used comparatively, since in nature there is always some water vapor included in air, and such water vapor, being a gas, is dry. *2.* A papermaking term used to describe "dry" pulp containing about 10% moisture.

air ejector A device for removing air or noncondensible gases from a confined space, such as the shell of a steam condenser, by eduction using a fluid jet.

air entrainment Artificial infusion of a semisolid mass such as concrete or a dense slurry with minute bubbles of air, especially by mechanical agitation.

air filter A device for removing solid particles such as dust or pollen from a stream of air, especially by causing the airstream to pass through layered porous material such as cloth, paper or screening.

airfoils Structures, pieces, or bodies, originally likened to foils or leaves in being wide and thin, designed to obtain a useful reaction on themselves in their motion through the air.

airframe(s) The assembled structural and aerodynamic components of an aircraft or rocket vehicle which support the different systems and subsystems integral to the vehicle.

air furnace Any furnace whose combustion air is supplied by natural draft, or whose internal atmosphere is predominantly heated air.

air gage *1.* A device for measuring air pressure. *2.* A device for precisely measuring physical dimensions by measuring the pressure or flow of air from a nozzle against a workpiece surface and relating the measurement to distance from the nozzle to the workpiece.

air gap The space between two ferromagnetic elements of a magnetic circuit.

air-hardening steel A type of tool steel containing sufficient alloying elements to permit it to harden fully through the cross section on cooling in air from a temperature above its transformation temperature. The term should be restricted to steels that are capable of being hardened by cooling in air in cross sections of at least two inches (50 mm).

air melting An expression normally referring to the process of melting through the use of carbon arc electrodes and an electric arc furnace. The atmosphere above the melt is air, and a slag is used to remove the impurities in the melt.

air meter A device for measuring the flow of air or other gas and expressing it as weight or volume per unit time.

air moisture The water vapor suspended in the air.

air permeability A method of measuring the fineness of powdered materials, such as portland cement, by determining the ease with which air passes through a defined mass or volume.

air purge The removal of undesired matter by replacement with air.

air regulator A device for controlling airflow—for example, a damper to control flow of air through a furnace, or a register to control flow of heated air into a room.

air resistance *1.* A measure of the "drag" on a vehicle moving through air. Air resistance increases as the square of the speed, thus power requirements increase much faster than vehicle speed. *2.* The opposition offered to the passage of air through any flow path.

air, saturated Air that contains the maximum amount of the vapor of water or other compound that it can hold at its temperature and pressure.

air separator A device for separating materials of different density, or particles of different sizes, by means of a flowing current of air.

airspace The atmosphere above a particular potion of the earth, usually defined by the boundaries of an area on the surface projected perpendicularly upward.

airspeed Speed of an airborne object with respect to the surrounding air mass. In calm air, airspeed is equal to ground speed; true airspeed is a calibrated airspeed that has been corrected for pressure and temperature effects due to altitude, and for compressibility effects at high airspeeds.

air thermometer A device for measuring temperature in a confined space by detecting variations in pressure or volume of air in a bulb inside the space.

airtight Sealed to prevent passage of air or other gas; impervious to leakage of gases across a boundary.

air viscosity (μ) $\mu = 1.7894 \times 10^{-5}\,Ns^2$ at standard day conditions.

airy disk The central bright spot produced by a theoretically perfect circular lens or mirror. The spot is surrounded by a series of dark and light rings, produced by diffraction effects.

AIT *See* autoignition temperature.

air zero gas Air containing less than 2 ppm hydrocarbon on a methane equivalence basis.

alarm *1.* An instrument, such as a bell, light, printer, or buzzer, that indicates when the value of a variable is out of limits. *2.* An abnormal process condition. *3.* The sequence state when an abnormal process condition occurs.

alarm point The level at which an environmental condition or process variable exceeds some predetermined value.

alarm severity A selection of levels of priority for the alarming of each input, output, or rate of change.

alarm system An integrated combination of detecting instruments and visible or audible warning devices that actuates when an environmental condition or process variable exceeds some predetermined value.

albedo The ratio of the amount of electromagnetic radiation reflected by a body to the amount incident upon it, often expressed as a percentage, for example, the albedo of the earth is 34%.

alclad A composite aluminum product with an anodic alloy coating metallurgically bonded to an aluminum core for corrosion protection.

alcohol A class of organic compounds characterized by the presence of a hydroxyl group attached to an alkyl hydrocarbon structure.

aldehyde(s) *1.* A class of chemical compounds having the general formula RCHO, where R is an alkyl (aliphatic) or aryl (aromatic) radical. *2.* Carbonyl groups to which a hydrogen atom is attached; the first stage of an alcohol; –CHO.

alert box In data processing, a window that appears on a computer screen to alert the user of an error condition.

algebraic adder An electronic or mechanical device that can automatically find the algebraic sum of two quantities.

algorithm *1.* A prescribed sequence of well-defined rules or operations for the solution of a problem in a specified number of steps. *2.* Detailed procedures for giving instructions to a computer. *3.* A recursive computational procedure. Contrast with heuristic and stochastic.

algorithmic language A language designed for expressing algorithms.

Algorithmic-Oriented Language An international procedure-oriented language.

alias Describes the relationship of one frequency to another when varying signals are sampled at equally spaced intervals, and the frequencies cannot be distinguished from each other by an analysis of their equally spaced values.

aliasing A peculiar problem in data sampling, where data are not sampled enough times per cycle, and the sampled data cannot be constructed.

alidade *1.* An instrument used in the plane-table method of topographic surveying and mapping. *2.* Any sighting device for making angular measurements.

aliphatic amines Hardeners used in the ambient curing of epoxy resins.

aliphatic hydrocarbons Hydrocarbons with an open-chain structure, for example, gasoline.

aliquot A sample of a larger item.

alkali A substance that is water soluble and ionizes in solution, providing hydroxyl ions.

alkali metals Metals in group IA of the periodic system, namely, lithium, sodium, potassium, rubidium, cesium, and francium.

alkaline Describes any solution having a pH greater than 7.

alkaline cleaner An alkali-based aqueous solution for removing soil from metal surfaces.

alkaline earth metals The metals in group IIA of the periodic table, namely, beryllium, magnesium, calcium, strontium, barium, and radium.

alkalinity Represents the amount of carbonates, bicarbonates, hydroxides, and silicates or phosphates in the water. Reported as grains per gallon or ppm as calcium carbonate.

alkali vapor lamps Lamps in which light is produced by an electric discharge between electrodes in an alkali vapor at low or high pressure.

alkoxysilanes A monomer from which silicone rubbers are produced.

alkyd Resin used in coatings. Reaction products of polyhydric alcohols and polybasic acids.

alkyd plastic Thermoset plastic based on resins composed principally of polymeric esters, in which the recurring ester groups are an integral part of the main polymer chain, and in which ester groups occur in most crosslinks that may be present between chains.

alkyd resins Polyester resins used as coating compounds and in the preparation of inks, varnishes and paints.

alkylsilicones Oily liquids used to form strong copolymer resins.

alkylurethane A flexible foam polyurethane with low heat resistance.

alligatoring *1.* Pronounced wide cracking over the entire surface of a coating having the appearance of alligator hide. Also known as crocodiling. *2.* Surface roughening of very coarse-grained sheet metal during forming. *3.* Longitudinal splitting of flat slabs in a plane parallel to the rolled surface that occurs during hot rolling. Also called fishmouthing.

allobar A form of an element having a distribution of isotopes that is different from the distribution in the naturally occurring form. Thus, an allobar has a different apparent atomic weight than the naturally occurring form of the element.

allophanate A product resulting from the reaction of an isocyanate and hydrogen atoms in a urethane.

allotropy The existence of a substance or element in two or more forms.

allowance Specified difference in limiting sizes—either minimum clearance or maximum interference between mating parts—computed mathematically from the specified dimensions and tolerances of both parts. *See also* machining allowance.

alloy *1.* (metal) A material composed of two or more chemical elements, one of which is a metal. *2.* (plastic) A material composed of two or more polymers or elastomers. Also called a polymer blend.

alloy cast irons The resulting product when alloying elements are added to cast iron to provide improved properties, such as strength, hardness and heat resistance.

alloy-depleted surface layer A loss of alloying elements at the surface sufficient to effect a near-surface microstructure difference using a suitable etchant.

alloy family A group of alloys having similar yet varying service performance characteristics, for example high speed steels, hot work steels.

alloying element The chemical substance added to a material to cause changes in its properties.

alloy steel An alloy of iron and carbon which also contains one or more additional elements intentionally added to increase hardenability or to enhance other properties.

all-pass network A network designed to introduce phase shift or delay into an electronic signal without appreciably reducing amplitude at any frequency.

allylics Polyester resins used in electrical components that are characterized by good dimensional stability and electrical properties as well as resistance to heat.

allyl plastic A thermoset plastic made up of polymerized monomers containing allyl groups.

Alnico Any of a series of commercial iron-base permanent magnet alloys containing varying amounts of aluminum, nickel and cobalt as the chief alloying elements.

alpha The allotrope of titanium with a hexagonal, close-packed crystal structure.

alpha-beta brass A brass containing about 40% zinc with a duplex structure of alpha and beta crystals, and used in the manufacture of castings and pipe.

alpha-beta structure A microstructure which contains both alpha and beta as the principal phases at a specific temperature.

alpha brass Brass containing up to 35% zinc with an alpha crystal structure with good ductility.

alpha case The oxygen, nitrogen, or carbon enriched, alpha-stabilized surface which results from elevated temperature exposure to environments containing these elements.

alpha counter *1.* A system for detecting and counting energetic alpha particles; it consists of an alpha counter tube, amplifier, pulse-height discriminator, scaler, and recording or indicating mechanism. *2.* An alpha counter tube and necessary auxiliary circuits alone. *3.* A term sometimes loosely used to describe just the alpha counter tube or chamber itself.

alpha decay The radioactive transformation of a nuclide by alpha-particle emission.

alpha double prime A supersaturated nonequilibrium orthorhombic phase formed by a diffusionless transformation of the beta phase in certain alloys. It occurs when cooling rates are too high to permit transformation by nucleation and growth. It may be strain induced during working operations and may be avoided by appropriate in-process annealing treatments.

alpha emitter A radionuclide that disintegrates by emitting an alpha particle from its nucleus.

alphanumeric code A code whose code set consists of letters, digits, and associated special characters.

alphanumeric display A display that can present numeric, common alphabetic characters, and sometimes other special symbols. Dot matrix and segmented (with more than 7 segments) displays are typical examples.

alpha particle A positively charged particle emitted from the nuclei of certain atoms during radioactive disintegration.

alpha particle induced soft errors Integrated circuit memory transient errors due to emission of alpha particles during radioactive decay of uranium or thorium contamination in the IC packaging material.

alpha prime In titanium, a supersaturated, acicular, nonequilibrium hexagonal phase formed by a diffusionless transformation of the beta phase. Alpha prime occurs when cooling rates are too high to permit transformation by nucleation and growth. It exhibits an aspect ratio of 10:1 or greater. Also known as martensite or martensite alpha.

alpha radiation *See* alpha particle.

alpha-ray spectrometer An instrument used to determine the energy distribution in a beam of alpha particles.

alpha 2 structure A structure consisting of an ordered alpha phase, such as Ti^3 (Al,SN) found in highly stabilized alpha. Defined by X-ray diffraction, not optical metallography.

altazimuth A sighting instrument having both horizontal and vertical graduated circles so that both azimuth and declination can be determined from a single reading. Also known as an astronomical theodolite or universal instrument.

alternate code complement In a frame synchronization scheme, refers to the process whereby a frame synchronization pattern is complemented on alternate frames to give better synchronization.

alternate immersion test A corrosion test in which a specimen is immersed and removed from a liquid after specific time intervals.

alternating copolymer A polymer with two repeating structures spaced alternately along a polymer chain.

alternating current (AC) An electrical current that reverses direction at regular intervals, with the rate expressed in hertz (cycles per second).

alternating-current bridge A bridge circuit that utilizes an a-c signal source and a-c null detector; generally, both in-phase (resistive) and quadrature (reactive) balance conditions must be established to balance the bridge.

17

alternating stress amplitude One-half the algebraic difference between the maximum and minimum stress for one cycle of a fatigue test.

alternator An electric generator that produces alternating current.

altigraph A recording pressure altimeter.

altimeter An instrument for determining height of an object above a fixed level or reference plane—sea level, for example; the aneroid altimeter and the radio altimeter are the most common types.

altitude Height above a specified reference plane, such as average sea level, usually given as a distance measurement in feet or meters regardless of the method of measurement.

alum A general name for a class of double sulfates containing aluminum and another cation such as potassium, ammonium, or iron.

alumel A nickel-base alloy with oxidation resistance used to make thermocouples for use at high temperatures.

alumina The oxide of aluminum—Al_2O_3.

aluminides Intermetallic compounds of aluminum and a transition metal.

aluminizing *1.* Applying a thin film of aluminum to a material such as glass. *2.* Forming a protective coating on metal by depositing aluminum on the surface, or reacting surface material with an aluminum compound, and diffusing the aluminum into the surface layer at elevated temperature.

aluminum A soft, white metal that in pure form exhibits excellent electrical conductivity and oxidation resistance. Aluminum is the base metal for an extensive series of lightweight structural alloys used in such diverse applications as aircraft frames and skin panels, automotive body panels and trim, lawn furniture, ladders, and domestic cookware.

aluminum arsenides Binary compounds of aluminum with negative, trivalent arsenic.

aluminum boron composites Structural materials composed of aluminum alloys reinforced with boron fibers (filaments).

aluminum brass An alpha-brass alloy containing about 2% aluminum for improved corrosion resistance.

aluminum bronze A copper alloy containing up to 10% aluminum, with good mechanical properties and corrosion resistance.

aluminum coated steel Material produced by hot-dip coating cold-rolled sheet steel on continuous lines. It provides a material with the superior strength of steel and the surface properties of aluminum.

aluminum diboride A refractory material with high strength used to reinforce epoxy resins.

aluminum gallium arsenides Compounds exhibiting characteristics suitable for use in laser devices, light-emitting diodes, solar cells, etc.

aluminum graphite composites Structural materials composed of aluminum alloys reinforced with graphite.

aluminum titanate A ceramic derived from the oxide of alumina and titania, used in applications requiring high thermal shock resistance.

aluminum-zinc coated steel Material produced by hot-dip coating cold-rolled sheet steel on continuous lines. It has the superior strength of steel and excellent corrosion resistance.

AM *See* modulation.

amalgam An alloy of metals containing mercury.

amalgamation *1.* Forming an alloy of any metal with mercury. *2.* A process for separating a metal from its ore by extracting it with mercury in the form of an amalgam. This process was formerly used to recover gold and silver, which are now extracted chiefly with the cyanide process.

ambient *1.* Used to denote surroundings, encompassing, or local conditions. Usually applied to environments, for example, ambient temperature or ambient pressure. *2.* A surrounding or prevailing condition, especially one that is not affected by a body or process contained in it.

ambient air The air that surrounds the equipment. The standard ambient air for performance calculations is air at 80°F, 60% relative humidity, and a barometric pressure of 29.921 inches Hg, giving a specific humidity of 0.013 pounds of water vapor per pound of air.

ambient air temperature/relative humidity Wet bulb and dry bulb readings that are recorded during the test, expressed in degrees Celsius (°C) or degrees Fahrenheit (°F).

ambient conditions The conditions (pressure, temperature, etc.) of the surrounding medium.

ambient level The levels of radiated and conducted signal and noise existing at a specified test location and time when the test sample is in operation. Atmospherics, interference from other sources, and circuit noise or other interference generated within the measuring set compose the ambient level.

ambient pressure error The maximum change in output, at any measurand value within the specified range, when the ambient pressure is changed between specified values.

ambient reference temperature The temperature to which rolling resistance data are referred. Variations in rolling resistance occur with changes in ambient temperature. Because precise control of room temperature is difficult, rolling resistance data must be referred to 24°C (75°F).

ambient temperature *1.* The temperature of the environment surrounding the test specimen. In many instances, the inferred ambient temperature is room temperature, typically 68–72°F. Unless otherwise specified, it is assumed that the sample is at the ambient temperature before being subjected to dynamic flexing. *2.* The environmental air temperature in which a unit is operating. In general, the temperature is measured in the shade (no solar radiation) and represents the air temperature for engine cooling performance measurement purposes. Air entering the radiator may or may not be the same as ambient, due to possible heating from other sources or recirculation.

ambient wind angle (v_a) The angle between the x-axis of the earth-fixed axis system and the ambient wind velocity vector.

ambient wind velocity (v_a) The horizontal component of the air mass velocity relative to the earth-fixed axis system in the vicinity of the vehicle.

Amici prism Also known as a roof prism. A right-angle prism in which the hypotenuse has been replaced by a roof, where two flat faces meet at a 90° angle. The prism performs image erection while deflecting the light by 90°.

amidopolyamines Compounds used in epoxy resins to react and create crosslinks within the resins.

amino Indicates the presence of NH_2 or NH group.

amino plastics Plastics based on resins made by the condensations of amines, such as urea and melamine, with aldehydes.

amino resins Resins resulting from polycondensation of a compound containing amino groups with an aldehyde.

amino-silane finish Finish applied to glass fibers to give a good bond to epoxide, phenolic, and melamine resins.

ammeter An instrument for determining the magnitude of an electric current in amperes.

ammonia A pungent, colorless, gaseous compound of hydrogen and nitrogen—NH_3.

amorphous Not having a crystal structure; having no determinate form.

amorphous film A film of material deposited on a substrate for corrosion protection, insulation, conductive properties, or a variety of other purposes. An amorphous film is non-crystalline and can be deposited by evaporation chemical deposition or by condensation.

amorphous plastic A plastic that has no crystalline component.

amorphous solid A rigid material with no patterned distribution of molecules or crystalline structure.

ampere The standard unit for measuring the strength of an electric current. A

rate of flow of a charge in a conductor or conducting medium of one coulomb per second.

ampere-hour A quantity of electricity equal to the amount of electrical energy passing a given point when a current of one ampere flows for one hour.

ampere per meter The SI unit of magnetic field strength; it equals the field strength developed in the interior of an elongated, uniformly wound coil excited with a linear current density in the winding of one ampere per meter of axial distance.

amperometry Chemical analysis of materials using measurements of electric current.

amphiboles A group of dark, rock-forming, ferromagnesian silicate minerals closely related in crystal form and composition.

amphoterics Substances having both acidic and basic properties.

amplification *1.* Increasing the amplitude of a signal by using a signal input to control the amplitude of a second signal supplied from another source. *2.* The ratio of the output-signal amplitude from an amplifier circuit to the input-signal amplitude from the control network, both expressed in the same units.

amplification factor The m factor for plate and control electrodes of an electron tube when the plate current is held constant.

amplifier *1.* Any device that can increase the magnitude of a physical quantity, such as mechanical force or electric current, without significant distortion of the wave shape of any variation with time associated with the quantity. *2.* A component used in electronic equipment to raise the level of an input signal so that the corresponding output signal has sufficient power to drive an output device such as a recorder or loudspeaker.

amplitude 1. (amplitude of displacement at a point in a vibrating system) The largest value of displacement that the point attains with reference to its equilibrium position. 2. The maximum value of the displacement of a wave or other periodic phenomenon from a reference position. Also, angular distance north or south of the prime vertical; the arc of the horizon; or the angle at the zenith between the prime vertical and a vertical circle, measured north or south from the prime vertical to the vertical circle.

amplitude distortion A condition in an amplifier or other device when the amplitude of the output signal is not an exact linear function of the input (control) signal.

amplitude modulation (AM) The process (or the results of the process) of varying the amplitude of the carrier in synchronism with and in proportion to the variation in the modulating signal.

amplitude response A measure of the time taken for a defined change of amplitude.

anaerobic Free of air or uncombined oxygen.

anaerobic adhesive An adhesive that dries in the absence of air.

analog 1. Of or pertaining to the general class of devices or circuits in which the output varies as a continuous function of the input. 2. The representation of numerical quantities by means of physical variables, such as translation, rotation, voltage, or resistance. 3. Contrasted with digital, describes a waveform that is continuous and varies over an arbitrary range.

analog channel A channel on which the information transmitted can take any value between the limits defined by the channel. Voice channels are analog channels.

analog data Data represented in a continuous form, as contrasted with digital data represented in a discrete, discontinuous form. Analog data are usually represented by means of physical variables, such as voltage, resistance, rotation, etc.

analog input 1. A continuously variable input. 2. A termination panel used to connect field wiring from the input device.

analog output Transducer output that is a continuous function of the measurand except as modified by the resolution of the transducer.

analog-to-digital (A/D) 1. A device, or sub-system, that changes real-world analog data (as from transducers) to a form compatible with binary (digital) processing, as done in a microprocessor. 2. The conversion of analog data to digital data. *See* analog-to-digital converter.

analog-to-digital converter (ADC) Any unit or device used to convert analog information to approximate corresponding digital information.

analysis 1. An evaluation based on decomposition into simple elements. 2. The part of mathematics that arises from calculus and deals primarily with functions. 3. Quantitative determination of the constituent parts.

analysis criticality A procedure by which each potential failure mode is ranked according to the combined influence of severity and probability of occurrence.

analysis, reliability The assessment of probabilities to determine satisfactory performance of an item under specified conditions of use over a given service period by means of statistical studies.

analysis system A system for measuring and collecting the coordinates of image points as a function of time.

analysis, ultimate Chemical analysis of solid, liquid, or gaseous fuels. In the case of coal or coke, determination of carbon, hydrogen, sulfur, nitrogen, oxygen, and ash.

analyte The substance that is being identified in an analysis.

analytical balance Any weighing device having a sensitivity of at least 0.1 mg.

analytical curve A graphical representation of some function of relative intensity in spectroscopic analysis plotted against some function of concentration.

analytical electron microscopy (AEM) Materials analysis using the electron microscope to recognize and quantify various responses from a specimen.

analytical gap The separation between the source electrodes in a spectrograph.

analytical line The spectral line of an element used to determine its concentration in spectroscopic analysis.

analytical train A general term that describes the entire system required to sample and analyze a particular constituent in exhaust gas. Typically, this train will include items such as tubing, condenser, particulate filter, sample pump, analytical instrument, and flow meter.

analytical wavelength In spectroscopy, the wavelength used for recognition or concentration of an element.

analyzer Any of several types of test instruments, ordinarily one that can measure several different variables either simultaneously or sequentially.

anchorite A zinc-iron phosphate coating.

AND A logic operator having the property that if P is an expression, Q is an expression, R is an expression ..., then the AND of P, Q, R ... is true if all expressions are true, false if any expression is false.

Anderson bridge A type of a-c bridge especially suited to measuring the characteristics of extremely low-Q coils.

andesite Volcanic rock composed essentially of andesine and one or more mafic constituents.

AND gate A combinational logic element such that the output channel is in its one state if and only if each input channel is in its one state.

anechoic chamber *1.* An acoustical device testing room in which all six of the surfaces absorbed at least 99% of the incident acoustic energy over the frequency range of interest. Also known as a dead room. *2.* A room lined with a material that absorbs radio waves of a particular frequency or band of frequencies; used chiefly for tests at microwave frequencies, such as a radar-beam cross section.

anelasticity A characteristic of some materials under stress whereby the loading curve does not coincide with the unloading curve.

anemometer A device for measuring wind speed; if it produces a recorded output, it is known as an anemograph.

anemoscope A device for indicating wind direction.

aneroid Describes a device or system that does not contain or use liquid.

angle (geometry) The inclination to each other of two intersecting lines, measured by the arc of a circle intercepted between the two lines forming the angle, the center of the circle being the point of intersection.

angle dekkor A testing instrument used for measuring angles by reflection between two surfaces.

angle modulation A type of modulation involving the variation of carrier-wave angle in accordance with some characteristic of a modulating wave; angle modulation can take the form of either phase modulation or frequency modulation.

angle of incidence The angle formed between the incident light ray and the normal at the point of incidence.

angle of refraction The angle formed at the point of emergence between the emergent ray and the normal.

angle of repose A characteristic of bulk solids equal to the maximum angle with the horizontal at which an object on an inclined plane will retain its position without tending to slide; the tangent of the angle of repose equals the coefficient of static friction.

angle-ply laminate A laminate with fibers of contiguous plies aligning at alternate angles.

angle valve A valve design in which one port is colinear with the valve stem or actuator, and the other port is at right angles to the valve stem.

angstrom A commonly used unit of length for light wavelength measurements equal to 10^{-10} meter or to 10^{-8} centimeter, or to about 0.00000004 inches. Nanometer is the preferred SI unit.

angular frequency A frequency expressed in radians per second; it equals 2π times the frequency in Hz.

angular momentum The product of a body's moment of inertia and its angular velocity.

anhydride *1.* A mixture from which water has been extracted. *2.* An oxide of a metal or nonmetal that forms an acid or base when mixed with water.

anhydrous Describes a chemical or other solid substance whose water of crystallization has been removed.

aniline An organic compound used in the production of aniline formaldehyde resins, with good resistance to moisture and chemical solvents.

anion A negatively charged ion of an electrolyte, which migrates toward the anode under the influence of a potential gradient.

anisotropic *1.* Having unequal responses to external stimuli. *2.* In physics, having properties, such as conductivity or speed of transmission of light, that vary according to the direction in which they are measured.

anisotropy The state of being anisotropic.

annealing A generic term denoting a treatment, consisting of heating to and holding at a suitable temperature followed by cooling at a suitable rate, used primarily to soften metallic materials, but also to simultaneously produce desired changes in other properties or in microstructure. The purpose of such change may be, but is not confined to, one or more of the following: improvement of machinability; facilitation of cold work; improvement of mechanical or electrical properties; or increase in stability of dimensions.

annealing-aluminum and aluminum alloys Annealing cycles are designed to: (a) remove part or all of the effects of cold working (where recrystallization may or may not be involved.); (b) cause substantially complete coalescence of precipitates from solid solution in relatively coarse form; or (c) both, depending on the composition and condition of the material. When the term is used without qualification, full annealing is implied. Specific process names in commercial use are: final annealing, full annealing, intermediate annealing, partial annealing, recrystallization annealing, and stress relief annealing.

annealing border *See* oxidized surface.

annealing-copper and copper alloys Depending on composition and condition, these materials are annealed by: (a) removal of the effects of cold work by recrystallization or recrystallization and grain growth; (b) substantially complete precipitation of the second phase in relatively coarse form in age (precipitation) hardened alloys; (c) solution heat treatment of age (precipitation) hardenable alloys; (d) relief of residual stress in castings. Specific process names in commercial use are: final annealing, full annealing, light annealing, soft annealing, and solution annealing.

annealing-ferrous The time-temperature cycles used vary widely in both maximum temperature attained and in cooling rate employed, depending on the composition of the material, its condition and the results desired. When applicable, the following more specific commercial process names should be used: black annealing, blue annealing, box annealing, bright annealing, cycle annealing, flame annealing, full annealing, graphitizing, in-process annealing, isothermal annealing, malleableizing, orientation annealing, process annealing, quench annealing, and spheroidizing. When the term is used without qualification, full annealing is implied. When applied only for the relief of stress, the process is properly called stress relieving.

annealing stain A discoloration on annealed material which may occur anywhere on the material. It results from residue, or oxidation, during annealing.

anneal orientation A final, high-temperature anneal applied principally to flat-rolled electrical steel to develop secondary grain growth and directionality of magnetic properties.

anneal to temper-copper and copper alloys A final anneal used to produce specified mechanical properties in a material.

annotate To add explanatory text to computer programming or any other instructions.

annular ducts Ring-shaped openings for the passage of fluids (gases, etc.) designed for optimum aerodynamic flow properties for the application involved.

annulus *1.* Any ring-shaped cavity or opening. *2.* A plate that protects or covers a machine.

annunciator An electromagnetic, electronic, or pneumatic signaling device that either displays or removes a signal light, metal flag, or similar indicator, or sounds an alarm, or both, when occurrence of a specific event is detected. In most cases, the display or alarm is single-acting, and must be reset after being tripped before it can indicate another occurrence of the event.

anode *1.* The electrode of an electrolyte cell at which oxidation occurs. *2.* The positive pole (+) in batteries, galvanic cells, or plating apparatus. In diodes, the positive lead. *3.* A more positive lead when conducting in the forward direction. *4.* The negative electrode in a storage battery, or the positive electrode in an electrochemical cell. *5.* The positive electrode in an x-ray tube or vacuum tube, where electrons leave the interelectrode space.

anodic coating An oxide film produced on a metal by treating it in an electrolytic cell with the metal as the cell anode.

anodic inhibitor A chemical substance or combination of substances that prevent or reduce the rate of the anodic or oxidation reaction by a physical, physico-chemical, or chemical action.

anodic potential Electronegative potential.

anodic protection Polarization to a more oxidizing potential to achieve a reduced corrosion rate by the promotion of passivity.

anodic stripping The removal of metallic coatings.

anodize To form a protective passive film (conversion coating) on a metal part, such as a film of Al_2O_3 on aluminum, by making the part an anode in an electrolytic cell and passing a controlled electric current through the cell.

anodizing Oxide coating formed on a metal surface (generally, aluminum) by an electrolytic process.

anomalies In general, deviations from the norm.

anorthosite A group of essentially mono-mineralic plutonic igneous rocks composed almost entirely of plagioclase feldspar.

antenna A device for sending or receiving radio waves, but not including the means of connecting the device to a transmitter or receiver. *See also* dipole antenna and horn antenna.

antenna array A single mounting containing two or more individual antennas coupled together to give specific directional characteristics.

anti-aliasing A technique used to provide a greater apparent resolution on the screen of a graphics display without the cost of a higher resolution display.

anticlines Geologic formations characterized by folds, the core of which contain stratigraphically older rocks; they convex upward.

anticorrosive Describes a substance, such as paint or grease, that contains a chemical that counteracts corrosion or produces a corrosion-resistant film by reacting with the underlying surface.

antidegradant An additive to polymers designed to prevent deterioration.

antifouling Measures taken to prevent corrosion or the accumulation of organic or other residues or growths on operating mechanisms, especially in underwater environments.

antifriction Describes a device, such as a bearing or other mechanism, that employs rolling contact with another part rather than sliding contact.

antigravity A hypothetical effect that would arise from cancellation by some energy field of the effect of the gravitational field of the earth or other body.

antimagnetic Describes a device that is made of nonmagnetic materials or employs magnetic shielding to avoid being influenced by magnetic fields during operation.

antimicrobial An additive to a polymer which prevents fungus or bacterial attack.

antimony A metallic element with atomic number 51, sometimes used as an alloying element.

antinodes The points, lines, or surfaces in a medium containing a standing wave where some characteristic of the wave field is at maximum amplitude.

antioxidant A substance that retards oxidation when added to another material.

antiozonant A substance used to prevent or retard the degradation of material through exposure to ozone.

antiparticles Particles with charges opposite in sign to the same particles in normal matter.

antireflective coating A coating designed to suppress reflections from an optical surface.

antiresonant Describes an electric, acoustic, or other dynamic system whose impedance is very high, approaching infinity.

antiskid Describes a material, surface, or coating that has been roughened or that contains abrasive particles to increase the coefficient of friction and prevent

sliding or slipping. Also known as antislip.

antislip *See* antiskid.

antistatic agent A substance that retards conductivity when mixed with or coated on another material.

antisurge control Control by which the unstable operating mode of compressors known as "surge" is avoided.

antisymmetric laminate *1.* A special laminate type that is balanced but unsymmetric. *2.* A laminate in which, for a given ply configuration in the lower half, there is an identical ply configuration in the upper half, but with an alternating ply angle.

anvil *1.* The stationary contact of a micrometer caliper or similar gaging device. *2.* In drop forging, the base of the hammer into which the lower die is set.

aperiodic Varying in a manner that is not periodically repeated.

aperiodically damped Reaching a constant value or steady state of change without introducing oscillation.

aperture A hole in a surface through which light is transmitted. When placed in the Fourier (focal) plane, apertures are sometimes called spatial filters, a more descriptive term.

apostilb (ASB) A unit of luminance equal to 1 candela/m^2 (1 unit). A surface of 1 m^2, all points of which are 1 m from a 1 candela source, has a luminance of 1 apostilb.

apparent candlepower Of an extended source measured at a specific distance, the candlepower of a point source that would produce the same illumination at the same distance.

apparent contaminant capacity The weight (grams) of contaminant introduced into a laboratory filter test system before the terminal pressure drop or specified reduction in flow is reached.

apparent density The density of loose or compacted particulate matter determined by dividing actual weight by volume occupied. Apparent density is always less than the true density of the material comprising the particulate matter because volume occupied includes the space devoted to pores or cavities between particles.

apparent porosity The measurement of void space within a material determined by dividing the measured exterior surface by volume of material.

apparent viscosity The resistance to continuous deformation (viscosity) in a non-Newtonian fluid subjected to shear stress.

appearance potential The minimum electron-beam energy required to produce ions of a particular type in the ion source of a mass spectrometer.

application The system or problem to which a computer is applied. Reference is often made to computation, data processing, and control as the three categories of application.

application time The time available for sealant application after mixing or after thawing a premixed and frozen cartridge of sealant. Acceptability limits established for class A brushable sealants are expressed in the time required for the viscosity to increase to a specified level at 77°F (25°C) and 50% relative humidity. The acceptability limits for class B extrudable sealants are expressed in terms of the extrusion rate of a sealant from a 6 fl. oz. (177 mL) cartridge through a nozzle with a 0.125 in. (3.18 mm) diameter orifice, using air pressure of 90 psi +5 (621 kPa +34) in a pneumatic sealant gun. The extrusion rate is expressed in grams/minute or (per some specifications) in cubic centimeters (cc)/minute. A minimum extrusion rate after the stated application time is given as the acceptable limit.

applied load *1.* Weight carried or force sustained by a structural member in service; in most cases the load includes the weight of the member itself. *2.* Material carried by the load-receiving member of a weighing scale, not including any load necessary to bring the scale into initial balance.

AQL *See* acceptable quality level.

aqueous vapor *See* water vapor.

A_{r1} Regarding steel, the temperature at which transformation from austenite to ferrite or to ferrite plus cementite is completed during cooling.

aragonite A white, yellowish, or gray orthorhombic mineral that contains calcium carbonate.

aramid A high-strength fiber derived from nylon with a significant benzene ring structure.

arbitration bar A test bar cast from molten metal at the same time as a lot of castings; it is used to determine mechanical properties in a standard tensile test, which are then evaluated to determine acceptability of the lot of castings.

arbor *1.* A piece of material usually cylindrical in shape, generally applied to locate a rotating part, or cutting tool, about a center. *2.* In machine grinding, the spindle for mounting and driving the grinding wheel. *3.* In machine cutting, such as milling, the shaft for holding and driving a rotating cutter. *4.* Generically, the principal spindle or axis of a rotating machine which transmits power and motion to other parts. *5.* In metal founding, a bar, rod, or other support embedded in a sand core to keep it from collapsing during pouring.

arbor press A mechanical or hydraulic machine for forcing arbors, mandrels, bushings, shafts, or pins into or out of drilled or bored holes.

arc *1.* A discharge of electricity across a gap between electrical conductors. *2.* A segment of the circumference of a circle. *3.* The graduated scale on an instrument for measuring angles. *4.* A narrow, curved pencil-like line in the coating running transversely approximately 2 inches from each edge of the strip.

arc furnace A furnace in which material is heated either directly by an electric arc between an electrode and the work or indirectly by an arc between two electrodes adjacent to the material.

arcing time With regard to fuses, the time measured from the point when element melt time ends to the point when current is interrupted and permanently becomes zero. If a mechanical indicator (not presently recommended) is utilized which incorporates a secondary element parallel to the fusible element, arcing time will commence from the point at which indicator melt time ends. With regard to breakers, arcing time is the time measured from the point when contacts first separate to the point when the current is interrupted and permanently becomes zero.

arc cutting Metal cutting with an arc between an electrode and the metal itself.

arc lamp A high-intensity lamp in which a direct-current electric discharge produces light that is continuous, as opposed to a flashlamp, which produces pulsed light.

arc line A spectral line in spectroscopy.

arc melting Raising the temperature of a metal to its melting point using heat generated by an electric arc; usually refers to melting in a specially designed arc furnace to refine a metal, produce an alloy, or prepare a metallic material for casting.

arc resistance The elapsed time during which an arc can contact a surface without causing conductivity.

arc strike *See* strike.

arc welding A group of welding processes that produce coalescence of metals by heating the metals with an arc, with or without pressure or the use of filler metal.

areal weight The weight of fiber per unit area of tape or fabric.

area meter A device for measuring the flow of fluid through a passage of fixed cross-sectional area, usually through the use of a weighted piston or float supported by the flowing fluid.

area, negative pressure Any region in which the static pressure is less than that of the static pressure of the undisturbed air stream.

area, positive pressure Any region in which the static pressure is greater than that of the static pressure of the undisturbed air stream.

argentometer A hydrometer used to find the concentration of a silver salt in water solution.

argon An inert gas with atomic number 18, often used to provide inert atmospheres during heat treating.

argument *1.* An independent variable, for example, in looking up a quantity in a table, the number or any of the numbers that identify the location of the desired value; or, in a mathematical function, the variable that, when a certain value is substituted for it, determines the value of the function. *2.* An operand in an operation on one or more variables. *See also* parameter and independent variable.

arithmetic and logical unit A component of the central processing unit in a computer, where data items are compared, arithmetic operations performed, and logical operations executed.

arithmetic check *See* mathematical check.

arithmetic element The portion of a mechanical calculator or electronic computer that performs arithmetic operations.

arithmetic expression An expression containing any combination of data names, numeric literals, and named constants, joined by one or more arithmetic operators in such a way that the expression as a whole can be reduced to a single numeric flue.

arithmetic operation A computer operation in which the ordinary elementary arithmetic operations are performed on numerical quantities. Contrast with logical operation.

arithmetic operator Any of the operators, + and –, or the infix operators, +, –, *, /, and **.

arithmetic shift *1.* A shift that does not affect the sign position. *2.* A shift that is equivalent to the multiplication of a number by a positive or negative integral power of the radix.

arm *1.* A rigid member that extends to support or provide contact beyond the perimeter of the basic item. *2.* Allows a hardware interrupt to be recognized and remembered. Contrast with disarm. *See* enable.

armature *1.* The core and windings of the rotor in an electric motor or generator. *2.* The portion of the moving element of an instrument that is acted upon by magnetic flux to produce torque.

aromatic compound An unsaturated hydrocarbon with benzene ring structures in the molecule.

aromatic polyester A polyester in which all hydroxyl and carboxyl groups are linked to benzene ring structures.

aromatics A hydrocarbon having a ring-type structure with the general formula C_nH_{2n-6} and containing three double bonds in the ring.

array *1.* An arrangement of elements in one or more dimensions. *See also* matrix and vector. *2.* In a computer program, a numbered, ordered collection of elements, all of which have identical data attributes. *3.* A group of detecting elements usually arranged in a straight line (linear array) or in two-dimensional matrix (imaging array). *4.* A series of data samples, all from the same measurement point. Typically, an array is assembled at the telemetry ground station for frequency analysis.

array dimension The number of subscripts needed to identify an element in the array.

array process A hardware device that processes data arrays; fast Fourier transforms (FFT) and power-spectral density (PSD) are typical processes.

array processor The capability of a computer to operate at a variety of data locations at the same time.

arrest marks *See* bench marks.

arrest points The temperature at which a rest occurs in heating or cooling curves.

Arrhenius acceleration factor (F) The factor by which the time to fail can be reduced by increased temperature. $F = \theta_1/\theta_2 = \exp (E/k) (1/T_2 - 1/T_1)$.

Arrhenius model A mathematic representation of the dependence of failure rate on absolute temperature and activation energy. The model assumes that degradation of a performance parameter is linear with time, with the failure rate a function of temperature stress.

arsenic A brittle, toxic metallic element with atomic number 33 which is used as an alloying element to improve strength and resistance to corrosion.

articulated structure A structure—either stationary or movable, such as a motor vehicle or train—which is permanently or semipermanently connected so that different sections of the structure can move relative to the others, usually involving pinned or sliding joints.

artificial aging Heat treating a metal at a moderately elevated temperature to hasten age hardening. Sometimes called accelerated aging.

artificial intelligence (AI) A subfield of computer science concerned with the concepts and methods of symbolic inference by a computer and the symbolic representation of the knowledge to be used in making inferences.

artificial language A language specifically designed for ease of communication in a particular area of endeavor, but one that is not yet natural to that area. This is contrasted with a natural language which has evolved through long usage.

ASA *See* acrylate styrene acrylonitrile.

ASB *See* apostilb.

asbestos A fibrous variety of the mineral hornblende, used extensively for its fire-resistant qualities to make insulation and fire barriers.

A-scan display A method in which results of ultrasonic tests are displayed on a CRT; echo signals from defects appear as a series of blips along a baseline.

ASCII (American Standard Code for Information Interchange) *1.* A widely-used code in which alphanumerics, punctuation marks, and certain special machine characters are represented by unique, 7-bit, binary numbers; 128 different binary combinations are possible ($2^7 = 128$), thus 128 characters may be represented. *2.* A protocol.

as-fabricated Describes the condition of a structure or material after assembly, and without any conditioning treatment such as a stress-relieving heat treatment.

ash The noncombustible inorganic matter in fuel.

aspect ratio *1.* The ratio of frame width to height for a television picture. This ratio is 4:3 in the United States, Canada, and United Kingdom. *2.* In any rectangular structure, such as the cross section of a duct or tubular beam, the ratio of the longer dimension to the shorter. *3.* A ratio used in calculating resistance to flow in a rectangular elbow; the ratio of width to depth.

asphalt A brown to black bituminous solid that melts on heating and is impervious to water but soluble in gasoline; used extensively in paving and roofing applications, and in paints and varnishes.

aspheric For optical elements, surfaces are aspheric if they are not spherical or flat. Lenses with aspheric surfaces are sometimes called aspheres.

aspiration Using a vacuum to draw up gas or granular material, often by passing a stream of water across the end of an open tube, or through the run of a tee joint, where the open tube or branch pipe extends into a reservoir containing the gas or granular material.

assembly language A computer programming language, similar to computer language, in which the instructions usually have a one-to-one correspondence with computer instructions in machine language, and which utilizes mnemonics for representing instructions.

assembly list A printed list which is the by-product of an assembly procedure. The assembly list lists in logical instruction sequence all details of a routine showing the coded and symbolic notation next to the actual notations established by the assembly procedure. This listing is highly useful in the debugging of a routine.

assessment *1.* A critical appraisal, including qualitative judgments about an item, such as the importance of analysis results, design criticality, and failure effect. *2.* The use of test data and/or operational service data to form estimates of population parameters and to evaluate the precision of these.

assign To designate a part of a system for a specific purpose.

assignment statement A program statement that calculates the value of an expression and assigns it a name.

association The combining of ions into larger ion clusters in concentrated solutions.

association reactions Gas phase chemical processes in which two molecular species A and B react to form a larger molecule AB.

associative processing (computers) Byte-variable computer processing with multifield search, arithmetic, and logic capability.

associative storage A storage device in which storage locations are identified by their contents, not by names or positions. Synonymous with content-addressed storage; contrast with parallel search storage.

assumptions Statements, principles, and/or premises offered without proof.

assurance The planned and systematic actions necessary to provide adequate confidence that a product or process satisfies given requirements.

A-stage An early stage of polymerization in which the material is fusible and still soluble in some liquids.

astatic Without polarity; independent of the earth's magnetic field.

astigmatism A defect in an optical element that causes rays from a single point in the outer portion of a field of view to fall on different points in the focused image.

astrakanite *See* bloedite.

astrochanite *See* bloedite.

astrodynamics The practical application of celestial mechanics, astroballistics, propulsion theory, and allied fields to the problem of planning and directing the trajectories of space vehicles.

astronomical theodolite *See* atlazimuth.

astronomy The science that treats the location, magnitudes, motions, and constitution of celestial bodies and structures.

astrophysics A branch of astronomy that treats the physical properties of celestial bodies, such as luminosity, size, mass, density, temperature, and chemical composition.

asymmetric rotor A rotating machine element whose axis of rotation is not the same as its axis of symmetry.

asymmetry potential The difference in potential between the inside and outside pH sensitive glass layers when they are both in contact with pH 7 solutions.

asymptotic properties Properties of any mathematical relation or corresponding physical system characterized by the value in question approaching but never reaching a particular magnitude.

asynchronous A mode of operation in which an operation is started by a signal before the operation on which this operation depends is completed.

atactic A molecular chain in which the methyl groups are more or less in random order.

atactic stereoisomerism A chain of molecules with the side chains or atoms positioned randomly.

atmidometer *See* atmometer.

atmometer A generic name for any instrument that measures evaporation rates. Also known as atmidometer, evaporimeter, or evaporation gage.

atmospheric air Air under the prevailing atmospheric conditions.

atmospheric chemistry Study of the production, transport, modification, and removal of atmospheric constituents in the troposphere and stratosphere.

atmospheric corrosion Corrosion that occurs naturally due to exposure to climatic conditions. Corrosion rates vary depending on specific global location.

atmospheric optics The study of the topical characteristics of the atmosphere and of the optical phenomena produced by the atmosphere's suspensoids and hydrometeors. Atmospheric optics embraces the study of refraction, reflection, diffraction, scattering, and polarization of light, but is not commonly regarded as including the study of any other kinds of radiation.

atmospheric pressure The barometric reading of pressure exerted by the atmosphere; at sea level, 14.7 pounds per square inch or 29.92 inches of mercury.

atmospheric radiation Infrared radiation emitted by or being propagated through the atmosphere.

atmospheric refraction Refraction resulting when a ray of radiant energy passes obliquely through an atmosphere.

atom The smallest particle of a chemical element composed of sub-atomic particles, electrons, protons, and neutrons.

atomic mass unit A unit for expressing atomic weights and other small masses; it equals, exactly, 1/12 the mass of the carbon-12 nuclide.

atomic number An integer that designates the position of an element in the periodic table of the elements; it equals the number of protons in the nucleus and the number of electrons in the electrically neutral atom.

atomic structure The arrangements of components of an atom.

atomic weight The weight of a single atom of any given chemical element; it is usually taken as the weighted average of the weights of the naturally occurring nuclides, expressed in atomic mass units.

atomization In powder metallurgy, the dispersion of molten metal into particles by a rapidly moving stream of gas or liquid.

atomizer A device by means of which a liquid is reduced to a very fine spray.

atom probe An instrument, consisting of a field-ion microscope with a probe hole in its screen that opens into a mass spectrometer, used to identify a single atom or molecule on a metal surface.

attemperation Regulating the temperature of a substance, for instance, passing superheated steam through a heat exchanger or injecting water mist into it to regulate final steam temperature.

attenuate To weaken or make thinner, for example, to reduce the intensity of sound or ultrasonic waves by passing them through an absorbing medium.

attenuation *1.* The lessening of energy due to geometrical spreading, adsorption, or scattering. *2.* The processing of molten glass to create thin fibers.

attenuator *1.* An optical device that reduces the intensity of a beam of light passing through it. *2.* An electrical component that reduces the amplitude of a signal in a controlled manner.

attitude The position of an object in space determined by the angles between its axes and a selected set of planes.

attraction *1.* In physics, the tendency or forces through which bodies or particles are attracted to each other. *2.* In chemistry, the force that is exerted between molecules not of the same kind, as when two molecules of hydrogen unite with one molecule of oxygen to form water. *3.* The force uniting molecules of the same nature—cohesion.

attribute A term used to designate a method of measurement whereby units are examined by noting the presence (or absence) of some characteristic or attribute in each of the units in the group under consideration and by counting how many units do (or do not) possess it. Inspection by attributes can be of two kinds: either the unit of product is classified simply as defective or nondefective, or the number of defects in the unit of product is counted with respect to a given requirement or set of requirements.

attribute sampling A type of sampling inspection in which an entire production lot is accepted or rejected depending on the number of items in a statistical sample that have at least one characteristic (attribute) that does not meet specifications.

attribute testing Testing to evaluate whether an item possesses a specified attribute.

audio Pertaining to audible sound—usually taken as sound frequencies in the range 20 to 20,000 Hz.

audiometer An instrument used to measure the ability of people to hear sounds; it consists of an oscillator, amplifier, and attenuator, and may be adapted to generate pure tones, speech, or bone-conducted vibrations.

audio signals Signals with a bandwidth of less than 20 kilohertz.

Auger electron An electron discharge from an atom with a vacancy in the inner shell.

Auger electron spectroscopy Chemical analysis of surface layers that recognizes the atoms within the surface layer by measuring the representative energies of their possessive Auger electrons.

auger spectroscopy Analytical technique in which the sample surface is irradiated with low-energy electrons, and the energy spectrum of electrons emitted from the surface is measured.

ausforming Mechanically working an appropriate high hardenability steel after quenching from above the upper critical temperature to a temperature between the lower critical and the M_s temperature, and isothermally transforming or quenching to produce the desired properties.

austempering Quenching a ferrous alloy from a temperature above the transformation range, in a medium having a rate of heat abstraction high enough to prevent the formation of high-temperature transformation products, and then holding the alloy, until transformation to bainite is complete, at a temperature below that of pearlite formation and above that of martensite formation.

austenite A face-centered cubic crystalline phase of iron-base alloys.

austenitic cast iron Cast irons containing up to 25% nickel that are nonmagnetic and resistant to heat and corrosion.

austenitic nitrocarburizing A lower-temperature variant of carbonitriding. Applied to ferrous materials at typical processing temperatures of 676°C to 774°C (1250°F to 1425°F). The process involves the diffusion of nitrogen and carbon into the surface of the work piece and the formation of a thin white layer of epsilon carbonitrides. Subsurface microstructure includes martensite and bainite, which improve the load-carrying capability when compared to ferritic nitrocarburizing.

austenitic stainless steel An alloy of iron containing at least 12% Cr plus sufficient Ni (or in some specialty stainless steels, Mn) to stabilize the face-centered cubic crystal structure of iron at room temperature. The structure is austenitic at room temperature.

austenitizing Forming austenite by heating a ferrous alloy into the transformation range (partial austenitizing) or above the transformation range (complete austenitizing). When used without qualification, the term implies complete austenitizing.

autocatalytic degradation The phenomenon whereby the breakdown products of the initial phase of degradation act to accelerate the rate at which subsequent degradation proceeds.

autoclave An airtight vessel in which the contents are heated and sometimes agitated; it usually uses high-pressure steam to perform processing, sterilizing, or cooking steps using moist or dry heat.

autocorrelation In statistics, the simple linear internal correlation of members of a time series (ordered in time or other domains).

autofrettage A process for manufacturing gun tubes and pressure vessels in which the inner surface layers of a plain tube are initially stressed by expansion beyond the yield strength so that residual compressive stresses are created.

autoignition temperature (AIT) Temperature at which the fluid flashes into flame without an external ignition source and continues burning. Actual value is to be determined by one of several approved test methods.

automate *1.* To apply the principles of automation. *2.* To operate or control by automation. *3.* To install automatic procedures, as for manufacturing, servicing, etc.

automatic *1.* Having the power of self-motion; self-moving; or self-acting; an automatic device. *2.* A machine that operates automatically. *3.* Pertaining to a process or device that, under specified conditions, functions without intervention by a human operator.

automatic control The type of control in which there is no direct action of man on the controlling device.

automatic control panel A panel of indicator lights and switches on which are displayed an indication of process conditions, and from which an operator can control the operation of the process.

automatic frequency control A device or circuit designed to maintain the frequency of an oscillator within a preselected band of frequencies. In an FM radio receiver, the circuitry that senses frequency drift and automatically controls an internal oscillator to compensate for the drift.

automatic gain control An auxiliary circuit that adjusts gain of the main circuit in a predetermined manner when the value of a selected input signal varies.

automatic self test Self-test to the degree of fault detection and isolation that can be achieved entirely under computer control, without human intervention.

automatic test Performance assessment, fault detection, diagnosis, isolation, and prognosis that is performed with a minimum of reliance on human intervention. This may include BIT.

automation *1.* The implementation of processes by automatic means. *2.* The theory, art, or technique of making a process more automatic. *3.* The investigation, design, development, and application of methods of rendering processes automatic, self-moving, or self-controlling. *4.* The conversion of a procedure, a process, or equipment to automatic operation.

autoradiography A technique for producing a radiographic image using ionizing radiation produced by radioactive decay of atoms within the test object itself.

autotrophs Organisms capable of synthesizing organic nutrients directly from simple inorganic substances such as carbon dioxide and inorganic nitrogen.

auto-zero logic module A component of a digital controller whose function is primarily to establish an arbitrary zero-reference value for each individual measurement.

auxiliary device Generally, any device that is separate from a main device but that is necessary or desirable for effective operation of the system.

auxiliary means A device or subsystem, usually placed ahead of the primary detector, which alters the magnitude of the measured quantity to make it more suitable for the primary detector without changing the nature of the measured quantity.

auxiliary panel *1.* A panel that is not in the main control room. The front of an auxiliary panel is normally accessible to an operator, but the rear is normally accessible only to maintenance personnel. *2.* Located at an auxiliary location.

availability The number of hours in the reporting period less the total downtime for the reporting period divided by the number of hours in the reporting period (expressed in percent).

availability factor The fraction of the time during which the unit is in operable condition.

available energy Energy that theoretically can be converted to mechanical power.

available heat In a thermodynamic working fluid, the amount of heat that could be transformed into mechanical work under ideal conditions by reducing the temperature of the working fluid to the lowest temperature available for heat discard.

available power An attribute of a linear source of electric power defined as $V_{rms}/4R$, where V_{rms} is the open circuit rms voltage of the power source and R is the resistive component of the internal impedance of the power source.

available power gain An attribute of a linear transducer defined as the ratio of power available from the output terminals of the transducer to the power available from the input circuit under specified conditions of input termination.

avalanche Production of a large number of ions by cascade action in which a single charged particle, accelerated by a strong electric field, collides with neutral gas molecules and ionizes them.

avalanche breakdown In a semiconductor diode, a nondestructive breakdown caused by the cumulative multiplication of carriers through field-induced impact ionization.

average The sum of a group of test values divided by the number of test values summed.

average outgoing quality limit The average percent of defective units that remain undetected in all lots that pass final inspection; it is a measure of the ability of sampling inspection to limit the probability of shipping defective product.

avionics Electrical and electronic equipment used in aviation, principally for navigation and communication.

Avogadro's number The number of molecules in a mole of a substance; 6.02252×10^{23}.

axes (coordinates) *See* coordinate.

axial fan Consists of a propeller or disc-type of wheel within a cylinder discharging the air parallel to the axis of the wheel.

axial-flow Describes a machine such as a pump or compressor in which the general direction of fluid flow is parallel to the axis of its rotating shaft.

axial modes Regimes of vibration along a given axis.

axial strain The linear strain in a plane that is parallel to the longitudinal axis of the test specimen.

axial surface roughness Surface roughness of a shaft measured in a direction parallel to the centerline axis.

axial tensile strength (related to fasteners) The maximum tensile stress that a fastener is capable of sustaining when axially loaded. Tensile strength is calculated from the original nominal tensile stress area of the threaded section and the maximum load occurring during a tensile test.

axial wound Filament-wound reinforced plastic where the filaments are parallel to the winding axis.

axle A rod, shaft, or other supporting member that carries wheels and either transmits rotating motion to the wheels or allows the wheels to rotate freely about it.

azeotrope A mixture whose evolved vapor composition is the same as the liquid it comes from. This phenomenon occurs at one fixed composition for a given system.

azeotropic distillation A distillation technique in which one of the product streams is an azeotrope. It is sometimes used to separate two components by adding a third, which forms an azeotrope with one of the original two components.

azimuth Horizontal direction or bearing.

azimuth angle An angular measurement in a horizontal plane about some arbitrary center point using true North or some other arbitrary direction as a reference direction (0°).

azimuth circle A ring scale graduated from 0° to 360°, and used with a compass, radar plan position indicator, direction finder, or other device to indicate compass direction, relative bearing, or azimuth angle.

azoles Compounds that contain a five-membered heterocyclic ring containing one or more nitrogen atoms.

B

B *See* bel.

babbitt Any of the white alloys composed principally of lead or tin which are used extensively to make linings for sliding bearings.

babble The composite signal resulting from cross talk among a large number of interfering channels.

backbone The trunk media of a multimedia LAN separated into sections by bridges, routers, or gateways.

backcoating Coating material that is deposited on the side opposite that which is being coated in the evaporated coating process, generally resulting in an objectionable appearing film or haze.

back extrusion *See* cold extrusion.

background In radiation counting, a low-level signal caused by radiation from sources other than the source of radiation being measured.

background discontinuities Small, scattered, subsurface discontinuities, such as shrink cavities, porosity, more-dense or less-dense foreign material, and gas holes, appearing over extended areas of the casting.

background discrimination The ability of a measuring instrument or detection circuit to distinguish an input signal from electronic noise or other background signals.

background fluorescence Fluorescent residues observed over the general surface of the test part.

background noise Undesired signals or other stimuli that are always present in a transducer output or electronic circuit, regardless of whether a desired signal or stimulus is also present.

background program A program of the lowest urgency with regard to time and which may be preempted by a program of higher urgency and priority. Contrast with foreground program.

backhand welding Laying down a weld bead with the back of the welder's principal hand (the one holding the torch or welding electrode) facing the direction of welding. In torch welding, this directs the flame backward against the weld bead to provide postheating. Contrast with forehand welding.

backing strip A piece of metal, asbestos, or other nonflammable material placed behind a joint prior to welding to enhance weld quality.

back reflection *1.* Indication of the echo from the far boundary of the product under test. *2.* An x-ray diffraction technique in which a narrow beam of polychromatic radiation normal to the surface of a material provides a diffraction pattern from the various crystal planes. The resulting pattern provides information on grain size, cold working, and any directionality from cold working.

backscattering The scattering of light in the direction opposite to the original direction in which it was traveling.

backstep sequence A method of laying down a weld bead in which a segment is welded in one direction, then the torch is moved in the opposite direction a distance approximately twice the length of the first segment, and another segment is welded back toward the first.

baffle *1.* A plate or vane, plain or perforated, used to regulate or direct the flow of fluid. *2.* A cabinet or partition used with a loudspeaker produced simultaneously by the front and rear surfaces of the diaphragm.

bag exhaust sampling A technique for collecting a sample of exhaust gas during a period of a test cycle and storing it for future analysis.

bainite A molecular structure formed in metal by isothermal transformation of austenite at a specific temperature. The properties of bainite resemble those of tempered martensite.

bake hardenable steels These steels that increase in strength while undergoing a paint-baking cycle.

bakelite A common name for a range of thermosets in the phenol formaldehyde group of plastics. A common use is to prepare microspecimens.

bakeout Heating the surfaces of a vacuum system during evacuation to degas them and aid the process of reaching a stable final vacuum level. *See also* degassing.

baking Heating to a low temperature usually to remove gases such as hydrogen. Aging may result from baking treatments.

balance *1.* Generically, a state of equilibrium—static, as when forces on a body exactly counteract each other, or dynamic, as when material flowing into and out of a pipeline or process has reached steady state and there is no discernible rate of change in process variables. *2.* An instrument for making precise measurements of mass or weight.

balanced construction A laminate construction in which reactions to loads of compression or tension result only in extension or compression deformation.

balanced design In filament-wound reinforced plastics, a winding pattern designed such that the stress in all filaments is equal.

balanced laminate A composite laminate in which all the laminae at angles, other than 0 and 90 degrees, occur only in plus and minus pairs and are symmetrical around the centerline.

balanced twist An arrangement of twists in a combination of two or more strands that does not cause them to kink or twisting on themselves when the yarn produced is held in the form of an open loop.

balance weight A mass positioned on the balance arms of a weighing device so that the arms can be brought to a predetermined position (null position) for all conditions of use.

ball burnishing *1.* Producing a smooth, dimensionally precise hole by forcing a slightly oversize tungsten-carbide ball through a slightly undersize hole at high speed. *2.* A method of producing a lustrous finish on small parts by tumbling them in a wood-lined barrel with burnishing soap, water, and hardened steel balls.

ball-type viscometer An apparatus for determining viscosity, especially of high-viscosity oils and other fluids, in which the time required for a ball to fall through liquid confined in a tube is measured.

Banbury mixer A heavy-duty batch mixer with two counterrotating rotors; it is designed for blending doughy material such as uncured rubber and plastics.

band *1.* The gamut or range of frequencies. *2.* The frequency spectrum between two defined limits. *3.* Frequencies that are within two definite limits and used for a different purpose. *4.* A group of channels. *See* channel. *5.* A group of recording tracks on a computer magnetic disk or drum.

band density In filament winding, the number of strands of reinforcement per inch of band width.

banded structure A microstructure found in wrought alloys in which the phases appear as parallel bands.

band spectrum A spectral distribution of light or other complex wave in which the wave components can be separated into a series of discrete bands of wavelengths. *See also* continuous spectrum.

bandwidth *1.* In filament winding, the width of the reinforcement strand. *2.* The difference, expressed in hertz, between the two boundaries of a frequency range. *3.* A group of consecutive frequencies constituting a band that exists between limits of stated frequency attenuation. A band is normally defined as more than 3.0 decibels greater than the mean attenuation across the band. *4.* A group of consecutive frequencies constituting a band that exists between limits of stated frequency delay. *5.* The

range of frequencies that can be transmitted in an electronic system.

bar *1.* A solid elongated piece of metal, usually having a simple cross section and usually produced by hot rolling or extrusion, which may or may not be followed by cold drawing. *2.* A unit of pressure. One bar equals 10^5 pascals.

Barba's Law Law stating that the value of percentage elongation on gage length from tensile tests is independent of testpiece dimensions when the ratio of gage length to cross-sectional area is constant.

Barcol hardness A value obtained by measuring the penetration of a spring loaded steel point into a material; most often used to measure the degree of cure of a plastic.

bare glass Reinforcement glasses from which sizing or finish has been removed; or such glasses before any addition of such sizing or finish.

barium A white alkaline earth metal with chemical properties similar to calcium.

barium titanate A ceramic composed of barium oxide and titanium used in capacitors due to its high dielectric constant.

barium yttrium copper oxide A superconducting ceramic made up of barium oxide, yttrium, and copper.

bark A decarburized layer on steel, just beneath the oxide scale, that results from heating the steel in an oxidizing atmosphere.

Barkhausien effect Changes in magnetic induction caused by varying the magnetizing force acting on a ferromagnetic material.

Barkometer scale A specific gravity scale used primarily in the tanning industry, in which specific gravity of a water solution is determined from the formula: sp gr = $1.000 \pm 0.001n$, where n is degrees Barkometer. On this scale, water has a specific gravity of zero Barkometer.

barn A unit of nuclear cross section where the probability of a specific nuclear interaction, such as neutron capture, is expressed as an apparent area; in this context, one barn equals 10^{-24} m^2.

baroclinity The state of stratification in a fluid in which surfaces of constant pressure (isobaric) intersect surfaces of constant density (isoteric). The number, per unit area, of isobaric-isoteric solenoids intersecting a given surface is a measure of baroclinity.

barograph An instrument for recording atmospheric pressure.

barometer An absolute pressure gage for determining atmospheric pressure; if it is a recording instrument, it is known as a barograph.

barometric pressure Atmospheric pressure as determined by a barometer, usually expressed in inches of mercury. *See also* atmospheric pressure.

barostat A device for maintaining constant pressure within a chamber.

barothermograph An instrument for automatically recording both atmospheric temperature and pressure.

barothermohygrograph An instrument for automatically recording atmospheric pressure, temperature, and humidity on the same chart.

barotropism The state of a fluid in which surfaces of constant density (or temperature) are coincident with surfaces of constant pressure; it is the state of zero baroclinity.

barrel *1.* A cylindrical component of an actuating cylinder, accumulator, etc. in which a piston or sealed separator moves. *2.* A unit of volume; for petroleum, one barrel equals 9702 in^3.

barrier Any material limiting passage through itself of solids, liquids, semi-solids, gases, or forms of energy such as ultraviolet light.

barrier coat Any protective coating applied to a composite structure.

barrier plastics A general term applied to a group of lightweight, transparent, impact-resistant plastics, usually rigid copolymers of high acrylonitrile content.

barrier protection A type of protection that relies on the coating preventing access of moisture or oxygen to the material being protected. Organic coatings often offer barrier protection to underlying substrates.

barrier resins Polymers developed for the packaging industry to provide a seal against oxygen and carbon dioxide to prevent spoilage of foodstuffs.

barytes Barium sulfate used in the manufacture of paint and paper, and as a flame retardant in foam products.

base A chemical substance that hydrolyzes to yield OH^- ions. Opposite of acid.

base address *1.* A number that appears as an address in a computer instruction, but serves as the base, index, initial, or starting point for subsequent addresses to be modified. Synonymous with presumptive address and reference address. *2.* A number used in symbolic coding in conjunction with a relative address; an address used as the basis for computing the value of some other relative address.

baseband *1.* A single channel signaling technique in which the digital signal is encoded and impressed on the physical medium. *2.* The frequencies starting at or near d-c.

baseline *1.* Generally, a reference set of data against which operating data or test results are compared to determine such characteristics as operating efficiency or system degradation with time. *2.* In navigation, the geodesic line between two stations operating in conjunction with each other.

baseline technique A method of measuring adsorbance, where the distance from the drawn baseline to the adsorption peak is the adsorbance.

base metal *1.* The metallic element present in greatest proportion in an alloy. *2.* The type of metal to be welded, brazed, cut or soldered. *3.* In a welded joint, metal that was not melted during welding. *4.* Metal to which a plated, sprayed, or conversion coating is applied. Also known as basis metal.

base metal catalyst A catalyst in which the active catalytic material is one or more non-noble metals such as copper or chromium.

base metal hardness (related to fasteners) Hardness at root diameter on a line bisecting the included angle of the thread.

base number Same as radix number.

BASIC *See* high-level language(s).

basic element A single component or subsystem that performs one necessary and distinct function in a measurement sequence. To be considered a basic element, the component must perform one and only one of the smallest steps into which the measuring sequence can be conveniently divided.

basic failure rate The basic failure rate of a product derived from the catastrophic failure rate of its parts, before the application of use and tolerance factors.

basic frequency In a waveform made up of several sinusoidal components of different frequencies, the single component having the largest amplitude or having some other characteristic that makes it the principal component of the composite wave.

basic input output system (BIOS) The part of a computer operating system that handles input and output.

basic recipe A generic, transportable recipe consisting of header information, equipment requirements, formula, and procedure.

basic reliability The duration or probability of failure-free performance under stated conditions.

basis metal The metal to which coatings are applied. *See* base metal.

basis weight For paper and certain other sheet products, the weight per unit area.

basketweave Basketweave, as it refers to titanium and a microstructure thereof: alpha platelets, with or without interweaved beta platelets, that occur in colonies. Forms during cooling through the beta transus at intermediate cooling rates.

batch *1.* The quantity of material required for or produced by a production operation at a single time. *2.* An amount of material that undergoes some unit chemical process or physical mixing operation to make the final product homogeneous or uniform. *3.* The amount of material developed during a process having the same characteristics throughout. Also called a lot. *4.* A group of similar computer transactions joined together for processing as a single unit.

batch annealing *See* box annealing.

batch contaminant addition A filter test condition under which a specified contaminant is added in batches of specified size at specified time intervals.

batch distillation A distillation process in which a fixed amount of a mixture is charged, followed by an increase in temperature to boil off the volatile components. This process differs from continuous distillation, in which the feed is charged continuously.

batch or grab sample A sample, liquid or solid, that is analyzed for composite composition.

batch mixer A type of mixer in which starting ingredients are fed all at once and the mixture removed all at once at some later time. Contrast with continuous mixer.

batch monitor *See* monitor software.

batch process A process that manufactures a finite quantity of material by subjecting measured quantities of raw materials to a time-sequential order of processing actions using one or more pieces of equipment.

batch processing *1.* Pertaining to the technique of executing a set of programs such that each is completed before the next program of the set is started. *2.* Loosely, the execution of programs serially.

bathochrome An agent or chemical group that causes the absorption band of a solution to shift to lower frequencies.

bathtub curve A plot of failure rate of an item (whether repairable or not) versus time. The failure rate initially decreases, then stays reasonably constant,

then begins to rise rather rapidly. It has the shape of a bathtub. Not all items have this behavior.

batt Felt fabric created by compressing organic fibers.

battery The cased assembly of interconnected electrochemical cells ready for installation.

baud *1.* A unit of signaling speed equal to the number of discrete conditions or signal events per second. (This is applied only to the actual signals on a communication line.) *2.* If each signal event represents only one bit condition, baud is the same as bits per second. *3.* When each signal event represents other than the logical state of only one bit; used for data entry only in the simplest of systems. *4.* A unit of signaling speed equal to the number of code elements per second. *5.* The unit of signal speed equal to twice the number of Morse code dots continuously sent per second. *See* rate, bit and capacity, and channel.

Baudot code A three-part teletype code consisting of a start pulse (always a space), five data pulses, and a stop pulse (1.42 times the length of the other pulses) for each character transmitted; various combinations of data pulses are used to designate letters of the alphabet, numerals 0 to 9, and certain standard symbols.

baud rate Any of the standard transmission rates for sending or receiving binary coded data; standard rates are generally between 50 and 19,200 bauds.

Baumé scale Either of two specific gravity scales devised by French chemist Antoine Baumé in 1768 and often used to express the specific gravity of acids, syrups, and other liquids; for light liquids the scale is determined from the

43

formula: °Bé = (140/sp gr) − 130. For heavy liquids it is determined from: °Bé = 145 − (145/sp gr). 60°F is the standard temperature used.

Bauschinger effect The phenomenon whereby plastic deformation in one direction results in reduction of yield strength when stress is applied in an opposite direction.

Bayard-Alpert ionization gage Ionization vacuum gage using a tube with an electrode structure designed to minimize x-ray-induced electron emission from the ion collector.

B-basis The mechanical property value "B" is the value above which 90% of the test values are expected to fall.

BCD *See* binary coded decimal.

BCOMP *See* buffer complete.

BDC *See* buffered data channel.

beach marks A term used to describe the characteristic fracture markings produced by fatigue crack propagation. Known also as clamshell, conchoidal, and arrest marks.

bead *1.* A rolled or folded seam along the edge of metal sheet. *2.* A projecting band or rim. *3.* A drop of precious metal produced during cupellation in fire assaying. *4.* An elongated seam produced by welding in a single pass.

beaded tube end The rounded exposed end of a rolled tube when the tube metal is formed against the sheet in which the tube is rolled.

beads Small lumps in the coating surface. Particles of dross picked up in the coating of iron oxide particles embedded in the strip surface from the furnace hearth rolls.

bead weld A weld deposit that rises above the surface.

beam *1.* An elongated structural member that carries lateral loads or bending moments. *2.* A confined or unidirectional ray of light, sound, electromagnetic radiation, or vibrational energy, usually of relatively small cross section.

beam divergence The increase in beam diameter with increase in distance from a laser's exit aperture. Divergence, expressed in milliradians, is measured at specified points across the beam's diameter.

beam splitter *1.* Partially reflecting mirror which permits some incident light to pass through and reflect the remainder. *2.* A device that separates a light beam into two beams. Some types affect polarization of the beam.

beam spread The angle of divergence of an acoustic or electromagnetic beam from its central axis as it travels through a material.

bearing *1.* The angle, usually expressed in deg (0–360, clockwise), between the direction from an observer to an object or point and a reference line. 2. A machine part that supports another machine part while the latter undergoes rotating, sliding, or oscillating motion. 3. The portion of a beam, truss, or other structural member that rests on the supports.

bearing area The cross-sectional area of the bearing joint.

bearing bronze A generic term for bronzes used in many bearing applications.

bearing strain The ratio of the deformation of the bearing hole, in the direction of the applied force, to the pin diameter. Also, the stretch or deformation strain for a sample under bearing load.

bearing strength The maximum stress that a bearing area can withstand without failure.

bearing stress The shear load divided by the bearing area.

bearing yield strength The bearing stress at which a material exhibits a specified limiting deviation from the proportionality of bearing stress to bearing strain.

beats Periodic pulsations in amplitude that are created when a wave of one frequency is combined with a wave of a different frequency.

Beer's law The law relating the absorption coefficient to the molar density.

behavior The way in which an organism, organ, body, or substance acts in an environment or responds to excitation, as the behavior of steel under stress, or the behavior of an animal in a test.

behavioral modeling Modeling a device or component directly in terms of its underlying mathematical equations.

behind the panel *1.* A term applied to a location that is within an area that contains (a) the instrument panel, (b) its associated rack-mounted hardware, or (c) is enclosed within the panel. *2.* Describes devices that are not accessible for the operator's normal use. 3. Describes devices that are not designated as local or front-of-panel-mounted.

Beilby layer An amorphous layer of surface metal remaining after mechanical working.

Beilstein test A chemical analysis test to determine the presence of chlorine in polymers.

bel A dimensionless unit for expressing the ratio of two power levels; the value in bels equals $\log (P_2/P_1)$, where P_1 and P_2 are the two power levels.

bellmouth *See* nozzle.

bellows *1.* A convoluted unit consisting of one or more convolutions, used to obtain flexibility. *2.* An enclosed chamber with pleated or corrugated walls so that its interior volume may be varied, either to alternately draw in and expel a gas or other fluid, or to expand and contract in response to variations in internal pressure.

bellows gage A pressure-measuring device in which variations in internal pressure within a flexible bellows causes movement of an end plate against spring force.

bellows meter A differential pressure-measuring instrument having a measuring element of opposed metal bellows, the motion of which positions the output actuator.

bench check A laboratory-type test of an assembly, component, or subassembly to verify its function or identify a source of malfunction, often done with the unit removed from its housing or system for service or repair. Also known as a bench test.

bench mark A natural or artificial object having a specific point marked to identify a reference location, such as a reference elevation.

benchmark program A routine used to determine the performance of a computer or software.

bench test *See* bench check.

bending Applying mechanical force or pressure to form a metal part by plastic deformation around an axis lying parallel to the metal surface. Commonly used to produce angular, curved, or flanged parts from sheet metal, rod, or wire.

bending moment The force times the distance from a reference point to the point the force is applied causing bending.

bending stiffness The sandwich property that resists bending deflections.

bending-twisting coupling A property of certain classes of laminates that exhibit twisting curvatures when subjected to bending moments.

bend loss Attenuation caused by high-order modes radiating from the side of a fiber. The two common types of bend losses are: (a) those occurring when the fiber is curved around a restrictive radius of curvature and (b) microbends caused by small distortions of the fiber imposed by externally induced perturbations, such as poor cabling techniques.

bend test Ductility test in which a metal or plastic specimen is bent through a specified arc around a support of known radius; used primarily to evaluate inherent formability of metal sheet, rod, wire, or plastic, and to evaluate weld quality produced with specific materials, joint design, and welding technique.

benzene An aromatic hydrocarbon with a ring structure of six carbon atoms.

BER *See* bit error rate.

Bernoulli coefficient Dimensionless coefficient that relates the change in the velocity and the corresponding change in static pressure, or "head," in a stream when the area is changed, as by a reducer. The pressure change is measured in units of velocity head.

Bernoulli theorem In aeronautics, a law or theorem stating that in a flow of incompressible fluid the sum of the static pressure and the dynamic pressure along a streamline is constant if gravity and frictional effects are disregarded.

beryllium A metal lighter than aluminum, non-magnetic, and characterized by good electrical conductivity and high thermal conductivity. Beryllium is used in alloys, especially beryllium copper alloy.

beryllium copper alloys Alloys that can be precipitation hardened to provide a very hard product; used in the manufacture of springs and nonsparking tools.

beryllium nickel A precipitation hardened alloy with similar characteristics to beryllium copper.

best straight line A line midway between the two parallel straight lines closest together and enclosing all output versus measurand values on a calibration curve.

best-straight-line linearity Also called independent linearity. An average of the deviation of all calibration points.

beta The allotrope of titanium with a body-centered cubic crystal structure occurring at temperatures between the solidification temperature of molten titanium and the beta transus.

beta cenidendorl An additive to acetals to prevent oxidation at elevated temperatures.

beta eucryptite Lithium aluminum silicate that is formed into glass-ceramic products.

beta eutectoid stabilizer An alloying element that dissolves preferentially in the beta phase, lowers the alpha-beta to beta transformation temperature, under equilibrium conditions, and results in the beta decomposition to alpha plus a compound.

beta factor In plasma physics, the ratio of the plasma kinetic pressure to the magnetic pressure.

beta fleck Transformed alpha-lean and/or beta-rich region in the alpha-beta microstructure. This area has a beta transus measurably below that of the matrix. Beta flecks have reduced amounts of primary alpha that may exhibit a morphology different from the primary alpha in the surrounding alpha/beta matrix.

beta isomorphous stabilizer An alloying element that is soluble in beta titanium in all proportions. It lowers the alpha-beta to beta transformation temperature without a eutectoid reaction and forms a continuous series of solid solutions with beta titanium.

beta particle An electron or positron emitted from the nucleus of a radioactive nuclide.

beta ratio The ratio of the diameter of the constriction to the pipe diameter, B = D_{const}/D_{pipe}.

beta ray A stream of beta particles.

beta-ray spectrometer An instrument used to measure the energy distribution in a stream of beta particles or secondary electrons.

beta spodumene Lithium aluminum silicate that is formed into glass-ceramic products resistant to thermal shock.

beta test The second stage of testing a new software program.

betatizing Forming beta constituent by heating a non-ferrous alloy into the temperature region in which the constituent forms.

beta transus In titanium alloys, the temperature that designates the phase boundary between the alpha plus beta and beta fields. Commercially pure grades transform in a range of 130 to 1760°F (890 to 960°C) depending on oxygen and iron content. In general, aircraft alloys vary in transformation temperature from 1380 to 1900°F (750 to 1040°C).

betatron Particle accelerator in which magnetic induction is used to accelerate electrons.

bezel A ring-shaped member surrounding a cover glass, window, cathode-ray tube face, or similar area to protect its edges and often to also provide a decorative appearance.

Bezold-Brucke effect Refers to the effect when luminance is increased, all chromatic colors, except a certain invariable blue, yellow, green, and red, appear increasingly like blue or yellow and decreasingly like green or red.

B-H meter An instrument used to determine the intrinsic hysteresis loop of a magnetic material.

bias *1.* To influence or dispose to one direction, for example, with a direct voltage or with a spring. *2.* A constant or systematic error as opposed to a random error. Bias manifests itself as a persistent positive or negative deviation of reference value. *3.* The departure from a reference value of the average of a set of values; thus, a measurement of the amount of unbalance of a set of measurements or conditions; error having an average value that is non-zero. *4.* The average d-c voltage or current maintained between a control electrode and the common electrode in a transistor or vacuum tube. *5.* A recurring error inherent in some measuring systems.

bias fabric Warp and fill fibers woven at an angle to the fabric length.

biaxle orientation In composites, the axes of reinforcing fibers lying in two direc-

47

tions at 90° to each other. In polymers, a polymer chain orientation in both longitudinal and transverse directions.

bidirectional laminate A plastic laminate with fibers aligned in two different directions in the same plane.

bidirectional load cell A column-type strain-gage load cell with female or male fittings at both ends for attaching load hardware; it can be used to measure either tension or compression loading. Also known as a universal load cell.

Bielby layer An amorphous layer at the surface of mechanically polished metal.

bilateral Of or having two sides; affecting two sides equally; reciprocal. 2. Fabric used in composite laminates which has substantially equal strength in both the warp and fill directions.

bilateral tolerance The amount of allowable variation about a given dimension, usually expressed as plus-or-minus a specific fraction or decimal.

billet *1.* A semifinished primary mill product ordinarily produced by hot rolling or pressing metal ingot to a cylinder or prism of simple cross-sectional shape and metallurgically appropriate cross-sectional area. *2.* A general term for the starting stock used to make forgings and extrusions.

bimetal *1.* A type of bearing construction in which a single layer of bearing material is bonded to a steel backing. A common example of this is a babbitt bearing. *2.* A bonded laminate consisting of two strips of dissimilar metals; the bond is usually a stable metallic bond produced by corolling or diffusion bonding; the composite material is used most often as an element for detecting temperature changes by means of differential thermal expansion in the two layers.

bimetallic corrosion A type of accelerated corrosion induced by differences in galvanic potential between dissimilar metals immersed in the same liquid medium (electrolyte) and also in electrical contact with each other.

bimetallic element A temperature-sensitive device composed of two materials having different thermal coefficients of expansion resulting in a proportional movement of the free segment of the device with changes in temperature.

bimetallic thermometer element A temperature-sensitive strip of metal (or other configuration) made by bonding or mechanically joining two dissimilar strips of metal together in such a manner that small changes in temperature will cause the composite assembly to distort elastically, and produce a predictable deflection; the element is designed to take advantage of the fact that different metals have different coefficients of thermal expansion.

bin activator A vibratory device sometimes installed in the discharge path of a mass-flow bin or storage hopper to promote steady discharge of dry granular material.

binary *1.* A characteristic or property involving a selection, choice, or condition in which there are but two possible alternatives. *2.* A computer numbering system that uses two as its base rather than ten. The binary system uses only 0 and 1 in its written form. *3.* A device that uses only two states or levels to perform its functions, such as a computer.

binary alloy A metallic material composed of only two chemical elements (neglecting minor impurities), at least one of which is a metal.

binary cell An information-storage element that can assume either of two stable conditions, and no others.

binary code A code that uses two distinct characters, usually 0 and 1.

binary coded decimal (BCD) Describes a decimal notation in which the individual decimal digits are represented by a group of binary bits. For example, in the 8-4-2-1 coded decimal notation, each decimal digit is represented by a group of four binary bits. The number twelve is represented as 0001 0010 for 1 and 2, respectively, whereas in binary notation it is represented as 1100. Related to binary.

binary-coded decimal system A system of number representation in which each digit in a decimal number is expressed as a binary number.

binary counter *1.* A counter that counts according to the binary number system. *2.* A counter whose basic counting elements are capable of assuming one of two stable states.

binary digit *1.* In binary notation, either of the characters 0 or 1. *2.* Same as bit. *3. See* equivalent binary digits.

binary distillation A distillation process that separates only two components.

binary notation A numbering system using the digits 0 and 1, with a base of 2.

binary number A number composed of the characters 0 and 1, in which each character represents a power of two. The number 2 is 10; the number 12 is 1100; the number 31 is 11111, etc.

binary point The radix point in a binary number system.

binary scaler A signal-modifying device (scaler) with a scaling factor of 2.

binary unit *1.* A binary digit. *2.* A unit of information content, equal to one binary decision, or the designation of one of two possible and equally likely values or states of anything used to store or convey information. *3. See* check bit and parity bit. *4.* Same as bit.

binary word A group of binary digits with place values in increasing powers of two.

binder *1.* In metal founding, a material other than water added to foundry sand to make the particles stick together. 2. In powder metallurgy, a substance added to the powder to increase green strength of the compact, or a material (usually of relatively low melting point) added to a powder mixture to bond particles together during sintering which otherwise would not bond into a strong sintered body. *3.* A resin that holds plastic components, such as a laminate, together.

Bingham body A non-Newtonian substance that exhibits true plastic behavior—that is, it flows when subjected to a continually increasing shear stress only after a definite yield point has been exceeded.

Bingham viscometer A time-of-discharge device for measuring fluid viscosity in which the fluid is discharged through a capillary tube instead of an orifice or nozzle.

biochemistry Chemistry dealing with the chemical processes and compounds of living organisms.

biocompatible In considering a material for use as a human prostheses, an expression meaning the material would not tend to be rejected by the body. Included among these materials are certain stainless steels, certain titanium alloys, and a special 35% Ni, 35% Co, 20% Cr, 10% Mo alloy.

bioconversion The transformation of algae and/or other biomass materials in successive stages to aliphatic organic acids to aliphatic hydrocarbons to diesel and/or other liquid fuels.

biodegradability The characteristic of a substance that can be decomposed by microorganisms.

biological corrosion Deterioration of metal surfaces due to the presence of plant or animal life; deterioration may be caused by chemicals excreted by the life form, or by concentration cells such as those under a barnacle, or by other interactions.

biomedical engineering The application of engineering principles to the solution of medical problems, including the design and fabrication of prostheses, diagnostic instrumentation, and surgical tools.

bionics The study of systems, particularly electronic systems, which function after the manner characteristic of, or resembling living systems.

bioreactors Biological processors to remove or produce certain chemicals or a particular chemical.

biotite A widely distributed and important rock-forming mineral of the mica group.

Biot number A standard heat-transfer dimensionless number.

bipolarity Capability of assuming negative or positive values.

bipolar technology Technology that uses two different polarity electrical signals to represent logic states of 1 and 0.

biquinary code A method of coding decimal digits in which each numeral is coded in two parts—the first being either 0 or 5, and the second any value from 0 to 4; the digit equals the sum of the two parts.

birefringence *1.* The difference between the two principal refractive indices. *2.* The ratio between the retardation and thickness of a material at a given point.

Birmingham wire gage A system of standard sizes used in the United States for brass wire, and for strips, bands, hoops, and wire made of ferrous and nonferrous metals; the decimal equivalent of standard Bwg sizes is generally larger than for the same gage number in both the American wire gage and U.S. steel wire gage systems.

biscuit *1.* A piece of pottery that has been fired but not glazed. *2.* An upset blank for drop forging. *3.* A small cake of primary metal, generally one produced by bomb reduction or a similar process.

bismaleimide (BMI) A polyimide that cures by an addition reaction, which avoids a formation of undesirable volatiles.

bisphenol A monomer used in the production of polycarbonates and epoxy resins.

bistable The capability of assuming either of two stable states, hence of storing one bit of information.

bit *1.* A cutting tool for drilling or boring. *2.* The blade of a cutting tool such as a plane or ax. *3.* A removable tooth of a saw, milling cutter, or carbide-tipped cutting tool. *4.* The heated tip of a soldering iron. *5.* An abbreviation of binary digit. *6.* A single character in a binary number. *7.* A single pulse in a group of pulses. *8.* A unit of information capacity of a storage device. The capacity in bits is the logarithm to the base two of the number of possible states of the device. Related to storage capacity. *9.* The smallest unit of information that can be recognized by a computer.

bit density A measure of the number of bits recorded per unit of length or area.

bit error rate The number of erroneous bits or characters received from some fixed number of bits transmitted.

bit error rate tester A system that measures the fraction of bits transmitted incorrectly by a digital communication system.

bit map A table that describes the state of each member of a related set; bit map is most often used to describe the allocation of storage space; each bit in the table indicates whether a particular block in the storage medium is occupied or free.

bit pattern A combination of n binary digits to represent 2^n possible choices, for example, a 3-bit pattern represents 8 possible combinations.

bit rate *1.* The speed at which bits are transmitted, usually expressed in bits per second. Compare with baud. *2.* The rate at which binary digits, or pulses representing them, pass a given point on a communications line or channel. Clarified by baud.

bits per second In a serial transmission, the instantaneous bit speed within one character, as transmitted by a machine or a channel. *See* baud.

bit stream A binary signal without regard to grouping by character.

bit string A string of binary digits in which each bit position is considered as an independent unit.

bit synchronizer A hardware device that establishes a series of clock pulses in synchronism with an incoming bit stream and identifies each bit.

bituminous Describes a substance that contains organic matter, mostly in the form of tarry hydrocarbons (described as bitumen).

bituminous coating Coal tar or asphalt based coating.

black annealing Box annealing or pot annealing ferrous alloy sheet, strip, or wire. *See* box annealing.

black body *1.* A body that has an absorption factor of 1 for all wavelengths of interest. The radiation from a black body is a function solely of its temperature and is useful as a standard for light measurements. 2. A physical object that absorbs incident radiation, regardless of spectral character or directional preference of the incident radiation; a perfect black body is most closely approximated by a hollow sphere with a small hole in its wall—the plane of the hole being the black body. *3.* Denotes a perfectly absorbing object, from which none of the incident energy is reflected.

black box A generic term used to describe an unspecified device which performs a special function or in which known inputs produce known outputs in a fixed relationship.

blackening *See* black oxide.

black lining Dark lines that appear along the tops of ridges in the sealer due to paint flowing down the sides of the ridges.

black liquor *1.* The solution remaining after cooking pulpwood in the soda or sulfite papermaking process. *2.* A black, iron-acetate solution containing 5 to 5.5% Fe, and sometimes tannin or copperas, used in dyes and printing inks.

black oxide A black finish produced by dipping metal in oxidizing salts. Also called blackening.

black panel thermometer A temperature device used to provide an estimation of the maximum temperature a specimen may attain during exposure to natural or artificial light. In practice, this device usually consists of a black coated or anodized metal panel with a platinum RTD, thermocouple, or other temperature sensor attached to the panel.

black spots Carbonaceous deposits caused by tightly wound areas in a coil not being exposed to the circulating gases during open coil type annealing. This condition is aggravated by poor strip shape.

black standard thermometer A temperature-measuring device that uses a resistance thermometer with good heat-conducting properties, fitted to the reverse side of a metal plate. The metal plate is fixed to a plastic plate so that it is thermally insulated. It is coated with a black layer.

blank *1.* A flat piece of sheet steel produced in blanking dies or by shearing for an identified part. The blank is usually subjected to further press operations. *2.* In computer programming, the character used to represent a space.

blank alarm point *See* alarm point.

blank carburizing Simulating the carburizing operation without introducing carbon. This is usually accomplished by using an inert material in place of the carburizing agent, or by applying a suitable protective coating to the ferrous alloy.

blank common *See* global common.

blanking *1.* Inserting a solid disc at a pipe joint or union to close off flow during maintenance, repair, or testing. *2.* Using a punch and die to cut a shaped piece from sheet metal or plastic for use in a subsequent forming operation. *3.* Using a punch and die to make a semifinished powder-metal compact.

blank nitriding Simulating the nitriding operation without introducing nitrogen. This is usually accomplished by using an inert material in place of a nitriding agent, or by applying a suitable protective coating to the ferrous alloy.

blasting *1.* Detonating an explosive. *2.* Using abrasive grit, sand, or shot carried in a strong stream of air or other medium to remove soil or scale from a surface.

bleeding *1.* Allowing a fluid to drain or escape to the atmosphere through a small valve or cock; used to provide controlled slow reduction of slight overpressure, to withdraw a sample for analysis, to drain condensation from compressed air lines, or to reduce the airspace above the liquid level in a pressurized tank. *2.* Withdrawing steam from an intermediate stage of a turbine to heat a process fluid or boiler feedwater. *3.* Natural separation of liquid from a semisolid mixture, such as oil from a lubricating grease or water from freshly poured concrete. *4.* The migration of color out of a material onto the surrounding surface.

blend *1.* To mix ingredients so that they are indistinguishable from each other in the mixture. *2.* To produce a smooth transition between two intersecting surfaces, such as at the edges of a radiused fillet between a shaft and an integral flange or collar.

blind sample A sample whose composition is not known to the analyst, but is known by those who submitted it.

blip *1.* On radar screens, a streak of light caused by an object, vehicle, or some electronic disturbance passing through the path of the radar beam. *2.* Any erratic signal on a computer screen.

blister *1.* A small raised area on the surface resulting from the expansion of gas concentrated at a subsurface inclusion. May occur as isolated spots, but often found in longitudinal streaks. *2.* A raised cavity or sack that deforms a surface of the seal material.

blistering Raised areas on a painted surface due to volatile substances coming out of the sealer after the topcoat has started to "cure."

block *1.* A piece of material such as wood, stone, or metal, usually with one or more plane or approximately plane faces, used to strengthen or sustain. *2.* A set of things, such as words, characters, or digits, handled as a unit. *3.* A collection of contiguous records recorded as a unit; blocks are separated by interblock gaps, and each block may contain one or more records. *4.* In data communication, a group of contiguous characters formed for transmission purposes. The groups are separated by interblock characters.

block coat Tie coat (adhesive) between noncompatible paints.

block copolymer A polymer formed by addition polymerization consisting of alternating hard and soft blocks along the length of the molecular chain.

block, data A set of associated characters or words handled as a unit.

block diagram A graphical representation of the hardware in a computer system. The primary purpose of a block diagram is to indicate the paths along which information or control flows between the various parts of a computer system. Not to be confused with the term flow chart.

blocker-type forging A shape forging designed for easy forging and extraction from the die through the use of generous radii, large draft angles, smooth contours, and generous machining allowances.

block flow The distance that freshly mixed, uncured sealant will sag while hanging from a vertical surface for approximately on-half hour at standard conditions.

blocking *1.* In forging, a preliminary operation performed in closed dies, usually hot, to position metal properly so that in the finish operation the dies will be filled correctly. *2.* Reducing the oxygen content of the bath in an open-hearth furnace. *3.* Undesired adhesion between plastics surfaces during storage or use. *4.* (of computer records) *See* grouping.

block sequence A welding sequence in which separated lengths of a continuous multiple-pass weld are built up to full cross section before gaps between the segments are filled in. Compare with cascade sequence.

bloedite A mineral consisting of hydrous sodium magnesium sulfate that is colorless. Also known as astrakanite or astrochanite.

bloom *1.* A semifinished metal bar of large cross section (usually a square or rectangle exceeding 36 square inches) hot rolled or sometimes forged from ingot. It might also be called a billet. *2.* Visible fluorescence on the surface of lubricating oil or an electroplating bath. *3.* A bluish fluorescent cast to a painted surface caused by a thin film of smoke, dust or oil. *4.* A loose, flower-like corrosion product formed when certain nonferrous metals are exposed in a moist environment. *5.* To apply an antireflection coating to glass. *6.* To hammer or roll metal to brighten its surface. *7.* An undesirable finish resulting on the surface of a plastic during manufacture.

blowhole A pocket of air or gas trapped during solidification of a cast metal.

blowing agent An additive that provides escaping gas in the production of foam molded products.

blue annealing Heating hot rolled ferrous sheet in an open furnace to a temperature within the transformation range and then cooling in air, in order to soften the metal. The formation of a bluish oxide on the surface is incidental.

blue brittleness In some steels, loss of ductility associated with tempering or service temperatures in the blue heat range, typically 400–700°F (200–370°C).

blueing *See* bluing.

blue vitriol A solution of copper sulfate sometimes applied to metal surfaces to make scribed layout lines more visible.

blue wool light fastness standard One of a group of dyed fabrics used to determine the amount of light, or combined light and heat, to which a specimen is exposed during fade/weathering testing.

bluing Also spelled blueing. *1.* Subjecting the scale-free surface of a ferrous alloy to the action of air, steam, or other agents at a suitable temperature, thus forming a thin blue film of oxide and improving the appearance and resistance to corrosion. *2.* Heating formed springs after fabrication to improve their properties and reduce residual stress.

blushing Whitening and loss of gloss of a coating due to moisture; blooming. Hue may be lightened due to moisture in the air, moisture in the sealer, or poor hiding power of the topcoat.

BMC *See* bulk molding compound.

BMI *See* bismaleimide.

board In computers, a flat sheet on which integrated circuits are mounted. *See* panel.

Bode diagram *See* Bode plot.

Bode plot A graph of transfer function versus frequency whereby the gain (often in decibels) and phase (in degrees) are plotted against the frequency on log scale. Also called bode diagram.

body block The test device used to apply the seat belt load to the seat system.

body-centered structure A crystalline structure where one atom is positioned at each cell corner and one atom positioned in the center of the cell.

body, encapsulated A body with all surfaces covered by a continuous surface layer of a different material, usually an elastomeric or polymeric material.

body leakage The static airflow leakage rate of a test vehicle at both $+$ and -1 in H_2O (250 Pa).

Bohr magneton A constant equivalent to the magnetic moment of an electron.

bolometer Instrument that measures the intensity of radiant energy by employing thermally sensitive electrical resistors; a type of actinometer.

Boltzmann distribution The probability that a molecule of gas in equilibrium will have position and momentum coordinates within a given range of values.

Boltzmann's constant (k) A constant that is equal to 1.380×10^{-23} J/K.

bomb calorimeter An apparatus for measuring the quantity of heat released by a chemical reaction.

bomb test A form of leak test in which enclosures are immersed in a fluid which is then pressurized for the purpose of driving it through possible leak passages and thus into the internal cavities where its presence will usually cause some form of electrical disturbance.

bond *1.* An interconnection that performs a permanent electrical and/or mechanical function. *2.* To join with adhesives. *3.* The adhesion established by vulcanization between two cured elastomer surfaces, or between one cured elastomer surface and one non-elastomer surface. *4.* A wire rope that attaches a load to a crane hook. *5.* In an adhesive bonded or diffusion bonded joint, the junction between faying surfaces. *6.* In welding, brazing, or soldering, the junction between assembled parts; where filler metal is used, it is the junction between fused metal and heat-affected base metal. *7.* In grinding wheels and other rigid abrasives, the material that holds abrasive grains together. *8.* Material added to molding sand to hold the grains together. *9.* The junction between base metal and cladding in a clad metal product.

bonded-phase chromatography (BPC) Liquid chromatography with a chemically bonded stationary phase.

bonded strain gage A device for measuring strain which consists of a fine-wire resistance element, usually in zigzag form, embedded in nonconductive backing material such as impregnated paper or plastic, which is cemented to the test surface or sensing element.

bond energy The potential energy of a bond between contiguous atoms.

bonding *See* bond.

bond ply The ply of a prepreg material that is placed against the fluted core of a radome.

bond strength In wire bonding, the pull force at rupture of the bond interface.

bondtester An electronic device used for testing the integrity of an adhesive bond.

bone dry A papermaking term used to describe pulp fibers or paper from which all water has been removed. Also known as oven dry or moisture free.

bonfire test The exposure of inflators and module assemblies to fire or associated high temperatures to confirm structural integrity when auto ignition occurs. Used for shipping/handling evaluation and simulated exposure to car fires.

Boolean Pertaining to logic quantities.

Boolean algebra The study of the manipulation of symbols representing operations according to the rules of logic. Boolean algebra corresponds to an algebra using only the number 0 and 1, therefore can be used in programming digital computers which operate on the binary principle.

Boolean functions A system of mathematical logic often executed in circuits to provide digital computations such as "OR," "AND," "NOR," "NOT," etc.

Boolean operator A logic operator each of whose operands and whose result have one of two values.

Boolean variable See logical variable.

bore probe A test probe for eddy current inspection of tubing and holes.

bore Reynolds number Calculated Reynolds number including R_d using V_{bore}, P_{bore}, μ_{bore}, d_{bore}; also $R_d = R_D/\beta$.

boric oxide An oxide of boron used in manufacture of glasses and glazes.

boron fibers Fibers produced by vapor deposition methods; used in various composite materials to impart a balance of strength and stiffness.

boron carbide A ceramic material with high abrasion resistance.

boron counter tube A type of radiation counter tube used to detect slow neutrons.

boron nitride A ceramic material used in grinding wheels and in the manufacture of metal cutting tools.

borosilicate glass A type of heat-resisting glass that contains at least 5% boric acid with low thermal expansion and high chemical resistance.

bort Industrial diamonds or diamond fragments.

boss *1.* A raised portion of metal or small area and limited thickness on flat or curved metal surfaces. *2.* A short projecting section of a casting, forging, or molded plastics part, often cylindrical in shape, used to add strength or to provide for alignment or fastening of assembled parts.

Bouguer law A relationship describing the rate of decrease of flux density of a plane-parallel beam of monochromatic radiation as it penetrates a medium that both scatters and absorbs at that wavelength.

bounce motion Translational motion of the vehicle in a direction parallel to its nominal plane of symmetry (i.e., relative to body fixed axes). Vertical translation if the vehicle is upright.

boundary In microstructures, the interface between two crystal grains or phases. Commonly called grain boundary.

boundary layer In a flowing fluid, a low-velocity region along a tube wall or other boundary surface.

boundary value problem Physical problem completely specified by a differential equation in an unknown and certain information (boundary condition) about the unknown. The differential equation is valid in a certain region of space. The information required to determine the solution depends completely and uniquely on the particular problem.

bound water In a moist solid to be dried, the portion of the water content that is chemically combined with the solid matter.

Bourdon tube A flattened tube, twisted or curved, and closed at one end, which is used as the pressure-sensing element in a mechanical pressure gage or recorder.

Boussinesq approximation The assumption (frequently used in the theory of convection) that the fluid is incompressible except insofar as the thermal expansion produces a buoyancy.

box annealing The term given to the process of softening steel by heating it in a prescribed manner in a closed container through which, in most cases, a controlled atmosphere is circulated to prevent oxidation during the heating and cooling cycle.

B power supply An electrical power supply connected in the plate circuit of a vacuum tube electronic device.

BPS *See* bits per second. *See also* baud.

Bragg curve A curve showing the average specific ionization of an ionizing particle of a particular kind as a function of its kinetic energy, velocity, or residual range.

Bragg's law A principle describing the apparent reflection of x-rays (and DeBroglie waves associated with certain particulate beams) from atomic planes in crystals.

braid An assembly of fibrous or metallic filaments woven to form a protective and/or conductive covering.

braiding Weaving fibers in a tubular, rather than flat, shape.

Brale A 120° conical diamond indenter used in Rockwell hardness testing of relatively hard metals. This indenter is used for the A, C, D, and N Rockwell scales.

branch The selection of one of two or more possible paths in the control of flow based on some criterion.

branched polymer A polymer structure containing branches or protrusions along the main chain on an irregular basis.

branch instruction An instruction that performs a branch.

branchpoint A point in a routine where one of two or more choices is selected under control of the routine. *See* conditional transfer.

brass Any of the many alloys based on the binary system copper-zinc; most brasses contain no more than 40 wt% zinc.

Brayton cycle A thermodynamic cycle consisting of two constant-pressure processes interspersed with two constant-entropy cycles.

braze welding A joining process similar to brazing, but in which the filler metal is not distributed in the joint by capillary action.

brazing A joining process whereby a filler material that melts at a temperature in excess of 800°F but at less than the melting point of the base metal, combines with the base metal.

breadboard model A preliminary assembly of parts to test the feasibility of an item or principle without regard to eventual design or form. Usually refers to a small collection of electronic parts.

break *1.* An interruption in computer processing. *2.* To interrupt the sending end and take control of the circuit of the receiving end.

breakdown *1.* Initial hot working of ingot-cast or slab-cast metal to reduce its size prior to final working to finished size. *2.* A preliminary press-forging operation.

breaking extension *1.* The elongation required to cause rupture of a test specimen. *2.* The tensile strain at the instant of rupture.

breaking factor The breaking load divided by the initial width of the test specimen.

breakout Fiber separation or fraying at drilled or machined edges of composites.

break point *1.* The point in the pressure-time curve that is preceded by a pressure drop of exactly 2 psi (13.8 kPa) within 15 min and succeeded by a drop of not less than 2 psi (13.8 kPa) in 15 min. *2.* The junction of two confluent straight-line segments of a plotted curve. *Note*: In the asymptotic approximation of a log-gain versus log-frequency relation in a Bode diagram, the value of the abscissa is called the corner frequency. *3.* (a) Pertaining to a type of instruction, instruction digit, or other condition used to interrupt or stop a computer at a particular place in a routine when manually requested. (b) A place in a routine where such an interruption occurs or can be made to occur.

breakpoint instruction *1.* An instruction that will cause a computer to stop or to transfer control, in some standard fashion, to a supervisory routine which can monitor the progress of the interrupted program. *2.* An instruction that, if some specified switch is set, will cause the computer to stop or take other special action.

breaks Creases or ridges usually in "untempered" or in aged material where the yield point has been exceeded. Depending on the origin of the break, it might be termed a crossbreak, a coil break, an edge break, a sticker break, etc.

Bremsstrahlung *1.* Electromagnetic radiation produced by the rapid change in the velocity of an electron or another fast, charged particle as it approaches an atomic nucleus and is deflected by it. In German, it means braking radiation. *2.* X-rays having a broad spectrum of wavelengths; formed due to deceleration of a beam of energetic electrons as they penetrate a target. Also known as white radiation.

Brewster's Law Law stating that if the electric vibration of the incident waves are parallel to the plane of incidence, and if the angle of incidence is such that the angle between the reflected and refracted beams equals 90 deg, the entire beam is transmitted, and none is reflected.

bridge A network device that interconnects two local area networks that use the same LLC (logical link control), but may use different MACs (media access controls).

bridging *1.* Premature solidification of metal across a mold section before adjacent metal solidifies. *See also* pipe. *2.* Welding or mechanical jamming of the charge in a downfeed furnace. *3.* Forming an arched cavity in a powder metal compact. *4.* Forming an unintended solder connection between two or more conductors, either a secure connection or merely an undesired electrical path without mechanical strength. Also known as a crossed joint or solder short.

bright annealing Annealing in a protective medium to prevent discoloration of the surface.

bright dipping Producing a bright surface on metal, such as by immersion in an acid bath.

bright finish A high-quality finish produced on sheets by rolls which have been ground and polished. Suitable for electroplating.

brightness A term used in nonquantitative statements with reference to sensations and perceptions of light; in quantified statements with reference to the description of brightness by photometric units.

brightness distribution The statistical distribution based on brightness, or the distribution of brightness over the surface of an object.

bright plating Electroplating to yield a highly reflective coated surface.

brine Water saturated or strongly impregnated with common salt.

Brinell hardness test A standard bulk hardness test in which a 10-mm-diameter ball is pressed into the surface of a test piece. A hardness number is determined by dividing applied load in kg by the area of the circular impression in sq mm.

briquetting Producing relatively small lumps or block of compressed granular material, often incorporating a binder to help hold the particles together.

British thermal unit 1/180 of the heat required to raise the temperature of 1 pound of water from 32°F to 212°F at a constant atmospheric pressure. It is about equal to the quantity of heat required to raise 1 pound of water 1°F. A Btu is essentially 252 calories.

brittle Describes materials that yield less than 0.5% before fracturing. Such materials (for example, ceramics) usually have a much greater compressive strength than tensile strength. The brittle material fails by fracture, not by yielding.

brittle fracture Separation of solid material with little or no evidence of macroscopic plastic deformation, usually by rapid crack propagation involving less energy than for ductile fracture of a similar structure.

brittleness The tendency of a material to fracture without apparent plastic deformation.

broaching Cutting a finished hole or contour in solid material by axially pulling or pushing a bar-shaped, toothed, tapered cutting tool across a workpiece surface or through a pilot hole.

bronze *1.* A copper-rich alloy of copper and tin, with or without small amounts of additional alloying elements. *2.* By extension, certain copper-base alloys containing less tin than other elements, such as manganese bronze and leaded tin bronze, and certain other copper-base alloys that do not contain tin, such as aluminum bronze, beryllium bronze and silicon bronze. 3. Trade names for certain copper-zinc alloys (brasses), such as architectural bronze (Cu-40Zn-3Pb) and commercial bronze (Cu-10Zn).

bronzing *1.* Plating a bronze alloy on another metal. *2.* Chemical finish of copper or bronze alloys to change the color.

brushes (electrical contacts) Conductive metal or carbon blocks used to make sliding electrical contact with a moving part, as in an electric motor.

brush plating An electroplating process in which the surface to be plated is not immersed, but rather rubbed with an electrode containing an absorbent pad or brush that holds (or is fed) a concentrated electrolyte solution or gel.

B-scan display A method in which results of ultrasonic tests are displayed. It provides a display of the side elevation of the test piece showing the position of any defects.

B-stage An intermediate stage in the reaction of thermal setting resins when the material can be softened by heating and is fusible.

Btu *See* British thermal unit.

bubble *1.* A small volume of steam enclosed within a surface film of water from which it was generated. *2.* Any small volume of gas or vapor surrounded by liquid; surface-tension effects tend to make all bubbles spherical unless they are acted upon by outside forces.

bubble gas Any gas selected to bubble from the end of a tube immersed in liquid for level measurement from the hydrostatic back pressure created in the tube.

bubble tube A length of pipe or tubing placed in a vessel at a specified depth to transport a gas injected into the liquid to measure level from hydrostatic back pressure in the tube.

buckle *1.* Localized waviness in a metal bar or sheet, usually transverse to the direction of rolling. *2.* An indentation in a casting due to expansion of molding sand into the cavity.

buckle line A line of collapsed honeycomb cells, two to three cells wide, with undistorted cells on either side.

Buckley gage A device that measures very low gas pressure by sensing the amount of ionization produced by a prescribed electric current.

buckling Producing a lateral bulge, bend, bow, kink, or wavy condition in a metal or composite beam, bar, column, plate, or sheet by applying compressive loading.

buffer *1.* A substance designed to minimize the chemical effects of other substances in a mixture. *2.* An isolating component designed to eliminate the reaction of a driven circuit on the circuits driving it, for example, a buffer amplifier.

buffer complete (BCOMP) In TELEVENT, the signal that indicates when the computer buffer is complete.

buffered computer A computing system with a storage device which permits input and output data to be stored temporarily in order to match the slow speeds of input and output devices.

buffered data channel A device that provides high-speed parallel data interfaces into and out of the computer memory.

buffer storage In computer operations, storage used to compensate for a difference in rate of flow or time of occurrence when transferring information from one device to another.

buffing Producing a very smooth and bright surface by rubbing it with a soft wheel, belt, or cloth impregnated with fine abrasive such as jeweler's rouge.

bug An error, defect, or malfunction in a computer program.

buildup *1.* Excessive electrodeposition on areas of high current density, such as at corners and edges. *2.* Small amounts of work metal that adhere to the cutting edge of a tool and reduce its cutting efficiency. *3.* Deposition of metal by electrodeposition or spraying to restore required dimensions of worn or undersize machine parts. *4.* Localized lineal areas showing a difference in cross-sectional contour during coiling. Usually occurs on the edges of the strip.

bulb *See* envelope.

bulge A local distortion of swelling outward caused by internal pressure on a tube wall or boiler shell caused by overheating or defective material.

bulk density Mass per unit volume of a bulk material, averaged over a relatively large number of samples.

bulk factor The ratio of the volume of a raw molding compound or powdered plastic to the volume of the finished solid piece produced from it.

bulk memory *See* secondary storage.

bulk modulus An elastic modulus determined by dividing hydrostatic stress by the associated volumetric strain (usually computed as the fractional change in volume).

bulk molding compound (BMC) A thermosetting resin molding compound reinforced with various fibers and used to manufacture high-strength components.

bulk sampling Testing a portion of a material that is typical of the whole lot.

bulk storage A hardware device in a computer system that supplements computer memory; typically, a magnetic tape or disk.

bulk volume The volume of a solid material including the open and closed internal pores.

bullion *1*. A semirefined alloy containing enough precious metal to make its recovery economically feasible. *2*. Refined gold or silver, ready for coining.

bull's eye structure The microstructure of cast iron when graphite nodules are encircled by a ferrite layer.

Buna-N A nitrile synthetic rubber known for resistance to oils and solvents. *See also* nitrile rubber.

bundle Parallel reinforcement fibers or filaments.

buoyancy The tendency of a fluid to lift any object submerged in the body of the fluid; the amount of force applied to the body equals the product of fluid density and volume of fluid displaced.

buret A tubular instrument used to deliver specific volumes of a liquid during titration or volumetric analysis.

burner *1*. Any device for producing a flame using liquid or gaseous fuel. *2*. A device in the firebox of a fossil-fuel-fired boiler that mixes and directs the flow of fuel and air to give rapid and complete combustion. *3*. A worker who cuts metal using an oxyfuel-gas torch.

burning Permanently damaging a metal or alloy by heating to cause either incipient melting or intergranular oxidation. *See* overheating.

burnish *1*. To polish or make shiny. *2*. Specifically, to produce a smooth, lustrous surface finish on metal parts by tumbling them with hardened metal balls or rubbing them with a hard metal pad.

burr *1*. A thin, turned over edge or fin produced by a grinding wheel, cutting tool,

or punch. *2*. A rotary tool having teeth similar to those on a hand file.

burst *1*. Surface or internal fissures or ruptures caused by metal movement during rolling or forging. *2*. A period in which data are present on the network which is preceded and succeeded by periods in which data are not present. *3*. A transient with multiple spikes.

burst strength *1*. Measurement of the ability of a formed component or test specimen to withstand internal pressure before rupture. *2*. The pressure required to rupture the component or test specimen.

burst test A hydrostatic test to establish the safety factor of structural housing (that is, inflator housing).

bus *1*. A group of wires or conductors, considered as a single entity, which interconnects parts of a system. *2*. In a computer, signal paths such as the address bus, the data bus, etc. *3*. A circuit over which data or power is transmitted; often one that acts as a common connection among a number of locations. Synonymous with trunk. *4*. A communications path between two switching points.

bus cycle The transfer of one word or byte between two devices.

bushing *1*. A removable piece of soft metal or impregnated sintered-metal sleeve used as a bearing or guide. *2*. A ring-shaped device made of ceramic or other nonconductive material used to support an electrical conductor while preventing it from becoming grounded to the support structure.

busing The joining of two or more circuits to provide a common electrical connection.

butadiene An unsaturated monomer which can be polymerized to produce synthetic rubber, and copolymerized with other substances to produce a wide range of thermoplastic molding compounds.

buttering Coating the faces of a weld joint prior to welding to preclude cross contamination of a weld metal and base metal.

butt joint A joint between two members lying approximately in the same plane; in welded joints, the edges may be machined or otherwise prepared to create any of several types of grooves prior to welding.

butt weld A weld that joins the edges or ends of two pieces of metal having similar cross sections, without overlap or offset along the joint line.

butylene An unsaturated hydrocarbon monomer which is copolymerized with isoprene to produce butyl rubber.

butyl rubber An elastomer with good abrasion and tear resistance and good resistance to acids and alkalis.

bypass component characteristic test A test that determines the performance of bypass relief valve with respect to its leakage rate, opening and reseat pressures, and resistance to flow.

bypass valve leakage test A test to determine the flow rate, expressed as volume per unit of time, through the closed valve, at specified static pressure head on the valve.

bypass valve opening pressure test A test to determine the pressure required to open the valve to permit flow at a specified minimum rate.

bypass valve reseat pressure test A test to determine the pressure at which the valve closes and restricts the flow rate to a level below a specified minimum rate as the pressure across the valve decreases.

byproduct Incidental or secondary output of a chemical production or manufacturing process that is obtained in addition to the principal product with little or no additional investment or allocation of resources.

byte *1.* Generally accepted as an eight-bit segment of a computer word. *2.* Eight contiguous bits starting on an addressable byte boundary; bits are numbered from the right, 0 through 7, with 0 the low-order bit. *3.* A collection of eight bits capable of representing an alphanumeric or special character.

C

C *1.* High-level programming language. *2.* Chemical symbol for carbon. *3.* The SI symbol for coulomb, a unit of electrical charge.

Ca Chemical symbol for calcium.

CA *See* constant amplitude. *See also* cellulose acetate.

CAB *See* cellulose acetate butyrate.

cable A composite electrical conductor consisting of one or more solid or stranded wires usually capable of carrying relatively large currents, covered with insulation and the entire assembly encased in a protective overwrap.

cadmium mercury A silvery white metal used as a plating material for corrosion protection of steel, and as an alloying element in some low melting alloys.

cadmium mercury tellurides *See* mercury-cadmium tellurides.

cadmium plating An electroplated coating of cadmium on a steel surface which resists atmospheric corrosion. Applications include nuts, bolts, screws, and many hardware items in addition to enclosures.

caesium A reactive alkali metal that reacts actively with air and water, and is used as a source of radium in radiography.

cage A circular frame for maintaining uniform separation between balls or rollers in a rolling-element bearing. Also known as separator.

cake A coalesced mass of metal powders.

caking Producing a solid mass from a slurry or mass of loose particles by any of several methods involving filtration, evaporation, heating, pressure, or a combination of these.

calcine *1.* To heat a material such as coke, limestone, or clay without fusing it, for the purpose of decomposing compounds such as carbonates and driving off volatiles such as moisture, trapped gases, and water of hydration. *2.* To heat a material under oxidizing conditions. *3.* The product of a calcining or roasting process.

calcite A colorless or white mineral form of calcium carbonate.

calcium Used as a reducing agent in the preparation of metals such as Th, V, Zr, as a deoxidizer, and as an alloying agent.

calcium silicates Silicates of calcium used in the manufacture of ceramics and tiles.

calculation A group of numbers and mathematical symbols that is executed according to a series of instructions.

calculus of variations The theory of maxima and minima of definite integrals whose integrand is a function of the dependent variables, the independent variables, and their derivatives.

calderas Large, basin-shaped volcanic depressions, more or less circular in form, the diameter of which is many times greater than that of the included vent or vents.

calefaction *1.* A warming process. *2.* The resulting warmed condition.

calender *1.* To pass a material such as rubber or paper between rollers or plates to make it into sheets or to make it smooth and glossy. *2.* A machine for performing such an operation.

calibrate To check or adjust the graduations of a quantitative measuring instrument.

calibrated flow In a unit that controls or limits rate or quantity of flow, the rate or quantity of flow for which the unit is calibrated or adjusted.

calibrating gas A precisely analyzed gas of known concentration, used to determine the response curve of an analytical instrument.

calibrating tank A liquid vessel of known capacity which is used to check the volumetric accuracy of positive-displacement meters.

calibration A test during which known values of input are applied to the device and corresponding output readings are recorded under specified conditions. To calibrate an instrument is to prepare it to provide meaningful measurements.

calibration curve A plot of indicated value versus true value used to adjust instrument readings for inherent error.

calibration cycle The frequency with which a device is due for calibration. This cycle could be dependent on time duration or cycles of use.

calibration gas A mixture of gases of specified and known composition used as the basis for interpreting analyzer response in terms of the concentration of the gas to which the analyzer is responding.

calibration hierarchy The chain of calibrations that links or traces a measuring instrument to a primary standard.

calibration interval, period, or cycle The time during which the instrument should remain within specific performance levels, with a specified probability, under normal conditions of handling and use.

calibration procedure The specific steps and operations to be followed by personnel in the performance of an instrument calibration

calibration record *1.* A record of the measured relationship of the transducer output to the applied input over the transducer range. *2.* A historical record of when calibration of an instrument was conducted.

calibration speed A specified speed for purposes of output voltage standardization and reference.

calibration value The value measured and read during the calibration of data channel.

caliper A gaging device with at least one adjustable jaw used to measure linear dimensions such as lengths, diameters, and thicknesses.

calk *See* caulk.

calorie 1/100 of the heat required to raise the temperature of 1 gram of water from 0°C to 100°C at a constant atmospheric pressure. It is about equal to the quantity of heat required to raise one gram of water 1°C. More recently, a calorie is defined 3600/860 joules, a joule being the amount of heat produced by a watt in one second.

calorific value The number of heat units liberated per unit of quantity of a fuel burned in a calorimeter under prescribed conditions.

calorimeter An instrument designed to measure heat evolved or absorbed.

calorimetric detector A detector that operates by measuring the amount of heat absorbed; incident radiation must be absorbed as heat to be detected.

calorize To produce a protective coating of aluminum and aluminum-iron alloys on iron or steel (or, less commonly, on brass, copper or nickel); the calorized coating is protective at temperatures up to about 1800°F.

CALS Computer-aided acquisition and logistic support.

cam A machine element that produces complex, repeating translational motion in a member known as a follower which slides or rolls along a shaped surface or in a groove that is an integral part of the cam.

camber *1.* Deviation from edge straightness usually referring to the greatest deviation of side edge from a straight line. *2.* Sometimes used to denote crown in rolls where the center diameter has been increased to compensate for deflection caused by the rolling pressure.

camber angle *1.* The inclination of the wheel plane to the vertical. Camber angle is considered positive when the wheel leans outward at the top and negative when it leans inward. *2.* The roll or tilt angle of the plane of a wheel, measured relative to vertical.

camera tube An electron-beam tube in which an optical image is converted to an electron-current or charge-density image, which is scanned in a predetermined pattern to provide an electrical output signal whose magnitude corresponds to the intensity of the scanned image.

cam follower The output link of a cam mechanism.

Campbell bridge A type of a-c bridge used to measure mutual inductance of coil or other inductor in terms of a mutual inductance standard.

camshaft The rotating member that drives a cam.

can A metal vessel or container, usually cylindrical, and usually having an open top or removable cover.

candela (cd) Equal to the obsolete units of candles and candlepower. Used to express the intensity of light visible to the human eye. Defined as 1/60 of the intensity of 1 square cm of a black-body radiator at the temperature of solidification of platinum (2046 K), and emitting one lumen per steradian.

candle The unit of luminous intensity. Defined as 1/60 of the intensity of 1 square cm of a black-body radiator at the temperature of solidification of platinum (2046 K). *See* candela.

candlepower *See* candela.

canned *1.* Describes a pump or motor enclosed within a watertight casing. *2.* Describes a composite billet or slab consisting of a reactive metal core encased in metal that is relatively inert so that the reactive metal may be hot worked in air by rolling, forging, or extrusion without excessive oxidation.

cannibalize To disassemble or remove parts from one assembly and use the parts to repair other, like assemblies.

canning A dish-like distortion on a flat surface.

cantilever A beam or other structural member fixed at one end and hanging free at the other end.

capability *1.* A measure of the ability of an item to achieve mission objectives given the conditions during the mission. *2.* The spread of performance of a process in a state of statistical control; the amount of variation from common causes identified after all special causes of variation have been eliminated.

capacitance The ability of a condensor to store a charge before the terminals reach a potential difference of one volt. The greater the capacitance, the greater the charge that can be stored.

capacitance meter An instrument for determining electrical capacitance of a circuit or circuit element. *See also* microfaradmeter.

capacitance-voltage characteristics The characteristics of a metal semiconductor contact or a semiconductor junction that manifests a measured capacitance as a function of a d-c bias voltage with small, superimposed a-c voltage applied to that junction or contact.

capacitive instrument A measuring device whose output signal is developed by varying the capacitive reactance of a sensitive element.

capacitor A device used for storing an electrical charge.

capacity factor The ratio of the average load carried to the maximum design capacity.

capacity lag In any process, the amount of time it takes to supply energy or material to a storage element at one point in the process from a storage point elsewhere in the process. Also known as transfer lag.

capillary *1.* Having a very small internal diameter. *2.* A tube with a very small diameter.

capillary action *1.* Spontaneous elevation or depression of a liquid level in a fine hair-like tube when it is dipped into a body of the liquid. *2.* Action induced by differences in surface energy between the liquid and the tube material.

capillary attraction The force of a liquid to infiltrate the pores of a compact.

capillary drying Progressive removal of moisture from a porous solid by evaporation at an exposed surface followed by movement of liquid from the interior to the surface by capillary action until the surface and core reach the same stable moisture concentration.

capillary tube A tube sufficiently fine that capillary action is significant.

capped steel A type of steel with characteristics similar to those of rimmed steels, but to a degree intermediate between those of rimmed and semikilled steels. It can be either mechanically capped or chemically capped when the ingot is cast, but in either case the full rimming action is stopped, resulting in a more uniform composition than rimmed steel.

capping Separation of a powder metal component during processing by cracks that originate on the surface.

caprolactam An amide compound with six carbon atoms which can be polymerized into nylon resins.

captive laboratory A laboratory that belongs to a material or product supplier, with systems that are dependent on those of the supplier, and with testing capabilities that exclusively provide those required by the supplier's material or product. Contrast with independent laboratory.

captive tests Holddown tests of a propulsive subsystem, rocket engine, or motor, as distinguished from a flight test.

captured sample (grab sample) A sample that has been taken into a container, or adsorbed onto a substrate, from the flowing stream, and that will be analyzed subsequently by an off-line measurement technique.

carbanion A negatively charged organic ion.

carbenes An organic radical containing divalent carbon.

carbides Compounds of carbon with one or more metallic elements. These are extremely hard compared to the matrix in which they exist.

carbide tool A cutting tool whose working edges and faces are made of tungsten, titanium, or tantalum carbide particles, compacted and sintered via powder metallurgy into a hard, heat-resistant and wear-resistant solid.

carbon *1.* An element that is the principal combustible constituent of all fuels. *2.* The principal element for organic polymers. *3.* Existing in varying amounts in steel, it is very influential with regard to metallurgical characteristics.

carbonaceous materials Substance composed of or containing carbon or carbon compounds.

carbon-carbon A composite made up of fibers of carbon or graphite in a carbon or graphite matrix.

carbon edge Carbonaceous deposits in a wavy pattern along the edges of a sheet or coil.

carbon electrode A carbon or graphite rod used in carbon-arc welding. Also used in electric furnace melting.

carbon equivalent An empirical relationship that is used to estimate the ability to produce gray cast iron, or one that is used to rate weldability of alloy steels.

carbon fibers Fibers produced by heating organic fibers, such as polyacrylonitrile, in an inert atmosphere.

carbonitriding A case hardening process in which a suitable ferrous material is heated above the lower transformation temperature in a gaseous atmosphere of such composition as to cause simultaneous absorption of carbon and nitrogen by the surface and, by diffusion, create a concentration gradient. The process is completed by cooling at a rate that produces the desired properties in the workpiece. Erroneously referred to as gas cyaniding.

carbonium ion Positively charged organic ions.

carbonization The process of converting coal to carbon by removing other ingredients.

carbon loss The loss representing the unliberated thermal energy occasioned by failure to oxidize some of the carbon in the fuel.

carbon monoxide (CO) A colorless, odorless, toxic gas usually resulting from combustion of carbonaceous compounds in an insufficient supply of oxygen.

carbon-pile pressure transducer A resistive-type pressure transducer that depends for its operation on the change in resistance that occurs when irregular carbon granules or smooth carbon disks are pressed together.

carbon potential A measure of the ability of an environment containing active carbon to alter or maintain, under prescribed conditions, the carbon content of the steel exposed to it. The carbon level attained by the steel will depend on such factors as temperature, time, and steel composition.

carbon restoration Replacing the carbon lost in the surface layer from previous processing by carburizing this layer to substantially the original carbon level. *See* recarburize.

carbon steel An alloy of carbon and iron that contains not more than 2% carbon, and that does not contain alloying elements other than a small amount of manganese.

carbon suboxides Colorless lacrimatory gases having unpleasant odors and boiling points of approximately –7°C.

carbon tetrachloride A nonflammable liquid used as an industrial cleaning agent.

carbonyl powder Metal powders created by the decomposition of various metal carbonyl compounds.

carburization In metallographic examination of steel, a darker shade of tempered martensite than that of the immediately adjacent base metal and harder by at least 30 points (Knoop or Vickers DPH [diamond -pyramid hardness]) than the hardness of the base metal.

carburizing A process in which an austenitized ferrous material is brought into contact with a carbonaceous atmosphere of sufficient carbon potential to cause absorption of carbon at the surface and, by diffusion, create a concentration gradient. The carburized material is then usually quench hardened.

Carnot cycle An idealized reversible thermodynamic cycle. The Carnot cycle consists of four stages: (a) an isothermal expansion of the gas at tempera-ture T_1; (b) an adiabatic expansion to temperature T_2; (c) an isothermal compression at temperature T_2; and (d) an adiabatic compression to the original state of the gas to complete the cycle.

carrier In spectrochemical analysis, a substance added to the sample to promote vaporization.

carrier band A single-channel signaling technique in which the digital signal is modulated on a carrier and transmitted.

carrier density The charge carrier concentration of holes and/or electrons in a semiconductor which determines its electronic characteristics and function.

carrier transport The mobility of conduction electrons or holes in semiconductors.

carryover The chemical solids and liquid entrained with the steam from a boiler.

Cartesian coordinates A coordinate system in which the locations of points in space are expressed in reference to three planes, called coordinate planes, no two of which are parallel. Also called rectangular coordinates.

cartridge *1.* A gas-generation device, packaged to include ignition, propellant, and other required items. *2.* A small unit used for storing computer programs or data values.

cascade sequence A welding sequence in which a continuous multiple-pass weld is built up by depositing weld beads in overlapping layers. Compare with block sequence.

cascading failure A failure for which the probability of occurrence is substantially increased by the existence of a previous failure.

case A hardened outer layer on a ferrous alloy produced by suitable heat treatment, which sometimes involves altering the chemical composition (for example, carburizing) of the outer layer before hardening.

case depth The distance from the surface of the hardened case to a point where the properties of the case, particularly hardness, are no longer discernible.

cased glass Glass composed of two or more layers of different glasses, usually a clear, transparent layer to which is added a layer of white or colored glass. The glass is sometimes referred to as flashed glass, multilayer glass, or polycased glass.

case hardened glass *See* tempered glass.

case hardening *1.* A generic term covering several processes applicable to steel which change the chemical composition of the surface layer by absorption of carbon, nitrogen, or a mixture of the two and, by diffusion, create a concentration gradient. The processes commonly used are: carburizing and quenching, cyaniding, nitriding, carbonitriding. The use of the applicable specific process name is preferred. *2.* Hardening accomplished by utilizing the steel's hardenability; in so doing the chemical composition is not altered.

casein A protein derived from milk used in the manufacture of plastics.

casing A covering of sheets of metal or other material, such as fire-resistant composition board, used to enclose all or a portion of a steam-generating unit.

cassette *1.* A light-tight container for holding photographic or radiographic film, or a photographic plate, and positioning it within a camera or other device for exposure. *2.* A small, compact container holding magnetic tape along with supply and takeup reels so that it can be inserted and removed as a unit for quick loading and unloading of a tape recorder or playback machine.

cassette tape Magnetic tape for digital data storage.

CASS test *See* copper-accelerated salt bath spray test.

cast *1.* To produce a solid shape from liquid or semisolid bulk material by allowing it to harden in a mold. *2.* A tinge of a specific color; a slight overtint of a color different from the main color, for instance, white with a bluish cast.

casting *1.* The process of making a solid shape by pouring molten metal into a cavity, or mold, and allowing it to cool and solidify. *2.* A near-net-shape object produced by this process.

casting alloy An alloy having suitable fluidity when molten and having suitable solidification characteristics to make it capable of producing shape castings.

casting resin A resin in liquid form which can be poured or otherwise introduced into a mold and shaped without pressure into solid articles.

casting shrinkage *1.* Total reduction in volume due to the three stages of shrinkage—during cooling from casting temperature to the liquidus, during solidification, and during cooling from the solidus to room temperature. *2.* Reduction in volume at each stage in the solidification of a casting.

casting slip A slurry of clay and additives suitable for casting into molds to make unfired ceramic products.

casting stress Stress that develops when the shape of the casting restricts shrinkage when the casting cools.

cast iron Any iron-carbon alloy containing at least 1.8% carbon and suitable for casting to shape.

cast plastics Liquid resins with hardeners that are poured into molds for curing.

cast steel Ferrous alloy steels that are in the cast form.

catalyst *1.* A substance, which by its mere presence, changes the rate of a reaction (decomposition) and may be recovered unaltered in nature or amount at the end of the reaction. *2.* The component of a two-part curing-type sealant that causes the prepolymer to polymerize.

cataphoresis Movement of suspended solid particles in a liquid medium due to the influence of electromotive force.

catastrophic failure *1.* A sudden failure that occurs without prior warning, as opposed to a failure that occurs gradually by degradation. *2.* Failure of a mechanism or component which renders an entire machine or system inoperable.

cathetometer An optical instrument for measuring small differences in height, for instance, the difference in height between two columns of mercury.

cathode *1.* A general name for any negative electrode. *2.* The electrode of an electrolytic cell at which reduction occurs.

cathode corrosion *1.* Corrosion of the cathode in an electrochemical circuit, usually involving the production of alkaline corrosion products. *2.* Corrosion of the cathodic member of a galvanic couple.

cathode ray In an electron tube or similar device, a stream of electrons emitted by the cathode.

cathode-ray oscilloscope An instrument that indicates the shape of a waveform by producing its graph on the screen of a cathode ray tube.

cathode ray tube (CRT) *1.* An electronic vacuum tube containing a screen on which information may be stored for visible display by means of a multigrid modulated beam of electrons from the thermionic emitter; storage is effected by means of charged or uncharged spots. *2.* A storage tube. *3.* An oscilloscope tube. *4.* A picture tube. *5.* A computer terminal using a cathode ray tube as a display device.

cathodic coating A mechanical plate or electrodeposit on a base metal, with the coating being cathodic to the underlying base metal.

cathodic corrosion Corrosion resulting from a cathodic condition of a structure usually caused by the reaction of an amphoteric metal with the alkaline products of electrolysis.

cathodic delamination Type of corrosion damage caused by loss of adhesion between the paint finish and the metal.

cathodic inhibitor A chemical substance or mixture that prevents or reduces the rate of cathodic or reduction reaction.

cathodic potential Electropositive potential.

cathodic protection Reduction of corrosion rate by shifting the corrosion potential of the electrode toward a less oxidizing potential by applying an external emf.

cation A positively charged ion of an electrolyte, which migrates toward the cathode under the influence of a potential gradient.

caulk *1.* A heavy paste, such as a mixture of a synthetic or rubber compound and a curing agent, or a natural product such as oakum, used to seal cracks or seams and make them airtight, steamtight, or watertight. Also known as caulking compound or calk. *2.* To seal a crack or seam with caulk.

caulking compound *See* caulk.

caul plates Smooth metal plates, free of surface defects, the same size and shape as a composite lay-up, used immediately in contact with the lay-up during the curing process to transmit normal pressure and temperature, and to provide a smooth surface on the finished laminate.

cause, basic The cause of a defect, failure, or damage which results in malfunctioning of an item when (a) being operated and maintained in a manner for which it was designed, and (b) the cause was not externally induced.

caustic cracking A type of stress-corrosion cracking common to some carbon and alloy steels when exposed to concentrated hydroxide solutions at high temperatures. *See* caustic embrittlement.

caustic dip A strongly alkaline solution for immersing metal parts to etch them, to neutralize an acid residue, or to remove organic materials such as grease or paint.

caustic embrittlement *1.* A historical term denoting stress corrosion cracking of steel exposed to alkaline solutions. *2.* Intergranular cracking of carbon steel or Fe-Cr-Ni alloy exposed to an aqueous caustic solution at a temperature of at least 150°F while stressed in tension; a form of stress-corrosion cracking. Also known as caustic cracking.

caustic soda The most important of the commercial caustic materials. It consists of sodium hydroxide containing 76 to 78% sodium oxide.

Cavendish balance A torsional instrument for determining the gravitational constant by measuring the displacement of two spheres of known small mass, mounted on opposite ends of a thin rod suspended on a fine wire, when two spheres of known large mass are brought near the small spheres.

cavitation Formation of cavities, either gaseous or vapor within a liquid stream, which occurs where the pressure is locally reduced to vapor pressure of the liquid.

cavitation corrosion Cavitation corrosion occurs on the low-pressure side of propellers and pump impellers where interruption in smooth flow causes vapor bubbles to form. When these bubbles collapse, they can destroy any protective coating and remove minute particles of metal.

cavitation damage Damage of a material associated with collapse of cavities in the liquid at a solid-liquid interface under conditions of severe turbulent flow. *See also* cavitation erosion.

cavitation erosion Progressive removal of surface material due to localized hydrodynamic impact forces associated with the formation and subsequent collapse of bubbles in a liquid in contact with the damaged surface. Also known as cavitation damage or liquid-erosion failure.

cavitation flow The formation of bubbles in a liquid, occurring whenever the static pressure at any point in the fluid flow becomes less than the fluid vapor pressure.

cavitons Density cavities created by localized oscillating electric fields.

cavity Formation of gaseous or vapor spaces within a liquid stream, which occurs where the pressure is locally reduced to vapor pressure of the liquid.

CBED *See* convergent-beam electron diffraction.

CBN Cubic boron nitride.

CBW *See* constant-bandwidth.

Cd Chemical symbol for cadmium.

cd *See* candela.

Ce Chemical symbol for cerium.

celestial mechanics The study of the theory of motions of celestial bodies under the influence of gravitational fields.

cell *1.* Electrochemical system consisting of an anode and a cathode immersed in an electrolyte. The anode and a cathode may be separate metals or dissimilar areas on the same metal. *2.* One of a series of chambers in which a chemical or electrochemical reaction takes place, for example, the chambers of a storage battery or electrolytic refining bath. *3.* One of the cavities in a honeycomb structure. *4.* The storage of one unit of information, usually one character or one word. *5.* A location specified by whole or part of the address and possessed of the faculty of store.

cellular plastic A porous plastic with numerous cells or bubbles dispersed throughout its bulk.

cellular rubbers A generic term for materials containing many cells (either open, closed, or both) dispersed throughout the mass.

cellulose acetate (CA) A strong thermoplastic material derived from an acetic acid ester of cellulose.

cellulose acetate butyrate (CAB) A strong, moisture-resistant thermoplastic derived from acetic and butyric acids reacting with cellulose.

cellulose ester A derivative of cellulose in which free hydroxyl groups attached to the cellulose chain are replaced by acetic groups. Used in the forming of thermoplastic molding compounds.

cellulose nitrate (CN) A nitric acid ester of cellulose.

cellulose propionate (CP) A derivative of cellulose made by the reaction of cellulose with propionic anhydride, producing a stiff glossy polymer.

cellulosic plastics Plastics derived from cellulose esters or ethers.

celsian Barium aluminum silicate that occurs naturally as feldspar.

Celsius A scale for temperature measurement based on the definition of 0°C and 100°C as the freezing point and boiling point, respectively, of pure water at standard pressure.

cement *1.* A dry, powdery mixture of silica, alumina, magnesia, lime, and iron oxide which hardens into a solid mass when mixed with water; it is one of the ingredients in concrete and mortar. *2.* An adhesive for bonding surfaces where intimate contact cannot be established and the adhesive must fill a gap over all or part of the faying surfaces.

cementation *1.* The introduction of one or more elements into the outer portion of a metal object by means of diffusion at high temperature. *2.* Conversion of wrought iron into steel by packing it in charcoal and heating it at about 1800°F for 7 to 10 days.

cemented carbide A powder-metallurgy product consisting of granular tungsten, titanium, or tantalum carbides in a temperature-resistant matrix, usually cobalt; used for high-performance cutting tools, punches, and dies.

cementite An intermetallic compound containing iron and carbon.

center gage A gage used to check angles, such as the angle of a cutting-tool point or screw thread.

centerless grinding Grinding the outside or inside of a workpiece mounted on rollers rather than on centers. The workpiece may be in the form of a cylinder or the frustum of a cone, and tolerances obtained by this technique are much better than tolerances obtained by hot rolling or cold drawing.

centerline shrinkage Reduction of a cast metal section along the central plane or axis.

center of gravity The center of mass of a system of masses, as the barycenter of the earth-moon system.

center of mass A point of a material body or system of bodies which moves as though the system's total mass existed at that point and all external forces were applied at the point.

centigrade A nonpreferred term formerly used to designate the temperature scale now referred to as the Celsius scale.

central integrated test system An on-line test system which processes, records, or displays at a central location, information gathered by test point datasensors at more than one remotely located equipment or system under test.

centrifugal casting A casting made by pouring metal into a mold which is rotated or revolved.

centrifugal force A force acting in a direction along and outward on the radius of turn for a mass in motion.

centrifugal separator A device that utilizes centrifugal force to separate materials of differing densities, such as water droplets or impurities from air.

centrifugal tachometer An instrument that measures the instantaneous angular speed of a rotating member such as a shaft by measuring the centrifugal force on a mass that rotates with it.

centrifuge A rotating device that separates suspended fine or colloidal particles from a liquid, or separates two liquids of different specific gravities, by means of centrifugal force.

cepstra The Fourier transformation of the logarithm of a power spectrum.

cepstral analysis The application of cepstral methods to wave or signal phenomena in seismology, speech analysis, echoes, underwater acoustics, etc.

ceramal protective coating. *See* cermet.

ceramic *1.* A heat-resistant natural or synthetic inorganic product made by firing a nonmetallic mineral. *2.* A shape made by baking or firing a ceramic material, such as brick, tile, or labware.

ceramic coating A protective coating made by thermal spraying a material such as aluminum or zirconium oxide, or by cementation of a material such as aluminum disilicide, on a metal substrate.

ceramic fiber Fiber composed of ceramic materials and usually used for reinforcement.

ceramic matrix composite Composite material consisting of a reinforced ceramic matrix.

ceramic tool A cutting tool made from fused, sintered, or cemented metallic oxides.

Cerenkov radiation Visible light produced when charged particles pass through a transparent medium at a speed exceeding the speed of light in the medium.

cerium A rare earth metal used as an alloying agent to improve strength and ductility.

cermet A body consisting of ceramic particles bonded with a metal; used in aircraft, rockets, and spacecraft for high-strength, high-temperature applications. The name is derived from a combination of CERamic and METal.

certificate of conformance Certificate issued by the laboratory which documents conformance of material to the material specification and which may describe the testing performed, but does not include the numerical values of results obtained.

certificate of test, test report Document issued by the laboratory describing the testing performed, the specific results obtained, and whether results conform to the material specification.

certification *1.* The act of certifying. *2.* The state of being certified. *3.* A means for individuals to indicate to the general public, co-workers, employers, and others that an impartial, nationally recognized organization has determined that they are qualified to perform specific technical tasks by virtue of their technical knowledge and experiences.

Cf Chemical symbol for californium.

C-factor *See* thermal conductance.

C-glass Glasses with a soda-lime-borosilicate composition for use in corrosive applications.

chafing Repeated motion between wires, cables, groups, harnesses, or bundles which results in wear.

chafing fatigue Fatigue caused by two materials rubbing against one another. *See also* fretting.

chain extensions A chemical reaction for increasing polymer chain length which results in greater hardness and stiffness of polymers.

chain reaction Chemical reactions consisting of similar or identical steps where each successive step is triggered by the previous one.

chain transfer agent A molecule from which an atom can be removed by a free radical.

chalking A defect of coated metals or plastics caused by formation of a layer of powdery material at the coating interface.

chamfer *1.* A beveled edge that relieves an otherwise sharp corner. *2.* A relieved angular cutting edge at a tooth corner on a milling cutter or similar tool.

chance failure Any failure whose occurrence is unpredictable in an absolute sense, but which is predictable only in a probabilistic or statistical sense.

changeover effect In ultrasonic inspection, effect whereby, when scanning from a good area onto a delaminated area, the back reflection will reduce in amplitude and will disappear completely as the delamination signal rises.

channel *1.* A path along which information, particularly a series of digits or characters, may flow. *2.* One or more parallel tracks treated as a unit. *3.* In a circulating storage, one recirculating path containing a fixed number of words stored serially by word. *4.* A path for electrical communication. *5.* A band of frequencies used for communication.

channel buffering A technique used to minimize the possibility of a failure in one channel from inducing a failure in another channel.

channel noise In communications, bursts of interruptive pulses caused mainly by contact closures in electromagnetic equipment or by transient voltages in electric cables during transmission of signals or data.

characteristic *1.* Specifically, distinguishing quality, property, feature, or capability of an entity. *2.* The integral part of a common logarithm, i.e., in the logarithm 2.5, the characteristic is 2, the mantissa is 0.5. *3.* Sometimes, that portion of a floating point number indicating the exponent.

characteristic curve A curve expressing a relation between two variable properties of a luminous source, such as candlepower and volts.

characteristic emulsion curve *See* characteristic curve.

characteristic impedance (Z_o) Of a uniform line, the ratio of an applied potential difference to the resultant current at the point where the potential difference is applied, when the line is of infinite length. This term is applied only to a uniform line. Coaxial cable is such a uniform line.

characteristic radiation Radiation produced by a particular element when its excitation potential is exceeded.

charge *1.* A defined quantity of an explosive. *2.* The starting stock loaded into a batch process. *3.* Material loaded into a furnace for melting or heat treating. *4.* A measure of the accumulation or depletion of electrons at any given point. *5.* The amount of substance loaded into a closed system, such as refrigerant into a refrigeration system. *6.* The quantity of excess protons (positive charge) or excess electrons (negative charge) in a physical body, usually expressed in coulombs.

chargeable failure A relevant, independent failure of equipment under test and any dependent failures caused thereby which are classified as one failure and used to determine contractual compliance with acceptance and rejection criteria.

Charpy test Measurement of the notch toughness or impact strength of a material. A specimen, usually notched and supported at both ends as a simple beam, is broken by a falling pendulum blow; values are determined by the height the pendulum striker rises in the absence of a specimen versus the height the striker rises after the fracture of a specimen. *See also* Izod test.

charring Determining the fiber content of a polymer matrix material by firing the matrix to ash.

chart A sheet or plate giving printed information in tabular and/or graphic form.

chart datum *See* datum plane.

chart recorder A device for automatically plotting a dependent variable against an independent variable.

chatter A series of lines uniformly spaced appearing transverse to the rolling direction, usually resulting from material being rolled on units having loose bearings. Results in a slight thickness variation where lines appear.

check *1.* A process of partial or complete testing of the correctness of machine operations. *2.* The existence of certain prescribed conditions within the computer, or the correctness of the results produced by a program.

check analysis (steel industry) An analysis by the purchaser of metal after it has been rolled or forged and sold. A check analysis is used either to verify the average composition of a heat or lot as represented by the producer's analysis, or to determine variations in the composition of a heat or lot. Materials specifications typically present check analysis limits; these are for use by the purchaser, not the producer, in determining conformance to specified limits. It is not used, as the term might imply, for a duplicate determination made to confirm a previous result. The results of analyses representing different locations in the same piece or taken from different pieces of a lot may differ from each other and from the producer's analysis due to segregation. Also, use of different analytical equipment, for example producer and purchaser, may result in different data.

check bit A binary check digit; often a parity bit. Related to parity check.

check digit In data transmission, one or more redundant digits appended to a machine word, and used in relation to the other digits in the word to detect errors in data transmission.

checked edges Sawtooth-like edges that can result from hot or cold rolling.

check, functional A quantitative check to determine if one or more functions of an item performs within specified limits.

checking *1*. Short axial cracks on the lip contact surface. *2*. Surface cracking in a checkerboard-like pattern; this may be in a surface layer (coating) or on the metal surface itself.

check, operational A task to determine if an item is fulfilling its intended purpose. The task does not require quantitative tolerances.

checkout *1*. Determination of the working condition of a system. *2*. A test or preliminary operation intended to determine whether a component or system is ready for service or ready for a new phase of operation.

checkpoint A point in time in a machine run at which processing is temporarily halted, to make a record of the condition of all the variables of the machine run, such as the status of input and output devices and a copy of working storage.

check problem A problem used to test the operation of a computer or to test a computer program; if the result given by the computer does not match the known result, it indicates an error in programming or operation.

checksum *1*. A routine for checking the accuracy of data transmission by dividing the data into small segments, such as a disk sector, and computing a sum for each segment. *2*. Entry at the end of a block of data corresponding to the binary sum of all information in the block. Used in error-checking procedures.

check, validity *See* validity check.

check valve A valve that allows flow in one direction only.

chelate A five-or-six-membered ring formation resulting from the molecular attraction of hydrogen, oxygen, or nitrogen atoms.

chelating agent A substance that combines with metal ions to form large molecules.

chemical affinity *1*. The relative ease with which two elements or compounds react with each other to form one or more specific compounds. *2*. The ability of two chemical elements to react to form a stable valence compound.

chemical analysis Determination of the chemical constituents.

chemical attack Damage to the resin matrix by accidental contact with, or unauthorized use of, chemical products.

chemical cleaning Processing using nitric, hydrofluoric, or nitric/hydrofluoric acid solutions, where material removal exceeds 0.0004 inch.

chemical clouds Artificial clouds of chemical compounds released in the ionosphere for observation of dispersion and other characteristics.

chemical conversion coating A protective or decorative coating produced in situ by chemical reaction of a metal with a chosen environment.

chemical core An object, composed of sodium chlorate or an analogous alkali metal chlorate or perchlorate, formulated with fuels, catalysts, and other modifiers and additives as required by the particular design, which evolves oxygen by a controlled chemical decomposition reaction, when actuated.

chemical defense All actions and counteractions designed for the protection of personnel and material against offensive chemical agents.

chemical degradation In the degradation of some materials (for example, rubber), accentuation of the process by the introduction of oxygen, resulting in significant crosslinks which cause the rubber to become hard and less deformable.

chemical deposition The plating of a metal to another metal from a solution of its salts.

chemical energy Energy produced or absorbed in the process of a chemical reaction.

chemical engineering A branch of engineering that deals with the design, operation, and maintenance of plants and equipment for chemically converting raw materials into bulk chemicals, fuels, and other similar products through the use of chemical reaction, often accompanied by a change in state or in physical form.

chemical fuels Fuels that depend on an oxidizer for combustion or for development of thrust, such as liquid or solid rocket fuel or internal combustion engine fuel; distinguished from nuclear fuel.

chemically foamed plastic A cellular plastic in which the cells are formed by gases generated from thermal decomposition or chemical reaction of the constituents.

chemical polishing Increasing the luster on the surface of a material by a chemical treatment.

chemical vapor deposited carbon Carbon deposited on a substrate by pyrolysis of a hydrocarbon, such as methane.

chemical vapor deposition (CVD) Metal deposited from a gaseous phase into a solid substrate.

chemiluminescence Any luminescence produced by chemical action.

chemiluminescent analyzer An analytical method for determining the NO_x concentration.

chemisorption The binding of a liquid or gas on the surface or in the interior of a solid by chemical bonds or forces.

chevron pattern A pattern of marks on the surface of a metal which looks like a series of letter 'V's nested together, such as a herringbone pattern.

chip *1.* Single large-scale integrated circuit. *2.* Material removed during the machining process.

chip breaker An attachment or a relieving channel behind the cutting edge of a lathe tool to cause removed stock to break up into pieces rather than to come off as long, unbroken curls.

chipping *1.* Using a manual or pneumatic chisel to remove seams, surface defects, or excess metal from semifinished mill products. *2.* Using a hand or pneumatic hammer with chisel-shaped or pointed faces to remove rust, scale, or other deposits from metal surfaces.

chirp An all-encompassing term for the various techniques of pulse expansion-pulse compression applied to pulse radar.

chlorinated hydrocarbon Any organic compound with chlorine atoms in its structure.

chlorine A toxic gaseous element used in the processing of materials.

chlorocarbons All compounds containing chlorine and carbon with or without other elements.

chlorofluorocarbon plastics Any plastic made with chlorine, fluorine, and carbon monomers.

chlorofluorohydrocarbon plastics Any plastic made with chlorine, fluorine, hydrogen, and carbon monomers.

chloroprene An elastomer derived from acetylene and hydrogen chloride with outstanding oil and chemical resistance.

chlorosulfonated polyethylene (CSP) An elastomer derived from polyethylene, chlorine, and sulfur dioxide used in the manufacture of hoses and seals.

choked flow The condition that exists when, with the upstream conditions remaining constant, the flow through a valve cannot be further increased by lowering the downstream pressure.

Cholesky factorization A numerical algorithm used to solve linear systems of equations.

chondrites Meteoritic stones character-ized by small rounded grains or spherules.

chord modulus The chord slope between points in the stress-strain curve.

chords Straight lines intersecting circles or other curves, or straight lines con-necting the ends of arcs.

Christiansen filter A device for admit-ting monochromatic radiation to a lens system.

chroma The attribute of color perception that expresses the degree of departure from gray, toward the pure hue, at the same value and hue.

chromadizing Improving paint adhesion on aluminum and its alloys by treating the surface with chromic acid.

chromated Material treated with chromic acid to improve its corrosion resistance.

chromate treatment Applying a solution of hexavalent chromic acid to produce a protective conversion coating of triva-lent and hexavalent chromium com-pounds.

chromatic aberration In a laser, the focus-ing of light rays of different wavelengths at different distances from the lens. This is not a significant effect with a single-wavelength laser source, but can be when working at different or multiple wavelengths.

chromatic contrast The difference between two areas, such as a symbol and its back-ground.

chromaticity *1.* The expression of color in terms of CIE codents. *2.* The color quality of a color stimulus definable by its chromaticity coordinates.

chromaticity coordinate The ratio of each of a set of three tristimulus values to their sum. They are given in ordered pairs, for example, $((x, y), u^1\,^71)$ etc.

chromaticity diagram Graphic representation of all possible colors on a two-dimensional diagram. Colors of similar dominant wavelength and excitation purity will plot close to one another.

chromaticity difference (CD) Distance between two color points.

chromatogram A plot of detector response against peak volume of solution emerging from the system for each of the constituents that have been separated.

chromatography The separation of chemical substances by making use of differences in the rates at which the substances travel through or along a stationary medium.

chromel A nickel-chromium alloy used in thermocouples.

chrome finish Finish applied to glass fibers to give good bonding to polyester and epoxy resins.

chrome magnesite Refractory material derived from chrome or calcined magnesite which is highly resistant to thermal shock.

chrome plating *See* chromium plating.

chromium A metallic element widely used as an electroplating material for oxidation resistance and for alloying steel. It is a grayish-white element which is hard and brittle and resistant to corrosion.

chromium coating *See* chromium plating.

chromium plating Electrodeposition of either a bright, reflective coating or a hard, less-reflective coating of chromium on a metal surface. Also known as chrome plating or chromium coating.

chromium steels Steels containing chromium as the main alloying element.

chromizing Forming a surface alloy of chromium onto a metal base.

chromophore The group of atoms within a molecule that contributes most heavily to its light-absorption qualities.

chronograph An instrument used to record the time at which an event occurs or the time interval between two events.

chronotron A device for measuring elapsed time between two events in which the time is determined by measuring the position of the superimposed loci of a pair of pulses initiated by the events. *See also* time lag.

Ci *See* curie.

CIL flowtest A method to determine the flow properties between various resins.

CIM *See* computer integrated manufacturing.

cinder A particle of gas-borne partially burned fuel larger than 100 microns in diameter.

circuit Any group of related electronic paths and components which electronic signals will pass to perform a specific function.

circuit analyzer A multipurpose assembly of several instruments or instrument circuits in one housing which are to be used in measuring two or more operating characteristics of an electronic circuit. *See* volt-ohm-milliammeter.

circuit breaker A resettable circuit-protective device.

circuit, fail-safe A circuit that has characteristics such that any probable malfunction will not adversely affect the safe operation of the aircraft or the safety of the passengers.

circuit-noise meter An instrument that uses frequency-weighting networks and other components to measure electronic noise in a circuit, giving approximately equal readings for noises that produce equal levels of interference.

circuit protection Automatic protection of a consequence-limiting nature used to minimize the danger of fire and/or smoke as well as the disturbance to the rest of the system, which may result from electrical faults or prolonged electrical overloads.

circular-chart recorder A type of recording instrument in which the input signal from a temperature, pressure, flow, or other transducer moves a pivoted pen over a circular piece of chart paper which rotates about its center at a fixed rate with time.

circularity In data processing, a warning message that the commands for two separate but interdependent cells in a program cannot proceed until a value for one of the cells is determined.

circularly polarized light Light in which the polarization vector rotates periodically, but does not change magnitude, describing a circle.

circular mil A wire-gage measurement equal to the cross-sectional area of a wire one mil (0.001 in.) in diameter; actual area is 7.8540×10^{-7} in.^{2}.

circular polarized wave An electromagnetic wave for which the electric field vector, magnetic field vector, or both, describe a circle.

circulating memory In an electronic memory device, a means of delaying information combined with a means for regenerating the information and reinserting it into the delaying means.

circulation The flow or motion of a fluid in or through a given area or volume. A precise measure of the average flow of a fluid along a given closed curve.

circumferential surface roughness Surface roughness of a shaft measured in a direction (plane) normal to the centerline axis.

CIS stereoisomer A polymer in which side atoms are arranged on the same side of a chain of double-bonded atoms.

Cl Chemical symbol for chlorine.

cladding *1.* Covering one piece of metal with a relatively thick layer of another metal and bonding them together; the bond may be produced by corolling or coextrusion at high temperature and pressure, or by explosive bonding. *2.* The low refractive index material that surrounds the core of a fiber and protects against surface contaminant scattering.

cladding strippers Chemicals or devices that remove the cladding from an optical fiber to expose the light-carrying core. The term might sometimes be misapplied to chemicals or devices that remove the protective coating applied over cladding to protect the fiber from the environmental stress.

clamp A device that, by rigid compression, holds a piece or part in position, or retains units in close proximity or parts in alignment.

clamping plate A plate for attaching a mold to a plastics-molding or die-casting machine.

clamping pressure In die casting, injection molding, and transfer molding, the force (or pressure) used to keep the mold closed while it is being filled.

clam shell *See* beach marks.

classification *1.* Separating a mixture into its constituents, such as by particle size or density. *2.* Segregating units of product into various adjoining categories, often by measuring characteristics of the individual units, thus forming a spectrum of quality. Also termed grading.

clay atmometer A simple device for determining evaporation rate to the atmosphere, which consists of a porous porcelain dish connected to a calibrated reservoir filled with distilled water.

clay filter A term used to describe a system of treating fuel to remove surface active agents.

clean fuels Energy sources from which pollutants and other impurities have been removed by refining, purification, and other means, to produce fuels less conducive to pollution.

cleanliness A term used in a general way to describe the presence of or freedom from nonmetallic inclusions. In specific instances, cleanliness might refer to the results of metallographic and/or magnetic analysis inspection.

cleanup *1.* Removing small amounts of stock by an imprecise machining operation, primarily to improve surface smoothness, flatness, or appearance. *See also* machining allowance. *2.* The time required for an electronic leak-testing instrument to reduce its output signal to 37% of the initial signal transmitted when tracer gas is first detected. *3.* The gradual disappearance of internal gases during operation of a discharge tube.

clearance *1.* The lineal distance between two adjacent parts that do not touch. *2.* Unobstructed space for insertion of tools or removal of parts during maintenance or repair.

clearance fit A type of mechanical fit in which the tolerance envelopes for mat-

ing parts always results in clearance when the parts are assembled.

clearcoat A paint without pigment applied over a color basecoat to enhance the appearance and durability of the total paint system.

cleavage factor A rupture where most of the grains have failed across a crystallographic plane of low index.

clinker A hard, compact congealed mass of fused furnace refuse, usually slag.

clinometer A divided-circle instrument for determining the angle between mutually inclined surfaces.

CLK *See* clock.

clock (CLK) *1.* A master timing device used to provide the basic sequencing pulses for the operation of a synchronous computer. *2.* A register that automatically records the progress of real time, or perhaps some approximation to it, records the number of operations performed, with its contents being available to a computer program. *3.* A timing pulse that coincides with or is phase-related to the occurrence of an event, such as bit rate or frame rate.

clock frequency The master frequency of periodic pulses which schedules the operation of the computer.

clock mode A system circuit that is synchronized with a clock pulse, that changes states only when the pulse occurs, and will change state no more than once for each clock pulse.

clock pulse A synchronization signal provided by a clock.

clock rate The time rate at which pulses are emitted from the clock. The clock rate determines the rate at which logical arithmetic gating is performed with a synchronous computer.

clock, real-time *See* real-time clock.

close annealing *See* box annealing.

closed bomb A fixed-volume chamber used for testing the pressure-time and chemical reaction characteristics of combustible materials.

closed-cell cellular plastic A cellular plastic in which almost all the cells are noninterconnecting.

closed circuit *1.* Any device or operation where all or part of the output is returned to the inlet for further processing. *2.* A type of television system that does not involve broadcast transmission, but rather involves transmission by cable, telephone lines, or similar method.

closed die A forming or forging operation in which metal flow takes place only within the die cavity. Contrast with open dies, in which there is little or no restriction to lateral flow.

closed loop *1.* A combination of control units in which the process variable is measured and compared with the desired value (or set point). *2.* A hydraulic or pneumatic system in which flow is recirculated following the power cycle. *3.* Pertaining to a system with feedback type of control, such that the output is used to modify the input.

closed-loop control *See* closed loop.

closed loop failure reporting system A controlled system assuring that all failures and faults are reported, analyzed, that positive corrective actions are identified to prevent recurrence, and that the adequacy of implemented corrective actions is verified by test.

closed pass A metal rolling arrangement in which a collar or flange on one roll fits into a groove on the opposing roll, thus permitting production of a flash-free shape.

closed pores Pores surrounded by solid material that is not accessible from the surface of the material.

close-grained Consisting of fine, closely spaced particles or crystals.

close-packed structure A crystalline structure in which the atoms are packed as close as possible.

close-tolerance forging Hot forging in which draft angles, forging tolerances, and cleanup allowances are considerably smaller than those used for commercial-grade forgings.

cloud chamber Device for observing the paths of ionizing particles, based on the principle that supersaturated vapor condenses more readily on ions than on neutral molecules.

clusec A unit of power used to express the pumping power of a vacuum pump; it equals about 1.333×10^{-6} watt, or the power associated with a leak rate of 10 mL/s at a pressure of 1 millitorr.

cluster analysis The analysis of data with the object of finding natural groupings within the data either by hand or with the aid of a computer.

CM *See* condition monitoring.

Cm Chemical symbol for curium.

CMOS Complementary metal oxide semiconductor. The combination of a PMOS (p-type channel metal oxide semiconductor) with an NMOS (n-type channel metal oxide semiconductor).

CMV *See* common mode voltage.

CN *See* cellulose nitrate.

CO *See* carbon monoxide.

Co Chemical symbol for cobalt.

CO_2 welding *See* gas metal-arc welding.

coagulation Bringing together tiny particles of an emulsion.

coal chemicals A group of chemicals used to make antiseptics, dyes, drugs, and solvents that are obtained initially as by-products of the conversion of coal to metallurgical coke.

coal derived gases The gases that are derived from various coal gasification processes.

coal derived liquids Fluid hydrocarbons derived from the liquefaction of coal.

coalescence A term used to describe the bonding of materials into one continuous body, with or without melting along the bond line, as in welding or diffusion bonding.

coal gas Gas formed by the destructive distillation of coal.

Coanda effect A phenomenon of fluid attachment to one wall in the presence of two walls.

coarse aggregate Crushed stone or gravel, used in making concrete, which will not pass through a sieve with 1/4-in. (6 mm) holes.

coarse grained *1.* Having a coarse texture. *2.* Having a grain size, in metals, larger than about ASTM No. 5.

coated materials A material that has a coating applied to the outer surface of the finished sample to impart some protective properties. Coating identification includes manufacturer's name, formulation designation (number), and recommendations for application.

coating A continuous film of some material on a surface.

coating (optics) A thin layer or layers applied to the surface of an optical component to enhance or suppress reflection of light, and/or to filter out certain wavelengths.

coating holes (voids) Areas devoid of coating. Coating holes or voids are caused by dust, dirt, lint, or improperly cleaned surfaces beneath the film.

coaxial cable Cable with a center conductor surrounded by a dielectric sheath and an external conductor. Has controlled impedance characteristics which make it valuable for data transmission.

cobalt A metallic element widely used as an alloying element in steel and nickel alloys.

COBOL *See* high-level language(s).

cobonding The curing together of two or more elements, of which at least one has already been fully cured and at least one is uncured.

cock A valve or other mechanism that starts, stops, or regulates the flow of liquid, especially into or out of a tank or other large-volume container.

co-curing The curing of a composite laminate or matrix while simultaneously bonding the respective components of the material structure.

COD *See* crack opening displacement.

CODAB *See* configuration data block.

code *1.* A system of symbols for meaningful communication. Related to instruction. *2.* A system of symbols for representing data or instructions in a computer or a tabulating machine. *3.* To translate the program for the solution of a problem on a given computer into a sequence of machine language, assembly language, or pseudo instructions and addresses acceptable to that computer. Related to encode. *4.* A machine language program.

coefficient of discharge The ratio of actual flow to theoretical flow. It includes the effects of jet contraction and turbulence.

coefficient of elasticity For tension tests, the reciprocal of Young's modulus.

coefficient of expansion The fractional change in dimension of a material for a unit change in temperature.

coefficient of friction Specifically, the ratio of the force at a common boundary of two bodies in contact that resists the motion or tendency to motion.

coefficient of influence Constant relating to the stress or strain behavior of a plastic laminate.

coefficient of rolling resistance The ratio of the rolling resistance force to the vertical load.

coefficient of thermal expansion (CTE) The change in length or volume of an item produced by a 1-degree temperature rise.

coefficient of variation The standard deviation divided by the mean, multiplied by 100 and expressed as a percentage.

coercimeter An instrument for measuring the magnetic intensity of a magnet or electromagnet.

coesite A polymorph of silicon dioxide.

coextrusion *1.* A process for bonding two metal or plastics materials by forcing them simultaneously through the same extrusion die. *2.* The bimetallic or bonded plastics shape produced by such a process.

Coffin-Manson law A relationship that enables one to estimate the fatigue life from the cyclic plastic strain range. The specific life of a given metal or alloy is determined by its tensile ductility.

cogeneration The generation of electricity or shaft power by an energy-conversion system and the concurrent use of the rejected thermal energy from the conversion system as an auxiliary energy source.

coherence A property of electromagnetic waves that are all the same wavelengths and precisely in phase with each other.

coherence length The distance over which light from a laser retains its coherence after it emerges from the laser.

coherent system of units A system of units of measurement in which a small number of base units, defined as dimensionally independent, are used to derive all other units in the system by rules of multiplication and division with no numerical factors other than unity. The SI base units, supplementary units, and derived units form a coherent set.

cohesion Refers to the internal forces holding a sealant together.

cohesive failure Failure of an adhesive joint occurring primarily in an adhesive layer.

cohesive strength *1.* The tensile strength required to fracture internal bonds with the force applied perpendicularly to the plane of the board. Also referred to as Z direction tensile or plybond. *2.* The stress related to the forces between atoms. *3.* The stress that can cause fracture without plastic deformation.

cohibitive effects The negative interaction of reinforcing elements in a composite causing reduced mechanical properties.

coil breaks Creases or ridges in metal sheet or strip which appear as parallel lines across the direction of rolling, generally extending the full width of the material.

coil weld A joint between two lengths of metal within a coil.

coincidence Existence of two phenomena or occurrence of two events simultaneously in time or space, or both.

coining Squeezing a metal blank between closed dies to form well-defined imprints on both front and back surfaces, or to compress a sintered powder-metal part to final shape.

coin test Tapping a composite laminate in different spots with a coin. A difference in sound could indicate a defect.

coke The solid residue remaining after most of the volatile constituents have been driven out by heating a carbonaceous material such as coal, pitch, or petroleum residues.

cold bend A test used to determine the effect of low temperatures on the insulation system of wire and cable when the wire or cable is flexed. Failure is characterized by the appearance of cracks or other defects in the insulation system.

cold cathodes Cathodes, the operation or which does not depend on their temperature being above the ambient temperature.

cold drawing *1.* Pulling rod, tubing, or wire through one or more dies that reduce its cross section, without applying heat either before or during reduction. *2.* Using equipment normally required for metalworking to form thermoplastic sheet.

cold extrusion Striking a cold metal slug in a punch-and-die operation so that metal is forced back around the die. Also known as cold forging, back extrusion, cold pressing, extrusion pressing, or impact extrusion.

cold-finished Refers to a primary-mill metal product, such as strip, bar, tubing, or wire, whose final shaping operation was performed cold; the material has more precise dimensions, and usually higher tensile and yield strength, than a comparable shape whose final shaping operation was performed hot.

cold finishing Processing by which final dimension and surface characteristics are produced below the recrystallization temperature.

cold flexibility Flexibility of a material during exposure to a predetermined low temperature for a specific length of time. *See* flexibility, cold.

cold flow Distortion in a material that takes place without a chemical or phase change.

cold forging *See* cold extrusion.

cold forming *1.* Any operation to shape metal that is performed cold. *2.* Shaping sheet metal, rod, or wire by bending, drawing, stretching, or other stamping operations without the application of heat.

cold galvanizing Painting a metal with a suspension of zinc particles in a solvent, so that a thin zinc coating remains after the organic solvent evaporates.

cold gas Gas at essentially room temperature, or at a temperature that is generally available from a pressure source without burning, decomposition, or external heating.

cold heading Cold working a metal by application of axial compressive forces that upset metal and increase the cross-sectional area over at least a portion of the length of the starting stock. Also known as upsetting.

cold joint In soldering, making a soldered connection without adequate heating, so that the solder does not flow to fill the spaces, but merely makes a mechanical bond.

cold pressing *1.* Forming powder metal shapes at room temperature or below. *2. See* cold extrusion.

cold resistance The ability of a material to withstand the effects of a low-temperature environment without loss of serviceability.

cold rolled sheets A product produced from a hot-rolled pickled coil which has been given substantial cold reduction at room temperature. The resulting product usually requires further processing to make it suitable for most common applications. The usual end product is characterized by improved surface, greater uniformity in thickness, and improved mechanical properties compared to hot-rolled sheet.

cold rolling Rolling metal at about room temperature; the process reduces thickness, increases tensile and yield strengths, improves fatigue resistance, and produces a smooth, lustrous or semilustrous finish.

cold room A facility capable of maintaining a temperature of $0 \pm 5°F$ ($-18 \pm 2.5°C$) and of sufficient size to contain the test set-up.

cold-setting adhesive A synthetic resin adhesive capable of hardening at normal room temperature in the presence of a hardener.

cold shut *1.* A defect that appears on the surface of a casting due to the failure of molten metal to unite. *2.* A defect on the surface of a forging due to lack of fusion during deformation.

cold slug flow Condition whereby insufficient heating of plasticizing cylinder results in unmelted pellets appearing in the molded part.

cold stretch Elongating filaments to increase tensile strength.

cold treatment Exposing to suitable sub-zero temperatures for the purpose of obtaining desired conditions or properties, such as dimensional or microstructural stability. When the treatment involves the transformation of retained austenite, it is usually followed by a tempering treatment.

cold welding Joining two materials by pressure at room temperature or below.

cold working Working of metals below the recrystallization temperature resulting in strain hardening.

collapse The breakdown of cell structure in a cellular material during a manufacturing process.

collector *1.* Any of a class of instruments for determining electrical potential at a point in the atmosphere, and ultimately the atmospheric electric field. *2.* A device used for removing gas-borne solids from flue gas. *3.* One of the functional regions in a transistor. *See also* accumulator.

colligative properties Properties based on the number of molecules present in a particular material.

collimation Producing a beam of light or other electromagnetic radiation whose rays are essentially parallel.

collision A close approach of two or more bodies (including energetic particles) that results in an interchange of energy, momentum, or charge. *See also* elastic collision and inelastic collision.

colloid A dispersion of particles of one phase in a second phase, where the particles are so small that surface phenomena play a dominant role in their chemical behavior.

colloidal suspension A suspension of finely divided particles approximately 5 to 10,000 angstroms (0.5 nm to 1 μm) in size dispersed in a continuous medium (for example, a gaseous, liquid, or solid substance), which do not settle or settle very slowly, and are not readily filtered.

Colmonoy A series of high nickel alloys used for hard facing of surfaces subject to erosion.

colonies Regions within prior beta grains with alpha platelets having nearly identical orientations. In commercially pure titanium, colonies often have serrated boundaries. Colonies arise as transformation products during cooling from the beta field at cooling rates slow enough to allow platelet nucleation and growth.

colorant Any substance such as a dye, pigment, or paint used to produce color in an object.

color attributes The four basic attributes used to describe the color of a material: hue, value, chroma, and metallic brilliance.

color change As used in fade/weathering testing, a change in color of any kind (whether a change in hue, saturation or lightness).

color code Any system of colors used to identify a specific type or class of objects from other, similar objects, for example, to differentiate steel bars of different grades in a warehouse.

color difference A color change in a sample or in a fabric, in which color has been altered or transferred from the sample as a result of testing.

colorfastness The resistance of a material to change in any of its color characteristics, to transfer its colorant(s) to adjacent materials, or both, as a result of the exposure of the material to any environment that might be encountered during testing.

colorimetry Any analytical process that uses absorption of selected bands of visible light, or sometimes ultraviolet radiation, to determine a chemical property.

color infrared photography A representation of temperature differences using false colors.

color temperature The temperature of a black body whose emitted light is a match for the given color.

column A vertical structural member of substantial length designed to bear axial compressive loads.

columnar structure A course crystalline structure of parallel columns of grains which have their long axis perpendicular to the casting surface.

column average stress (S_{ra}) The direct compression stress in a column or the average stress computed from the several gages located at the section.

column maximum stress (S_{rm}) The maximum compression stress in a column computed from the plane of buckling as established from the several gages located at the section.

combination scale An instrument scale consisting of two or more concentric or colinear scales, each graduated in equivalent values with two or more units of measure.

combinatorial logic system Digital system not utilizing memory elements.

combined carbon The portion of carbon in cast iron or steel other than free carbon.

combined life and particle retention test A filter life test in which specified amounts of inorganic classified contaminant are admitted to the influent stream at specified intervals during the test, to determine the particle separation efficiency of the filter at various stages of clogging. An organic clogging contaminant may be used in this test, in combination with the inorganic classified contaminant.

combined pressure pulse and vibration fatigue test A more sophisticated test method on which the laboratory test simulates simultaneously the service condition of impulse and fatigue.

combustibility *See* flammability.

combustible The heat-producing constituents of a fuel.

combustible loss The loss representing the unliberated thermal energy occasioned by failure to oxidize completely some of the combustible matter in the fuel.

combustion The rapid chemical combination of oxygen with the combustible elements of a fuel resulting in the production of heat.

combustion chamber Any chamber or enclosure designed to confine and control the generation of heat and power from burning fuels.

combustion rate The quantity of fuel fired per unit of time, as pounds of coal per hour, or cubic feet of gas per minute.

command *1.* A signal or input whereby functions are performed as the result of a transmitted signal. *2.* A signal that causes a computer to start, stop, or to continue a specific operation.

comment An expression that explains or identifies a particular step in a routine, but that has no effect on the operation of the computer in performing the instructions for the routine.

commonality The factors that are common in equipment or systems.

common cause analysis Generic term encompassing zonal analysis, particular risk analysis, and common mode analysis.

common field A field that can be accessed by two or more independent routines.

common mode In analog data, an interfering voltage from both sides of a differential input pair (in common) to ground.

common mode interference A form of interference that appears between the terminals of any measuring circuit and ground. *See* common mode voltage.

common mode voltage (CMV) In-phase, equal-amplitude signals that are applied to both inputs of a differential amplifier, usually referred to as a guard shield or chassis ground.

communication link *1.* The physical means of connecting one location to another for the purpose of transmitting and receiving information. *2.* The physical realization of a specified means by which stations communicate with each other. The specification normally cover the interfaces and some aspects of functional capability. *3.* A link may provide multiple channels for communications.

commutation Cyclic sequential sampling on a time-division basis of multiple data sources.

compact *1.* A powder-metallurgy part made by pressing metal powder, with or without a binder or other additives; prior to sintering it is known as a green compact, and after sintering as a sintered compact or simply a compact. *2.* To consolidate earth or paving materials by weight, vibration, impact, or kneading so that the consolidated material can sustain more load than prior to consolidation.

compacted graphite cast iron Cast iron with a graphite shape intermediate between a flake and spherical form.

compaction The process of producing powder metal parts.

compaction crack A crack produced in a compact during compression processing.

comparator *1.* In computer operations, device or circuit for comparing information from two sources. *2.* A device for inspecting a part to determine any deviation from a specific dimension by electrical, optical, pneumatic, or mechanical means.

compass bearing *See* bearing.

compatibility *1.* A characteristic ascribed to a major sub-system which indicates it functions well in the overall system. Also applied to the overall system with reference to how well its various sub-systems work together. *2.* The ability of two devices to communicate with each other in a manner that both understand. *3.* The ability of two or more substances to interact in a way to give a material desirable properties.

compatibility interface A point at which hardware, logic, and signal levels are defined to allow the interconnection of independently designed and manufactured components.

compatible The state in which different kinds of computers or equipment can use the same programs or data.

compensation *1.* A modifying or supplementary action (also, the effect of such action) intended to improve performance with respect to some specified characteristic. *2.* Provision of a supplemental device, circuit, or special materials to counteract known sources of systematic error.

compile *1.* A computer function that translates symbolic language into machine language. *2.* To prepare a machine-language program from a computer program written in another programming language.

compiler *1.* A program that translates a high-level source language (such as FORTRAN IV or BASIC) into a machine language suitable for a particular machine. *2.* A computer program more powerful than an assembler. In addition to its translating function, which is generally the same process as that used in an assembler, it is able to replace certain items of input with a series of instructions, usually called subroutines.

compile time In general, the time during which a source program is translated into an object program.

complement *1.* An angle equal to 90° minus a given angle. *2.* A quantity expressed to the base n, which is derived from a given quantity by a particular rule. Frequently used to represent the negative of the given quantity. *3.* A complement on n, obtained by subtracting each digit of the given quantity from n − 1, adding unity to the least significant digit, and performing all resultant carries.

complementary metal oxide semiconductor (CMOS) *1.* One type of computer semiconductor memory. The main feature of CMOS memory is its low power consumption. *2.* A type of semiconductor device not specifically memory.

complete decarburization *1.* Complete loss of carbon as determined by examination. *2.* *See* gross decarburization.

complete test *1.* A test composed of a baseline segment and a test segment. *2.* A test composed of two test segments.

complex compounds Chemical compounds in which part of the molecular boding is of the coordinate type.

complex frequency A complex number used to characterize exponential or damped sinusoidal waves in the same way as an ordinary frequency is used to characterize a simple harmonic wave.

complexing agent A substance that has an electron donor that will combine with a metal ion to form a complex ion.

complex modulus The ratio of stress to strain, where each vector is a complex number.

complex Young's modulus The sum of Young's modulus vectors and the loss modulus vectors.

compliance A term used in testing a material for stiffness or deflection. Tensile compliance is the reciprocal of Young's modulus; shear compliance is the reciprocal of shear modulus.

component An article that is a self-contained element of a complete operating unit and performs a function necessary to the operation of that unit.

composite A material or structure made up of physically distinct components that are mechanically, adhesively, or metallurgically bonded together; examples include filled plastics, laminates, filament-wound structures, cermets, and adhesive-bonded honeycomb-sandwich structures.

composite class A major subdivision of fibrous composite materials in which a class is defined by the geometric characteristic of the fiber arrangement.

composite compact A metal powder material with two or more layers or shapes, where each material can be uniquely identified.

composite electrode A welding electrode with two or more discriminate materials, one of which is a filler material.

composite joint A connection between two parts that involves both mechanical joining and welding or brazing, where both contribute to total joint strength.

composite materials Structural materials of metals, ceramics, or plastics with built-in strengthening agents. The strengthening agents may be in the form of filaments, foils, powders, or flakes of a different compatible material.

composite powder A metal powder where each particle contains two or more different materials.

composite propellants Solid rocket propellants consisting of a fuel and an oxidizer, neither of which would burn without the presence of the other.

composite solid propellants Propellants in which a granular inorganic oxidizer is suspended in an organic fuel binder, neither of which would burn without the presence of the other.

composite structure Any shape that is created by bonding together two or more discriminate components.

compound 1. An intimate admixture of a polymer with all the materials necessary for the finished article. 2. A substance, composed of two or more different chemicals having properties that are different from those of its constituent elements.

compound compact A compact consisting of more than one metal powder, where the metals are joined by centering and/or pressure.

compound layer The surface layer of the steel converted to essentially pure nitrides of metal, primarily iron.

compound semiconductor A semiconductor, such as gallium arsenide, that is made up of two or more materials, in contrast with simple, single-element materials, such as silicon and germanium.

compressed gas in solution A nonliquefied gas that is dissolved in a solvent.

compressibility *1*. The property of a substance, such as air, by virtue of which its density increases with increase in pressure. *2*. Volumetric strain per unit change in hydrostatic pressure. *3*. The ability of a metal powder to be formed into well-defined shapes. *4*. The ratio of a compact density.

compressibility factor (Z) A factor used to compensate for deviation from the laws of perfect gases. If the gas laws are used to compute the specific weight of a gas, the computed value must be adjusted by the compressibility factor, Z, to obtain the true specific weight.

compressibility test A test to determine the density and cohesiveness of a metal powder compact as a function of pressure.

compressible Capable of being compressed. Gas and vapor are compressible fluids.

compressible flow Fluid flow under conditions that cause significant changes in density.

compression *1*. Reduction of dimension from an external force. *2*. The relative displacement of sprung and unsprung masses in the suspension system in which the distance between the masses decreases from that at static condition.

compression, adiabatic Compression of a gas or mixture of gases without transmission of heat to or from it.

compression failure Buckling, collapse or fracture of a structural member that is loaded in compression.

compression, isothermal Compression of a gas or mixture of gases with subtraction of sufficient heat to maintain a constant temperature.

compression mold A type of plastics mold that is opened to introduce starting material, closed to shape the part, and reopened to remove the part and restart the cycle; pressure to shape the part is supplied by closing the mold.

compression ratio *1*. In internal combustion engines, the ratio between the volume displaced by the piston plus the clearance space, to the volume of the clearance space. *2*. In powder metallurgy, the ratio of the volume of loose powder used to make a part to the volume of the pressed compact.

compression set The deformation that remains in rubber after it has been subjected to and released from a specific percent compression for a definite period of time at a prescribed temperature. Compression set measurements are for the purpose of evaluating creep and stress relaxation properties of rubber.

compression spring An elastic member, usually made by bending metal wire into a helical coil, that resists a force tending to compress it.

compression test A destructive test for determining fracture strength, yield strength, ductility, and elastic modulus by progressively loading a short-column specimen in compression.

compressive modulus The ratio of compressive stress to strain; equal to Young's modulus for tensile test.

compressive strength The amount of pressure a material can withstand before shattering (brittle material), or buckling (ductile material).

compressive stress Stress that causes a material to be reduced in dimension. Contrast with tensile stress.

compressor *1.* A machine—usually a reciprocating-piston, centrifugal, or axial-flow design—that is used to increase pressure in a gas or vapor. *2.* A hardware or software process for removing redundant or otherwise uninteresting words from a stream, thereby "compressing" the data quantity.

Compton scattering A form of interaction between x-rays and loosely-bound electrons in which a collision between them results in deflection of the radiation from its previous path, accompanied by random phase shift and slight increase in wavelength.

computation The numerical solution of complex equations.

computational chemistry A complementary method for determining properties of gases, solids, and their interactions from first principle calculations. It extends testing capabilities to realms that are too dangerous or too costly to obtain experimentally.

computational fluid dynamics The application of large computer systems for the numerical solutions of complex fluid dynamics equations.

computational stability The degree to which a computational process remains valid when subjected to effects such as errors, mistakes, or malfunctions.

computer *1.* A data processor that can perform substantial computation, including numerous arithmetic or logic operations, without intervention by a human operator during the run. *2.* A device capable of solving problems by accepting data, performing described operations on the data, and supplying the results of these operations. Various types of computers are calculators, digital computers, and analog computers.

computer graphics *1.* The technique of combining computer calculations with various display devices, printers, plotters, etc. to render information in graphical or pictorial format. *2.* Any display in pictorial form on a computer monitor that can be printed.

computer integrated manufacturing (CIM) A system in which a central computer gathers all types of data, provides information stored in the database for decisions, and controls production input and output.

computerized simulation Computer-calculated representation of a process, device, or concept in mathematical form.

computer-limited Pertaining to a situation in which the time required for computation exceeds the time available.

computer simulation A logical-mathematical representation of a simulation concept, system, or operation programmed for solution on an analog or digital computer. *See also* computerized simulation.

computer systems simulation Forecasting of computer requirements by the use of predictive modeling and estimating computer workloads.

computer vision Capability of computers to analyze and act on visual input.

concatenate To combine several files into one file, or several strings of characters into one string, by appending one file or string after another.

concatenated codes Two or more codes that are encoded and decoded in series.

concave Describes a surface whose central region is depressed with respect to a projected flat plane approximately passing through its periphery.

concentrate *1.* To separate metal-bearing minerals from the gangue in an ore. *2.* The enriched product resulting from an ore-separation process. *3.* An enriched substance that must be diluted, usually with water, before it is used.

concentration *1.* Quantity of solid, liquid, or gaseous material related to that of another material in which it is found in the form of a mixture, suspension, or solution. *2.* The number of times that the dissolved solids have increased from the original amount in water due to evaporation.

concentricity The quality of two or more geometric shapes having the same center; usually, the term is applied to plane shapes or cross sections of solid shapes that are approximately circular.

conchoidal *See* beach marks.

conceptual phase The exploration of alternative solutions to satisfy a need.

concrete A mixture of aggregate, water, and a binder, usually portland cement, that cures as it dries and becomes rock hard.

concurrent processing Two (or more) computer operations that appear to be processed simultaneously, when in fact the CPU is rapidly switching between them.

cond *See* conductivity.

condensate *1.* The liquid product of a condensing cycle. Also known as condensate liquid. *2.* A light hydrocarbon mixture formed by expanding and cooling gas in a gas-recycling plant to produce a liquid output.

condensation The physical process whereby a vapor becomes a liquid or solid; the opposite of evaporation.

condensation nuclei Liquid or solid particles upon which condensation of water begins in the atmosphere.

condensation polymerization A chemical reaction in which two or more molecules combine by the separation of water or another substance. When a polymer is formed, the process is called polycondensation.

condensation reaction A chemical reaction in which two different molecules react to form a new compound of greater complexity, with the formation of water, alcohol, ammonia etc., as a byproduct.

condensation resin A resin derived from polycondensation.

condensation test A test used to evaluate the rust preventive properties of greases covering steel objects under alternate low-temperature and moderate-temperature, high-humidity conditions.

condenser The heat exchanger, located at the top of the column, that condenses overhead vapors. For distillation, the common condenser cooling media are water, air, and refrigerants such as propane.

conditional stability *1.* Describes a linear system that is stable for a certain interval of values of the open-loop gain, and unstable for certain lower and higher values. 2. The property of a controlled process whereby it can function in either a stable or unstable mode, depending on conditions imposed.

conditional transfer An instruction that, if a specified condition or set of conditions is satisfied, is interpreted as an unconditional transfer. If the conditions are not satisfied, the instruction causes the computer to proceed in its normal sequence of control.

conditioning *1.* Subjecting a material to a prescribed environmental and/or stress history before testing. *2.* A maintenance procedure consisting of deep discharge, short, and constant-current charge used to correct cell imbalance which may have been acquired during battery use.

conditioning heat treatment A preliminary heat treatment used to prepare a material for a desired reaction to a subsequent heat treatment. For the term to be meaningful, the treatment used must be specified.

conditioning of steel Refers to a method of removing scale, laps, and shallow defects from the surface of bars and billets by grinding with an abrasive wheel in a direction parallel to the length of the workpiece. Stock removal by this process will leave shallow grinding marks on the metal. Conditioning is to be differentiated from centerless grinding, rough turning, and cold drawing.

conditioning time With regard to bonding, the time interval between removal of the joint from conditions of heat and/ or pressure used to accomplish the bonding and the attainment of approximately maximum-bond strength.

condition monitoring (CM) A data gathering process which allows failures to occur, and relies on analysis of operating experience information to indicate the need for appropriate action.

conductance A measure of the ability of any material to conduct an electric charge. *See also* resistance.

conducting *See* conduction.

conducting media *See* conductors.

conducting polymer A plastics material having electrical conductivity approaching that of metals.

conduction *1.* The transfer of heat from one part of a body to another part of the same body or between bodies in physical contact. *2.* The transfer of energy within and through a conductor by means of internal particles or molecular activity and without any net external motion.

conduction band(s) A partially filled or empty energy band in which electrons are free to move easily.

conductive Capable of conducting electricity.

conductive elastomer An elastomeric material that conducts electricity; usually made by mixing powdered metal into a silicone before it is cured.

conductivity *1.* The ability to transmit heat or electricity; the reciprocal of resistivity. Electrical conductivity is expressed in terms of the current per unit of applied voltage. *2.* The amount of heat (Btu) transmitted in one hour through one square foot of a homogeneous material 1 inch thick for a difference in temperature of 1°F between the two surfaces of the material. *3.* The electrical conductance, at a specified temperature, between the opposite faces of a unit cube; usually expressed as $ohm^{-1} cm^{-1}$.

conductivity-type moisture sensor An instrument for measuring moisture content of fibrous organic materials such as wood, paper, textiles, and grain at moisture contents up to saturation.

conductometer An instrument that measures thermal conductivity, especially one that does so by comparing the rates at which different rods conduct heat.

conductor *1.* Substance or entity that transmits electricity, heat, or sound. *2.* Any material through which electrical current can flow.

conduit *1.* Any channel, duct, pipe, or tube for transmitting fluid along a defined flow path. *2.* A thin-wall pipe used to enclose wiring.

cone-plate viscometer An instrument for routinely determining the absolute viscosity of fluids in small sample volumes.

cones Geometric configurations having a circular bottom and sides tapering off to an apex (as in nose cone).

confidence A specialized statistical term referring to the reliance to be placed on an assertion about the value of a parameter of a probability distribution.

confidence coefficient *1.* A measure of assurance that a statement based on statistical data are correct. *2.* The probability that an unknown parameter lies within a stated interval or is greater or less than some stated value.

confidence factor The percentage figure that expresses confidence level.

confidence interval The interval within which it is asserted that the parameter of a probability distribution lies.

confidence level *1.* The probability that a given statement is correct, usually associated with statistical predictions. *2.* In acceptance sampling, the probability that accepted lots will be better than a specific value known as the rejectable quality level (RQL); a confidence level of 90% indicates that 90 out of every 100 lots accepted will have a quality better than the RQL. *3.* In statistical work, the degree of assurance that a

particular probability applies to a specific circumstance.

confidence limit A bound of confidence interval.

configuration *1.* The arrangement of the parts or elements of something. *2.* A particular selection of hardware devices or software routines and/or programs that function together. *3.* A term applied to a device or system whose functional characteristics can be selected or rearranged through programming or other methods. *4.* Chemical structures that share some relationship.

configuration data block (CODAB) In TELEVENT, the data section that identifies the "personality" of a hardware-software combination.

configuration interaction In physical chemistry, the interaction between two different possible arrangements of the electrons in an atom or molecule.

confluence *See* convergence.

conformance Describes the situation when a device conforms to the manufacturers specifications. *See* accuracy and error band.

conformations Varied shapes of molecules in a polymer chain resulting from molecular rotation around covalent bonds.

conjugate gradient method An interactive method for solving a system of linear equations of dimension N, which terminates at most N steps if no rounding errors are encountered.

connect To establish linkage between an interrupt and a designated interrupt servicing program. *See* disconnect.

connector *1.* Any detachable device for providing electrical continuity between two conductors. *2.* In fiber optics, a device that joins the ends of two optical fibers together temporarily.

consecutive access A method of data access that is characterized by the sequential nature of the I/O device involved; for example, a card reader is an example of a consecutive access device.

consistency *1.* A statistical term relating to the behavior of an estimator as the sample size becomes very large. An estimator is consistent if it converges to the population value as the sample size becomes large. *2.* A qualitative means of classifying substances, especially semisolids, according to their resistance to dynamic changes in shape.

console *1.* A main control desk for an integrated assemblage of electronic equipment. Also known as control desk. *2.* A grouping of control devices, instrument indicators, recorders, and alarms, housed in a freestanding cabinet or enclosure, to create an operator's work station.

consolidation A processing step in the manufacture of composites where the fiber and matrix are compressed to eliminate voids and increase density.

constant *1.* A value that remains the same throughout the distinct operation; opposite of variable. *2.* A data item that takes as its value its name (hence, its value is fixed during program execution).

constant amplitude Amplitude that is invariant in specified circumstances.

constantan An alloy of 55% copper and 45% nickel, used with copper in thermocouples.

constant bandwidth (CBW) The spacing of FM subcarriers equally with relation to each other. *See* proportional bandwidth.

constant failure period The period during which failures of some items occur at an approximately uniform rate.

constant failure rate *1.* A term characterizing the instantaneous failure rate in the middle or "useful life" period of the bathtub curve model of item life. *2.* Characterizes the hazard rate, h(t), of an item having an exponential reliability function.

constant-load balance A single-pan weighting device, having a constant load, in which the sample weight is determined by hanging precision weights from a counterpoised beam.

constant pressure (variable rate of fluid flow) A filter performance test condition which specifies that pressure difference across the filter shall be held constant throughout the test, which results in decreasing flow rate as the filter becomes clogged.

constant rate of fluid flow A filter performance test condition which specifies that test flow rate through the filter is held constant throughout the test, usually by varying the inlet pressure to compensate for increases of pressure drop across the filter.

constant-volume gas thermometer A device for detecting and indicating temperature based on Charles' Law—the pressure of a confined gas varies directly with absolute temperature.

constituent *1.* One of the elements that makes up a chemical system. *2.* A phase that occurs as a recurring configuration in a material microstructure. *3.* The elements of a composite, such as the fiber and the matrix.

constraint *1.* The limit of normal operating range. *2.* Anything that keeps a member under longitudinal tension from contracting laterally.

consumable electrode *1.* An arc-welding electrode that melts during welding to provide the filler metal. *2.* This expression is also used as part of "consumable electrode remelting," which is a special process for producing specialty steels and alloys; the electrode is consumed to produce an ingot.

consumable insert In welding, a piece of metal that is placed in the root of a weld prior to welding, and that melts during welding to supply part of the filler metal.

contact *1.* To establish communications with a facility. *2.* In hardware, a set of conductors that can be brought into contact by electromechanical action and thereby produce switching. In software, a symbolic set of points whose open or closed condition depends on the logic status assigned to them by internal or external conditions.

contact adhesive A dry-to-the-touch adhesive that will stick to itself or another smooth surface upon contact.

contact corrosion A term mostly used in Europe to describe galvanic corrosion between dissimilar metals.

contact element A temperature-measuring element used in intimate contact with the solid state body to be measured.

contact fatigue Cracking and pitting on surfaces that are subjected to alternating high stress, such as roller bearings or gears.

contact infiltration A process whereby an initially solid, but meltable material is placed in contact with a compact. When the solid material is melted, capillary action draws the liquid into the pores of the compact.

contact inspection In ultrasonic testing, a method of scanning a test piece which involves placing a search unit directly on a test piece surface covered with a thin film of couplant.

contact loads Dynamic loading by contact between two bodies.

contact material A composite of a wear and arc transfer-resistant metal with a soft and conductive metal.

contactor A mechanical or electromechanical device for repeatedly making and breaking electrical continuity between two branches of a power circuit, thereby establishing or interrupting current flow.

contact potential The difference in potential at the junction of two different materials.

contact pressure resins Resins that require little or no pressure when used for bonding.

contact scanning In ultrasonic testing, an organized movement of the beam relative to the object being tested.

contact testing Testing with transducer assembly in direct contact with material through a thin layer of couplant. *See* contact inspection.

contact thermography A method of measuring surface temperature in which the surface of an object is covered with a thin layer of luminescent material and then viewed under ultraviolet light in a darkened room; the brightness viewed indicates surface temperature.

contact tube In gas metal arc welding and flux cored arc welding, a metal part with a hole in it which provides electrical contact between the welding machine and the wire electrode, which is fed continuously through the hole.

contact-wear allowance The thickness that may be lost due to wear from either of a pair of mating electrical contacts before they cease to adequately perform their intended function.

contaminant *1.* A foreign material present on or in the contact surface. *2.* An impurity or foreign substance present in or on a material which affects one or more properties of the material.

contaminate That which contaminates to make impure or corrupt by contact or mixing.

contamination A general term used to describe an unwanted material that adversely affects the physical or electrical characteristics of an item.

content-addressed storage *See* associative storage.

context The composition, structure, or manner in which something is put together. Also refers to the situation or environment of an event.

contiguous file A file consisting of physically adjacent blocks on a mass-storage device.

continuous annealing The process of passing material such as steel through a controlled atmosphere furnace that has both heating and cooling zones. Temperatures, line speeds, and cooling rates are varied to obtain the desired properties for the product being heat treated. When processing fine wire or sheet, the time required for continuous annealing does not usually exceed a few minutes.

continuous casting A casting technique in which a cast cross-sectional shape is continuously withdrawn from a mold as it solidifies so that the length is independent of mold dimensions. The rapid solidification inherent in this process promotes minimal chemical segregation of the product.

continuous contaminant addition A filter test condition under which a specified contaminant is added to the test fluid at a continuous specified rate for the duration of the test.

continuous mixer A type of mixer in which starting ingredients are fed continuously and the final mixture is withdrawn continuously, without stopping or interrupting the mixing process. Generally, unmixed ingredients are fed at one end of the machine and blended progressively as they move toward the other end, where the mixture is discharged. Contrast with batch mixer.

continuous operation A process that operates on the basis of continuous flow, as opposed to batch, intermittent, or sequenced operations.

continuous rating The rating applicable to specified operation for a specific uninterrupted length of time.

continuous sampling The presentation of a flowing sample to the analytical analyzer so as to obtain continuous measurement of concentrations of the components of interest.

continuous sampling plan In acceptance sampling, a plan intended for application to a continuous flow of individual units of product that involves acceptance or rejection on a unit-by-unit inspection and sampling. Continuous sampling plans are usually characterized by requiring that each period of 100% inspection be continued until a specified number of consecutively inspected units are found clear of defects.

continuous spectrum A distribution of wavelengths in a beam of electromagnetic radiation in which the intensity varies continuously with wavelength. A continuous spectrum exhibits no characteristic structure, such as a series of bands, in which the intensity does not abruptly change at discrete wavelengths. *See also* band spectrum.

continuous weld A welded joint where the fusion zone is continuous along the entire length of the joint.

continuum Things that are continuous, that have no discrete parts. For example, the continuum of real numbers as opposed to the sequence of discrete integers.

contrast *1.* The subjective assessment of the difference in two parts of a field of view seen simultaneously or successively. (Hence, luminosity contrast, lightness contrast, color contrast, simultaneous contrast, successive contrast.) *2.* In a photographic or radiographic image, the ability to record small differences in light or x-ray intensity as discernible differences in photographic density.

contrast ratio (contrast) The quotient of the luminous intensity of an activated area of a display and the luminous intensity of the same point in the inactivated state. Contrast ratio (relative contrast) is also used to mean the quotient between the active and background area luminous intensities of a display. A complete definition requires that the ambient light conditions be specified for a given contrast ratio value.

control *1.* Frequently, one or more of the components in any mechanism responsible for interpreting and carrying out manually initiated directions. *2.* In some applications, a mathematical check.

control accuracy The degree to which a controlled process variable corresponds to the desired value or setpoint.

control action, derivative *See* derivative control.

control agent The energy or material comprising the process element that is controlled by manipulating one or more of its attributes—the attribute(s) commonly termed the controlled variable(s).

control bench Facility for testing of an engine control using a model or other artificial input in lieu of a hardware engine.

control block A storage area through which a particular type of information required for control of the operating system is communicated among its parts.

control board A panel that contains control devices, instrument indicators, and sometimes recorders which display the status of a system or subsystem.

control calculations Installation-dependent calculations that determine output signals from the computer to operate the process plant. These may or may not use generalized equation forms such as PID (proportional, integral, and derivative) forms.

control chart A plot of some measured quantity, which can be used to determine a quality trend or to make adjustments in process controls as necessary to keep the measured quantity within prescribed limits.

control circuit *1.* A circuit in a control apparatus which carries the electrical signal used to determine the magnitude or duration of control action. *2.* A circuit in a digital computer which performs any of the following functions: directs the sequencing of program commands, interprets program commands, or controls operation of the arithmetic element and other computer circuits in accordance with the interpretation.

control computer A process which directly controls all or part of the elements in the process. *See* process computer.

control counter A physical or logical device in a computer which records the storage locations of one or more instruction words which are to be used in sequence, unless a transfer or special instruction is encountered.

control desk *See* console.

control device Any device—such as a heater, valve, electron tube, contactor, pump, or actuator—used to directly effect a change in some process attribute.

controlled cooling Cooling from an elevated temperature in a predetermined manner, to avoid hardening, cracking, or internal damage, or to produce a desired microstructure or mechanical properties. The term applies to cooling following hot working.

controlled test A test designed to control or balance out the effects of environmental differences and to minimize the chance of bias in the selection, treatment, and analysis of test samples.

controlled variable *1.* The variable that the control system attempts to keep at the setpoint value. The setpoint may be constant or variable. *2.* The part of a process over which control is desired (flow, level, temperature, pressure, etc.). *3.* A process variable that is to be controlled at some desired value by means of manipulating another process variable.

controller *1.* A device for interfacing a peripheral unit or subsystem in a computer; for example, a tape controller or a disk controller. *2.* Device that con-

tains all the circuitry needed for receiving data from external devices, both analog and digital, processes the data according to pre-selected algorithms, then provides the results to external devices.

control limits In statistical quality control, the upper and lower values of a measured quantity that establish the range of acceptability.

control logic The sequence of steps or events necessary to perform a particular function. Each step or event is defined to be either a single arithmetic or a single Boolean expression.

control loop A combination of two or more instruments or control functions arranged so that signals pass from one to another for the purpose of measurement and/or control of a process variable. *See* closed loop, open loop, and loop.

control output module A device that stores commands from the computer and translates them into signals that can be used for control purposes.

control panel A part of a computer console that contains manual controls. *See* console.

control program *1.* A group of programs that provides such functions as the handling of input/output operations, error detection and recovery, program loading, and communication between the program and the operator. *2.* Specific programs that control an industrial process.

control programming Writing a user program for a computer which will control a process in the sense of reacting to random disturbances in time to prevent impairment of yield, or dangerous conditions.

control system A system in which deliberate guidance or manipulation is used to achieve a prescribed value of a variable.

control unit *1.* The portion of a computer that directs the sequence of operations, interprets the coded instructions, and initiates the proper commands to the computer circuits preparatory to execution. *2.* A device designed to regulate the fuel, air, water, or electrical supply to the controlled equipment.

control variable *1.* The variable that the control system attempts to keep at the setpoint value. *2.* The part of a process over which control is desired (flow, level, temperature, pressure, etc.) *3.* A process variable that is to be controlled at some desired value by means of manipulating another process variable.

convection The transfer of heat by movement of the heated and/or cooled particles of a fluid medium.

convection cooling Removing heat from a body by means of heat transfer using a moving fluid as the transfer medium.

conventional nitriding *See* nitriding.

convergence *1.* Approach to a limit, for example by an infinite sequence. *2.* The condition in which all the electron beams of a multibeam (color) cathode ray tube intersect at a specific point. *See also* misconvergence.

convergent-beam electron diffraction (CBED) Impinging electron beams on a crystal to produce a diffraction pattern made up of discs of different intensity.

conversion coating A protective surface layer on a metal that is created by chemical reaction between the metal and a chemical solution.

conversion time The time required for a complete measurement by an analog-to-digital converter.

conversion transducer Any transducer whose output-signal frequency is different from its input-signal frequency.

converter *1.* Rotary device for changing alternating current to direct current. Transducer whose output is a different frequency from its input. *2.* A type of refining furnace in which impurities are oxidized and removed by blowing air or oxygen through the molten metal.

convex A term describing a surface whose central region is raised with respect to a projected flat plane approximately passing through its periphery.

convex programming In operations research, a particular case of nonlinear programming in which the function to be maximized or minimized and the constraints are appropriately convex or concave functions of the controllable variables. Contrast with dynamic programming, integer programming, linear programming, mathematical programming, and quadratic programming.

coolant *1.* Liquid used for heat transfer composed of 50% glycol and 50% water by volume. *2.* A chemical that may be added to the gas generant to lower gas temperature. *3.* In a machining operation, any cutting fluid whose chief function is to keep the tool and workpiece cool.

cooling curve A curve showing the relation between time and temperature during the cooling of a material.

cooling stresses Residual stresses resulting from nonuniform distribution of temperature during cooling.

coordinate Set of measures defining points in space.

coordinate system *See* coordinate.

coordination number In a crystalline structure, the number of atoms or ions with which a single atom or ion is in contact.

cope The top part of the flask for a casting mold.

copolymer A compound resulting from the polymerization of two or more different monomers.

copper A ductile metal with excellent electrical and thermal conductivity and resistance to corrosion.

copper-accelerated salt-spray test (CASS test) A corrosion test for anodic aluminum coatings and other electrodeposits.

copper alloy An alloy in which copper is the predominant element. Generally, the addition of sulfur, lead, or tellurium improves machinability. Cadmium improves tensile strength and wearing qualities. Chromium gives very good mechanical properties at temperatures well above 200°C. Zirconium provides harness, ductility, strength, and relatively high electrical conductivity at temperatures where copper, and common high-conductivity copper alloys, tend to weaken. Nickel improves corrosion resistance, while silicon offers much higher mechanical properties. Beryllium, when present in an approximate 2% content in copper alloys, permits maximum strength, while about 0.5% content offers high conductivity.

copper and copper alloy annealing Depending on composition and condition, the materials are annealed by: (a) removal of the effects of cold work by recrystallization or recrystallization and grain growth; (b) substantially complete precipitation of the second phase in relatively coarse form in age (precipitation) hardened alloys; (c) solution heat treatment of age (precipitation) hardenable alloys; (d) relief of residual stress in castings. Specific process names in commercial use are final annealing, full annealing, light annealing, soft annealing, solution annealing, and stress relief annealing.

copper, ETP Electrolytic tough pitch copper (ETPC) has a minimum copper content of 99.9%. Annealed conductivity averages 101% with a 100% minimum.

copper-manganese A copper alloy with improved resistance to softening on heating.

copper, OFHC Oxygen-free high conductivity copper with a 99.95% minimum copper content, with an average annealed conductivity of 101%. It is suitable for an apparatus that is welded or exposed to reducing gasses at high temperatures. This copper has no residual deoxidant.

copper, silver bearing Silver-bearing copper with a 99.9% copper content that provides nearly the same electrical conductivity as ETP copper, but offers a higher softening point, greater resistance to creep, and higher strength at elevated temperatures. It also offers higher resistance to wear and oxidation, and improved machinability.

cord Textile, steel wire strands, and the like, forming the plies in the tire.

core *1.* A strongly ferromagnetic material used to concentrate and direct lines of flux produced by an electromagnetic coil. *2.* The inner layer in a composite material or structure. *3.* The central portion of a case-hardened part, which supports the hard outer case and gives the part its toughness and shock resistance. *4.* An insert placed in a casting mold to form a cavity, recess, or hole in the finished part. *5.* Magnetic memory elements; typically the main memory in a computer system.

core corrosion Oxidation or other chemical or electrolytic attack that adversely affects the core.

core crush A collapse, distortion, or compression of the core.

core depression A localized indentation or gouge in the core.

cored solder Wire solder having a flux-filled central cavity.

core iron A grade of soft steel suitable for making cores used in electromagnetic devices, such as chokes, relays, and transformers.

core nodes The points at which honeycomb cells are bonded to each other.

core splice A joint between two sections of honeycomb core, formed by bonding with foaming adhesive or crushing the core together to interlock the cells.

core stabilization A process to rigidize honeycomb core materials to prevent distortion during machining.

core wire Copper wire having a steel core, often used to make antennas.

coring A metallurgical condition where individual grains or dendrites vary in composition from center to grain boundary due to nonequilibrium cooling during solidification in an alloy that solidifies over a range of temperatures.

corpuscular radiation Nonelectromagnetic radiation consisting of energetic charged or neutral particles.

correction A quantity, equal in absolute magnitude to the error, added to a calculated or observed value to obtain a true value.

corrective action *1.* A documented design, material, or process change to correct the true cause of a failure. Part replacement with a like item does not constitute appropriate corrective action. Rather, the action should make it impossible for that failure to happen again. *2.* The change produced in a controlled variable in response to a control signal.

correlation *1.* A form of statistical dependence between two variables. Unless stated otherwise, linear correlation is implied. *2.* Measurement of the degree of similarity of two images as a function of detail and relative position of the images.

correlation coefficient A number between -1 and $+1$, which indicates the degree of linear relationship between two sets of numbers. Coefficients of -1 and $+1$ represent perfect linear agreements between two variables, while a coefficient of zero implies none.

corresponding states A principle that states that two substances should have similar properties at corresponding conditions with reference to some basic properties, for example critical pressure and critical temperature.

corrode To dissolve or wear away gradually, especially by chemical action.

Corrodekote test An accelerated corrosion test for electrodeposits in which a specimen is coated with a slurry of clay in a salt solution, and then is exposed for a specified time in a high-humidity environment.

corrosion The deterioration of a material, usually a metal, by reaction with its environment.

corrosion coating A material applied to integral tanks to coat the surface and supply protection to the metal, preventing corrosion. The term is used widely, but actually means, a corrosion preventive coating.

corrosion cracking Cracks within a metallic material that follow the paths of grain boundaries.

corrosion embrittlement The corrosive attack on metal causing the loss of ductility.

corrosion fatigue Effect of the application of repeated or fluctuating stresses in a corrosive environment characterized by shorter life than would be encountered as a result of either the repeated or fluctuating stresses alone or the corrosive environment alone.

corrosion potential The potential that a corroding metal exhibits under specific conditions of concentration, time, temperature, aeration, velocity, etc.

corrosion protection Preventing corrosion or reducing the rate of corrosive attack by any of several means including coating a metal surface with a paint, electroplate, rust-preventive oil, anodized coating, or conversion coating; adding a corrosion-inhibiting chemical to the environment; using a sacrificial anode; or using an impressed electric current.

corrosion rate The speed (usually an average) with which corrosion progresses; often expressed as though it were linear, in units of mdd (milligrams per square decimeter per day) for weight change, or mpy (mils per year) for thickness changes. Corrosion rates may be reported as: (a) A weight loss per area divided by the time (a milligram is 1/1000th of a gram, there are 453.6 g/lb,

and a decimeter is 3.937 in.; or (b) The depth of metal corroded, divided by the time (a "mil" is 1/1000th of an inch).

corrosion resistance The ability of a material, either in its natural form or after some conditioning, to withstand exposure to corrosive elements.

corrosive Any substance or environment that causes corrosion.

corrosive flux A soldering flux that removes oxides from the base metal when the joint is heated to apply solder.

corrosiveness The degree to which a substance causes corrosion.

corrosive wear Wear in which the reaction to an environment is notable.

corrugating Forming sheet metal into a series of alternating parallel ridges and grooves.

corrugations Transverse ripples caused by a variation in strip shape during hot or cold reduction.

corundum An aluminum oxide abrasive with higher purity than emery.

cosmetic corrosion Corrosion typically characterized by blistering and/or rusting which is aesthetically displeasing, but does not result in catastrophic failure of the item.

cosmetic defect A variation from the conventional appearance of an item, such as a slight deviation from its usual color, which is not detrimental to the performance of the item.

cosmetic irregularities Surface conditions that do not affect the integrity of the component and require no further action.

cosmic dust Finely divided solid matter with particle sizes smaller than a micrometeorite, thus with diameters much smaller than a millimeter, moving in interplanetary space.

Cottrell process The extraction of solid particles from gases by electrostatic precipitation.

coulometry An electromechanical measurement method in which the number of columns used in electrolysis determines the amount of substance electrolyzed.

coulomb Metric unit for quantity of electricity.

count *1.* (yarn) Size based on length and weight. *2.* (fabric) Number of yarns per woven inch. *3.* In computer programming, the total number of times a given instruction is performed.

counter *1.* A device or register in a digital processor for determining and displaying the total number of occurrences of a specific event. *2.* In the opposite direction. *3.* Device or PC program element that can total binary events and perform ON/OFF actions based on the value of the total. *4.* A device, register, or location in storage for storing numbers or number representations in a manner that permits these numbers to be increased or decreased by the value of another number, or to be changed or reset to zero or to an arbitrary value.

counting rate The average number of ionizing events that occur per unit time, as determined by a counting tube or similar device.

counts *1.* An alternate form of representing raw data corresponding to the numerical representation of a signal received from or applied to external hardware. *2.* The accumulated total of a series of discrete inputs to a counter. *3.* The discrete inputs to an accumulating counter. *See* digitized signal.

couple A cell developed in an electrolyte resulting from electrical contact between two dissimilar metals.

coupler *1.* In data processing, a device that joins similar items. *2.* In fiber optics, a device that joins together three or more fiber ends—splitting the signal from one fiber so it can be transmitted to two or more other fibers. Directional, star, and tee couplers are the most common.

coupling A means or a device for transferring power between systems.

coupling agent A substance that reacts with the matrix and reinforcement portions of a composite material to form a stronger bond.

coupon A material specimen for a test.

covalent bond The strongest primary atomic bond that is based on electron sharing.

covering power The ability of an electroplating solution to give a satisfactory plate at low current densities, such as occur in recesses, but not necessarily to build up a uniform coating. Contrast with throwing power.

cover plate *1.* Any flat metal or glass plate used to cover an opening. *2.* Specifically, a piece of glass used to protect the tinted glass in a welder's helmet or goggles from being damaged by weld spatter.

CP *See* cellulose propionate.

cP Abbreviation for centipoise. *See* poise.

CPU-bound A state of program execution in which all operations are dependent on the activity of the central processor, for example, when a large number of calculations are being performed; compare to I/O-bound.

crack *1.* A sharp break or fissure in the sealing element. *2.* Evidence of a full or partial break without separation of parts. *3.* The fissure or chink between adjacent components of a mechanical assembly. *4.* To incompletely sever a solid material, usually by overstressing it. *5.* To open a valve, hatch, door, or other similar device a very slight amount.

crack closure Phenomenon that occurs when the cyclic plasticity of a material gives rise to the development of residual plastic deformations in the vicinity of a crack tip, causing the fatigue crack to close at positive load.

crack geometry The shape and size of partial fractures or flaws in materials.

crack growth The spreading or progressing rate of a crack in a material due to an increased load.

cracking *1.* Separations in the topcoat film before or after baking. Cracking can be caused by shrinkage of the paint film during the bake cycle, swelling of the sealer after the topcoat has started to cure, or sliding of the uncured paint film on the uncured sealer due to incompatibility. *2.* Parting within the tread, sidewall, or inner liner of the tire extending to cord material. *3.* The interruption of a surface due to environment and/or stress. *4.* Apparent fine cracks at or under the surface of a plastic part.

cracking, edge Short, linear cracks that originate along the polished transition edge of outer panes and have propagated radially inward.

cracking star A group of several roughly linear surface micro-cracks that originate at a point and propagate outward to form a star-like pattern.

crack opening displacement (COD) The measurement of the opening of a notch in the direction perpendicular to the plane of the notch in the crack.

crack tip The boundary between cracked and uncracked material.

Crank-Nicholson method A method for solving parabolic partial differential equations, whose main feature is an implicit method that avoids the need for using very small time steps.

cratering Round depressions in a painted surface, usually due to contaminants in the paint or on the surface being painted.

craze A network of fine cracks in the surface.

crazing Apparent fine cracks at or under the surface of a plastic part.

creep Time-dependent plastic strain occurring in a metal or other material under stress which is lower than that required to cause failure in a tensile test, usually at elevated temperature.

creepback The undercutting or the separation of paint from the substrate at an edge, damage site, or a scribe line.

creep limit The maximum stress in a specific time that will cause a specific change in dimension.

creep rate The slope of a creep-time curve over a specific time.

creep resistance *See* creep strength.

creep-rupture strength The stress that will result in fracture at a specific time in a creep test.

creep rupture test *See* stress rupture test.

creep strain The total strain over a specific time produced by stress during a creep test.

creep strength The constant nominal stress that will cause a specified quantity of creep in a given time at constant temperature. Also referred to as creep resistance.

creep stress The stress load divided by the initial cross-sectional area of the specimen.

creep test A test to determine the creep characteristics of a material under a specific load at elevated temperatures.

crest *1.* The top of a screw thread. *2.* The bottom edge of a weir notch, sometimes referred to as the sill.

crevice corrosion *1.* A form of accelerated localized corrosion occurring at locations where easy access to the bulk environment is prevented, such as at mating surfaces of metal assemblies. *2.* A type of concentration-cell corrosion associated with the stagnant conditions in crevices, fissures, pockets, and recesses away from the flow of a principal fluid stream, where concentration or depletion of dissolved salts, ions or gases such as oxygen leads to deep pitting.

crimp anvil (nest) The portion of a crimping die that supports a barrel or ferrule during crimping.

crimping *1.* Forming small corrugations in order to set down and lock a seam, create an arc in a metal strip, or reduce the radius of an existing arc or circle. *2.* Causing something to become wavy, crinkled, or warped. *3.* Pinching or pressing together to seal or unite, especially the longitudinal seam of a tube or cylinder.

critical aspect ratio The ratio of fiber length to fiber diameter in fiber-reinforced composites.

critical buckling stress The average stress that produces an incipient buckling condition in column-type members.

critical cooling rate The minimum rate of continuous cooling to prevent undesirable transformations. For steel, it is the minimum rate at which austenite must be continuously cooled to suppress transformations above the M_s temperature.

critical damping *1.* The minimum amount of viscous damping required in a linear system to prevent the displacement of the system from passing the equilibrium position upon returning from an initial displacement. *2.* The value of damping that provides the most rapid transient response without overshoot. Operation is between underdamping and overdamping.

critical dimension *1.* Generally, any physical measurement whose value or accuracy is considered vital to the function of the involved component or assembly. *2.* In a waveguide, the cross-sectional dimension that determines the critical frequency of the waveguide.

critical fiber length The minimum length of a reinforced composite fiber that will produce the maximum possible load on the fibers.

critical flow A somewhat ambiguous term that signifies a point at which the characteristics of flow suffer a finite change.

critical fusion frequency Frequency of intermittent light stimulation at which flicker is no longer perceived.

critical humidity The relative humidity above which the atmospheric corrosion rate of a given metal increases sharply.

criticality A relative measure of the consequences of failure.

criticality analysis A procedure by which each potential failure modes is evaluated and ranked according to the combined influences of severity and probability of occurrence.

critical mass The amount of concentrated fissionable material that can just support a self-sustaining fission reaction.

critical pitting potential (E_{cp}, E_p, E_{pp}) The lowest value of oxidizing potential at which pits nucleate and grow. It is dependent on the test method used.

critical point The thermodynamic state in which liquid and gas phases of a substance coexist in equilibrium at the highest possible temperature.

critical pressure The equilibrium pressure of a fluid that is at its critical temperature.

critical-pressure ratio The ratio of downstream pressure to upstream pressure that corresponds to the onset of turbulent flow in a moving stream of fluid.

critical resolved shear stress The amount of stress required to cause slip to occur on a plane within a metal crystal.

critical strain The amount of prior plastic strain that is just sufficient to trigger recrystallization when a deformed metal is heated.

critical stress intensity factor The measurement of a powder metal part toughness.

critical temperature The temperature above which a substance cannot exist in the liquid state regardless of the pressure. As applied to materials, the temperature at which a change in phase takes place causing an appreciable change in the properties of the material.

critical temperature ranges Synonymous with transformation ranges, which is the preferred term. *See* transformation ranges.

critical velocity *1.* In rocketry, the speed of sound at the conditions prevailing at the nozzle throat. *2.* For a given fluid, the average linear velocity marking the upper limit of streamline flow and the lower limit of turbulent flow, at a given temperature and pressure, in a given confined flowpath.

cross-assembler An assembler program run on a larger host computer and used for producing machine code to be executed on another, usually smaller, computer.

cross axis sensitivity *See* transverse sensitivity.

cross-compiler A computer program run on a larger host computer and used for translating a high-level language program into the machine code to be executed on another computer.

crossed joint *See* bridging.

crosshair An inscribed line or a thin hair, wire, or thread used in the optical path of a telescope, microscope, or other optical device to obtain accurate sightings or measurements.

cross laminate A composite where some layers of reinforcement material are placed at right angles to other layers of the reinforcement material.

cross-linking The establishing of chemical links between the molecular chains in polymers through the process of chemical reaction, electron bombardment, or vulcanization.

crosslinking agent A substance that promotes or regulates intermolecular covalent or ionic bonding between polymer chains.

crossover frequency *1.* The frequency at which a dividing network delivers equal power to upper-band and lower-band channels. *2.* For an acoustic recording system, the frequency at which the asymptotes to the constant-amplitude and constant-velocity portions of the frequency-response curve intersect.

Also known as the transition frequency or the turnover frequency.

cross-ply laminate A composite where plies are placed at 0 and 90 degrees.

cross polarization The component of the electric-field vector normal to the desired polarization component.

cross section *1.* Measure of the effectiveness of a particular process expressed either as the area (geometric cross section) that would produce the observed results, or as the ratio. *2.* For a given confined flowpath or a given elongated structural member, the dimensions, shape, or area determined by its intersection with a plane perpendicular to its longitudinal axis. *3.* In characterizing interactions between moving atomic particles, the probability per unit flux and per unit time that a given interaction will occur.

cross-wire weld A resistance weld made by passing a controlled electric current through the junction of a pair of crossed wires or bars; used extensively to make mesh or screening.

crosswise direction Refers to the cutting of specimens and to the application of load.

crowfoot A type of woven reinforcement fabric where a filling thread is woven over three warped threads and then under one warped thread; the resulting fabric is more pliable and thus more readily formed around curves.

crown *1.* The domed top of a furnace or kiln. *2.* The central portion of sheet or plate material which is slightly thicker than at the edges. *3.* Any raised central portion of a nominally flat surface.

crown glass An optical glass of alkali-lime-silica composition with index refraction usually 1.5 to 1.6.

CRT *See* cathode ray tube.

crucible A pot or vessel made of a high-melting-point material, such as a ceramic or refractory metal, used for melting metals and other materials.

crude oil Unrefined petroleum.

crush *1.* A casting defect caused by displacement of sand as the mold is closed. *2.* Buckling or breaking of a section of a casting mold caused by incorrect register as the mold is closed.

crushing test *1.* The application of a vertical static load through a beam placed laterally across the uppermost members of the ROPS. *2.* A compression test for bearings or tubing to determine the maximum compression load.

cryochemistry The study of chemical phenomena in very low-temperature environment.

cryogenic Any process carried out at very low temperature, usually considered to be −60°F (−50°C) or lower.

cryogenic cooling Use of cryogenic fluids to reach temperatures low enough to conduct a cryogenic process.

cryogenic fluid A liquid that boils below −123°Kelvin (−238°F, −150°C) at one atmosphere absolute pressure.

cryogenic liquid Liquefied gas at very low temperature, such as liquid oxygen, nitrogen, or argon.

cryogenics The study of the methods of producing very low temperatures. The study of the behavior of materials and processes at cryogenic temperatures.

cryogenic treatments Same as cold treatment only specifically related to temperatures at or below that of liquid nitrogen (−196°C).

cryometer A thermometer for measuring very low temperatures.

cryoscope A device for determining the freezing point of a liquid.

cryostat An apparatus for establishing the very low-temperature environment needed for carrying out a cryogenic operation.

cryotron Device based on the principle that superconductivity established at temperatures near absolute zero is destroyed by the application of a magnetic field.

crystal A material form in which the atoms or molecules are aligned in a symmetrical pattern.

crystal dislocations Types of lattice imperfections whose existence in metals is postulated in order to account for the phenomenon of crystal growth and of slip, particularly for the low value of shear stress required to initiate slip.

crystal lattices Three-dimensional, recurring patterns in which the atoms of crystals are arranged.

crystalline fracture A type of fracture surface appearance characterized by numerous brightly reflecting facets resulting from cleavage fracture of a polycrystalline material.

crystalline plastic A polymeric material where the atom structure is arranged in an orderly three-dimensional manner.

crystallinity In polymers, describes a microstructure in which the linear molecular chains are arranged in an orderly fashion.

crystallization Crystal formation by solidification of a liquid or crystal growth from a solution.

crystallography The study of crystalline arrangements and external morphologies of crystal types by means of x-ray and electron diffraction.

crystal oscillator A device for generating an a-c signal whose frequency is determined by the properties of a piezoelectric crystal.

crystal spectrometer An instrument that uses diffraction from a crystal to determine the component wavelengths in a beam of x-rays or gamma rays.

C-scan Scanning a material specimen with ultrasonics to identify the presence of defects such as voids, foreign contamination, or delamination.

CSMA/CD Carrier sense multiple access with collision detection.

CSP *See* chlorosulfonated polyethylene.

C-stage The final, fully cured stage of a thermoset plastic in which the material is virtually insoluble and infusible.

CTE *See* coefficient of thermal expansion.

cumulative damage laws Laws, such as Miner's law, used to estimate the fatigue life of a particular component when subjected to a series of variable loading cycles.

cumulative dose The total amount of penetrating radiation absorbed by the whole body, or by a specific region of the body, during repeated exposures.

cup-cone fracture *See* cup fracture.

cup fracture A mixed mode fracture in ductile metals, usually observed in round tensile specimens, in which part of the fracture occurs under plane-strain conditions and the remainder under plane-stress conditions. Also called a cup-cone fracture.

cupping *1.* The first step in deep drawing. *2.* The fracture of severely worked rod or wire, where one of the fracture surfaces is roughly conical and the other cup-shaped.

cupping test A test to determine the ductility and drawing ability of a material by drawing it into a deep cup shape.

cure To change the physical properties of a material by chemical reaction, by the action of heat and catalysts.

cure cycle The time/temperature/pressure cycle used to cure a thermosetting resin system or prepreg.

cure-date The date the compounded, uncured elastomer is vulcanized to produce an elastomeric product.

cure monitoring, electrical Use of electrical techniques to detect changes in the electrical properties and/or mobility of the resin molecules during cure.

cure rate A measure of the rate of polymerization based on the increasing hardness of the sealant with time. The time required to reach the specified hardness varies widely with the type of sealant and with the temperature and relative humidity. Testing is performed at standard conditions.

cure shrinkage Shrinkage of composites during curing which results in stresses on the reinforcement fibers.

cure stress An internal stress that occurs in composites during the curing cycle.

cure time The time required to produce vulcanization at a given temperature.

Curie Abbreviated Ci. The standard unit of measure of radioactivity of a substance; it is defined as the quantity of a radioactive nuclide that is disintegrating at the rate of 3.7×10^{10} disintegrations per second.

Curie temperature The temperature in a ferromagnetic material above which the material becomes substantially nonmagnetic.

curing *1.* Allowing a substance such as a polymeric adhesive or poured concrete to rest under controlled conditions, which may include clamping, heating, or providing residual moisture, until it undergoes a slow chemical reaction to reach final bond strength or hardness. *2.* In thermoplastics molding, stopping all movement for an interval prior to releasing mold pressure so that the molded part has sufficient time to stabilize.

curing agent A catalytic substance that causes or speeds polymerization when added to a resin. *See also* hardener.

curing temperature The temperature at which the elastomeric product is vulcanized.

current The rate of flow of an electrical charge in an electric circuit analogous to the rate of flow of water in a pipe.

curtains An uneven pattern of a coating resulting from run-back of the applied material. Also called flowlines.

curvature of field A defect in an optical lens or system which causes the focused image of a plane field to lie along a curved surface rather than a flat plane.

curve-fit The process of determining the coefficients in a curve by mathematically fitting a given set of data to that curve class.

cut A deep discontinuity in the seal material whereby no material is removed.

cut edge A mechanically sheared edge obtained by slitting, shearing, or blanking.

cutoff *1.* An act or instance of shuffling something off. *2.* The parting line on a compression-molded plastics part. Also known as flash groove or pinchoff. *3.* The point in the stroke of an engine where admission of the working fluid to the cylinder is shut off. *4.* The time required to shut off the flow of working fluid into a cylinder.

CVD *See* chemical vapor deposition.

cyanate esters *See* cyanate resins.

cyanate resins Thermosetting resins derived from bisphenols or polyphenols. Also referred to as cyanate esters and triazine resins.

cyaniding A case hardening process in which a ferrous material is heated above the lower transformation range in a molten salt containing cyanide to cause simultaneous absorption of carbon and nitrogen at the surface and, by diffusion, create a concentration gradient. Quench hardening completes the process.

cybernetics The branch of learning that brings together theories and studies on communication and control in living organisms and machines.

cycle *1.* An interval of space or time in which one set of events or phenomena is completed. *2.* Any set of operations that is repeated regularly in the same sequence. *3.* To run a machine through a complete set of operating steps. *4.* The fundamental time interval for operations inside the computer.

cycle annealing An annealing process employing a predetermined and closely controlled time-temperature cycle to produce specific properties or microstructure.

cycle *n* One complete cycle of values of a load that is repeated.

cycle redundancy check (CRC) An error-detection scheme, usually hardware implemented, in which a check character is generated by taking the remainder after dividing all the serialized bits in a block of data by a predetermined binary number. This remainder is then appended to the transmitted date and recalculated and compared at the receiving point to verify data accuracy.

cycle stealing Refers to the process whereby data is transferred over the data bus during a direct memory access while little disruption occurs to the normal operation of the microprocessor.

cycle time *1.* The time required by a computer to read from or write into the system memory. If system memory is core, the read cycle time includes a write-after-read (restore) subcycle. *2.* Cycle time is often used as a measure of computer performance, since this is a measure of the time required to fetch an instruction.

cyclic A condition of either steady-state or transient oscillation of a signal about the nominal value.

cyclic code A form of gray code, used for expressing numbers in which, when coded values are arranged in the numeric order of real values, each digit of the coded value assumes its entire range of values alternately in ascending and descending order.

cyclic compounds In organic chemistry, compounds containing a ring of atoms.

cyclic load *1.* Repeated loading that can lead to fracture. *2.* Repeated loads that change in value at a specific point.

cyclic redundancy check (CRC) An error-checking technique in which a checking number is generated by taking the remainder after dividing all the bits in a block (in serial form) by a predetermined binary number.

cyclic redundancy check character A character used in a modified cyclic code for error detection and correction.

cyclic shift A shift in which the data moved out of one end of the storing register are re-entered into the other end, as in a closed loop.

cyclic strain hardening exponent (n´) The power to which "true" plastic strain amplitude must be raised to be proportional to "true" stress amplitude.

cyclic strength coefficient (K´) The "true" stress at a "true" plastic strain.

cyclic testing (without applying stress) Accelerated testing and simulation of service conditions by the use of controlled alternating exposure to at least two corrosive environments, such as salt or other chemical exposure, water immersion, temperature variations, humidity variations, ultraviolet (UV) light exposure, mud or clay contamination, gravel or shot blasting.

current density The current per unit area; generally expressed as amps per sq cm.

cyclic yield strength (0.2% σ_{ys}) The stress to cause 0.2% inelastic strain as measured on a cyclic stress-strain curve. It is determined by constructing a line parallel to the slope of the cyclic stress-strain curve at zero stress through 0.2% strain and zero stress. The stress where the constructed line intercepts the cyclic stress-strain curve is taken as the 0.2% cyclic yield strength.

cycling Periodic repeated variation in a controlled variable or process action. *See also* cycle.

cyclodextrin Degradation products of starch that have six to eight glucose residues.

cyclograph A device for electromagnetically sorting or testing metal parts by means of the pattern produced on a cathode-ray tube when a sample part is placed in an electromagnetic sensing coil; the CRT pattern is different in shape for different values of carbon content, case depth, core hardness, or other metallurgical properties.

cycloparaffins (naphthenes) Hydrocarbons having a ring-type structure with only single bonds between the carbon atoms.

cyclotron A device that utilizes an alternating electric field between electrodes positioned in a constant magnetic field to accelerate ions or charged subatomic particles to high energies.

cyclotron frequency Frequency at which a charged particle orbits in a uniform magnetic field.

cyclotron radiation The electromagnetic radiation emitted by charged particles as they orbit in a magnetic field.

cyclotron resonance Energy transfer to charged particles in a magnetic field from an alternating-current electric field whose frequency is equal to the cyclotron frequency.

cyclotron resonance device Microwave amplifier based on the interaction between electromagnetic waves and transverse electron streams moving along helical trajectories.

D

DA *See* double amplitude displacement.

D/A *See* digital-to-analog converter.

DAC *See* digital-to-analog converter.

DAIP *See* diallyl isophthalate.

daisy chain *1.* A serial interconnection of devices. Signals are passed from one device to another, generally in the order of high priority to low priority. *2.* A method of propagating signals along a bus, often used in applications in which devices are connected in series.

Dalton's law The empirical generalization that for many so-called perfect gases, a mixture of these gases will have a pressure equal to the sum of the partial pressures that each of the gases would have as a sole component with the same volume and temperature, provided there is no chemical interaction.

dam Boundary support or ridge used to prevent excessive edge bleeding or resin runout of a laminate and to prevent crowning of the bag during cure.

damage, cavitation *See* cavitation damage.

damage resistance Increases in resistance due to damage to the load.

damp To suppress oscillations or disturbances.

damped systems Systems in which energy is dissipated by forces opposing the vibratory motion. Any means associated with a vibrating system to balance or modulate exciting forces will reduce the vibratory motion, but are not considered to be in the same category as damping. The latter term is applied to an inherent characteristic of the system without reference to the nature of the excitation.

dampener A device for progressively reducing the amplitude of spring oscillations after abrupt application or removal of a load.

damper *1.* A device or system for the purpose of suppressing oscillations or disturbances. *2.* A device for introducing a variable resistance for regulating the volumetric flow of gas or air.

damping *1.* The dissipation of energy with time or distance. *2.* Decreasing the time of vibrations in the motion of a body subject to influences that cause vibration. *3.* The transitory decay of the amplitude of a free oscillation of a system, associated with energy loss from the system. *4.* Dissipation of energy in a vibrating or oscillating system.

damping capacity The proficiency of a material to assimilate vibration by converting internal friction into heat.

damping coefficient The proportionality factor between the component of the applied force vector that is in phase with velocity and the velocity vector.

damping constant The component of applied force that is 90 deg out of phase with the deformation, divided by the velocity of deformation.

damping factor In any damped oscillation, the ratio of the amplitude of any given half-cycle to the amplitude of the succeeding half-cycle.

damping ratio *1.* A measure of the amount of damping for oscillatory systems. *2.* The ratio of the degree of actual damping to the degree of damping required for critical damping. *3.* The ratio of the deviations of the indicator following an abrupt change in the measurand in two consecutive swings from the position of equilibrium, the greater deviation being divided by the lesser. The deviations are expressed in angular measure.

damping, structural Also called solid damping. Damping due to internal friction within the material itself. It is independent of frequency and proportional to the maximum stress of the vibration cycle.

DAP *See* diallyl phthalate.

d'Arsonval galvanometer *See* light-beam galvanometer.

d'Arsonval movement The mechanism of a permanent-magnet moving-coil instrument such as a d'Arsonval galvanometer.

DAS *See* data acquisition system.

data *1.* A representation of facts, concepts, or instructions in a formalized manner suitable for communications, interpretation, or processing by humans or automatic means. *2.* Any representations such as characters or analog quantities to which meaning is or might be assigned. *3.* Information of any type. *4.* A common term used to indicate the basic elements that can be processed or produced by a computer.

data acquisition The function of obtaining data from sources external to a microprocess or a computer system, converting it to binary form, and processing it.

data acquisition system A system used for acquiring data from sensors via amplifiers and multiplexers and any necessary analog to digital converters.

data analysis *See* data processing and data reduction.

data bank A comprehensive collection of data, for example, several automated files, a library, or a set of loaded disks. Synonymous with database.

database *1.* Any body of information. *2.* A specific set of information available to a computer. *3.* A collection of interrelated data stored together with controlled redundancy to serve one or more applications. *4. See* data bank.

data capture (logging) The systematic collection of data to use in a particular data processing routine, such as monitoring and recording temperature changes over a period of time.

data channel A bidirectional data path between I/O devices and the main memory of a digital computer. Data channels permit one or more I/O operations to proceed concurrently with computation thereby enhancing computer performance.

data code A structured set of characters used to represent the data items of a data element, for example, the data codes 1, 2, ... , 7 may be used to represent the data items Sunday, Monday , ... , Saturday.

data converter Any of numerous devices for transforming analog signals to digital signals, or vice versa.

data display module A device that stores computer output and translates this output into signals which are distributed to a program-determined group of lights, annunciators, numerical indicators, and cathode ray tubes in operator consoles and remote stations.

data input/output unit (DI/OU) A device that interfaces to a process for the sole purpose of acquiring or sending data.

data link *1.* Equipment that permits the transmission of information in data format. *2.* Facility for transmission of information. *3.* A fiber-optic signal transmission system which carries information in digital or analog form.

data logger *1.* A system or sub-system with a primary function of acquisition and storage of data in a form that is suitable for later reduction and analysis, such as computer-language tape. *2.* A computer system designed to obtain data from process sensors and to provide a log of the data.

data processing The execution of a systematic sequence of operations performed upon data.

data protection Any method to preserve computer data from destruction or misuse. Backing up computer files is one example.

data reduction The process of transforming masses of raw test or experimentally obtained data, usually gathered by automatic recording equipment, into useful, condensed, or simplified intelligence.

datum *1.* A point, direction, or level used as a convenient reference for measuring angles, distances, heights, speeds, or similar attributes. *2.* Any value that serves as a reference for measuring other values of the same quantity.

datum level *See* datum plane.

datum plane A permanently established reference level, usually average sea level, used for determining the value of a specific altitude, depth sounding, ground elevation or water-surface elevation. Also known as a chart datum, datum level, reference level, or reference plane.

daughter A nuclide formed as a result of nuclear fission or radioactive decay.

dawsonite A mineral consisting of aluminum sodium carbonate.

dB *See* decibel.

DC *See* direct current.

DDC *See* direct digital control.

dead band *1.* The range through which the measured signal can be varied without initiating response. *2. See* dead zone.

dead load stress (S_1) The stress computed as defined under "stress" by using the difference in the readings obtained under dead load stress condition and "initial reference test condition" for each gage.

dead load stress condition (DL) The completely assembled crane structure on the test site and in the position or attitude, ready to apply the specified live load at the specified radius. Under this condition, the second reading for each gage is obtained, N_2. Note: Although the hook, block, slings, etc, are considered part of the suspended load, for purposes of safety and practicality they may be supported by the crane when this reading is taken.

dead mild steel Very low carbon content steel usually produced in sheet and strip.

dead soft A condition of lowest hardness and tensile strength created by full annealing. This expression relates in a relative sense compared to the lowest hardness to which a particular material can be annealed; it has no absolute hardness value.

dead time *1.* The interval of time between initiation of an input change or stimulus and the start of the resulting response. *2.* Any definite delay deliberately placed between two related actions in order to avoid overlap that might cause confusion or to permit a particular different event, such as a control decision, switching event, or similar action to take place. *3.* The time during which an instrument or system is processing information and thus cannot accept input.

deadweight gage A device used to generate accurate pressures for the purpose of calibrating pressure gages.

dead zone *1.* In ultrasonic inspection, the zone(s) within which a defect cannot be observed. *2.* Also called dead band. A range of values around the set point. When the controlled variable is within this range, no control action takes place.

deaeration Removing a gas—air, oxygen, or carbon dioxide, for example—from a liquid or semisolid substance.

debond *1.* An unbonded or nonadhering area in a composite laminate. *2.* Separating a bonded joint for repair work.

debug *1.* To locate and correct any errors in a computer program. *2.* To detect and correct malfunctions in the computer itself. Related to diagnostic routine. *3.* To submit a newly designed process, mechanism or computer program to simulated or actual operating conditions for the purpose of detecting and eliminating flaws or inefficiencies.

debuggers System programs that enable computer programs to be debugged.

debugging A process to detect and remedy inadequacies. With respect to software, it is the development process to locate, identify, and correct programming mistakes.

debugging aid routine A routine to aid programmers in the debugging of their routines. Some typical aid routines are storage printout, tape printout, and drum printout routines.

debulking Compressing a composite laminate under heat to remove air and prevent wrinkles.

deburr To remove burrs, fins, sharp edges, and the like from corners and edges of parts or from around holes, by any of several methods, often involving the use of abrasives.

Debye length A theoretical length that describes the maximum separation at which a given electron will be influenced by the electric field of a given positive ion.

decade *1.* A group or assembly of ten units, for example, a counter that counts to ten in one column, or a resistor box that inserts resistance quantities in multiples of powers of 10. *2.* A period of ten years.

decalescence A phenomenon associated with the transformation of alpha iron to gamma iron on the heating (superheating) of iron or steel, revealed by the darkening of the metal surface owing to the sudden decrease in temperature caused by the fast absorption of the latent heat of transformation. Contrast with recalescence.

decanting Boiling or pouring off liquid near the top of a vessel that contains two immiscible liquids or a liquid-solid mixture that has separated by sedimentation, without disturbing the heavier liquid or settled solid.

decarburization The loss of carbon from the surface layer of a carbon-containing alloy due to reaction with one or more chemical substances in a medium that contacts the surface.

decarburizing Removing carbon from the surface layer of a steel or other ferrous alloy by heating it in an atmosphere that reacts selectively with carbon; atmospheres that are relatively rich in water vapor or carbon dioxide function in this manner.

decay Decrease of a radioactive substance because of nuclear emission of alpha or beta particles, positrons, or gamma rays.

decay time The time in which a voltage or current pulse will decrease to one tenth of its maximum value. Decay time is proportional to the time constant of the circuit.

decelerate To reduce in speed.

decelerating electrode An intermediate electrode in an electron tube which is maintained at a potential that induces decelerating forces on a beam of electrons.

deceleration The act or process of moving, or of causing to move, with decreasing speed. Sometimes called negative acceleration.

decelerometer An instrument for measuring the rate at which speed decreases.

decibel (dB) A unit for measuring relative strength of a signal parameter, such as power, voltage, etc. The number of decibels is twenty (ten for power ratio) times the logarithm (base 10) of the ratio of the measured quantity to the reference level.

decibel meter An instrument calibrated in logarithmic steps and used for measuring power levels, in decibel units, of audio or communication circuits.

decimal *1.* Pertaining to a characteristic or property involving a selection, choice, or condition in which there are ten possibilities. *2.* Pertaining to the numeration system with a radix of ten. *3. See* binary code decimal.

decimal balance A type of balance having one arm ten times as long as the other, so that heavy objects can be balanced with light weights.

decimal coded digit A digit or character defined by a set of decimal digits, such as a pair of decimal digits specifying a letter or special character in a system of notation.

decimal digit In decimal notation, one of the characters 0 through 9.

decimal notation A fixed radix notation, where the radix is ten; for example, in decimal notation, the numeral 576.2 represents the number 5×10^2 plus 7×10^1, plus 6×10, plus 2×10^{-1}.

decimal number A number, usually of more than one figure, representing a sum, in which the quantity represented by each figure is based on the radix of ten. The figures used are 0, 1, 2, 3, 4, 5, 6, 7, 8, and 9.

decimal point The radix point in decimal representation.

decimal-to-binary conversion The process of converting a number written to the base ten, or decimal, into the equivalent number written to the base two, or binary.

decision instruction An instruction that effects the selection of a branch in a program, for example, a conditional jump instruction.

decision table A table of all contingencies that are to be considered in the description of a problem, together with the actions to be taken. Decision tables are sometimes used in place of flow charts for problem description and documentation.

decode 1. To apply a code so as to reverse some previous encoding. 2. To determine the meaning of individual characters or groups of characters in a message. 3. To determine the meaning of instructions from the status of bits which describe the instruction, command, or operation to be performed.

decoder 1. A device that determines the meaning of a set of signals and initiates a computer operation based on them. 2. A matrix of switching elements that selects one or more output channels according to the combination of input signals present. Contrast with encoder. Clarified by matrix. 3. A device used to change computer data from one coded format to another.

decollate The separation of multi-part computer forms.

decomposition The separation of a compound into its individual chemical elements.

decompression Any method for relieving pressure.

decontamination Removing or neutralizing an unwanted chemical, biological, or radiological substance.

decoration aging treatment A relatively low-temperature, short duration aging treatment used to determine the degree of recrystallization in cold- or hot-worked metastable beta titanium alloys that have been solution heat treated.

decoupling control A technique in which interacting control loops are automatically compensated when any one control loop takes a control action.

decrement 1. The quantity by which a variable is decreased. 2. A specific part of an instruction word in some binary computers, thus a set of digits.

decremeter An instrument for measuring the damping of a train of waves by determining its logarithmic decrement.

deep drawing A press operation for forming cup-shaped or deeply recessed parts from sheet metal by forcing the metal to undergo plastic deformation between dies without substantial thinning.

default 1. The value of an argument, operand, or field assumed by a program if a specific assignment is not supplied by the user. 2. The alternative assumed when an identifier has not been declared to have one of two or more alternative attributes.

defect A deviation of an item or material from some ideal state. The ideal state is usually described in a formal specification. Such is generally thought or known to adversely influence service performance. See also imperfection.

defect, critical A defect that could result in hazardous or unsafe conditions for individuals using, maintaining, or depending on the item or material.

defective 1. An item or material with at least one defect. 2. A material with a variety of small defects in sufficient number to render the item or material unusable.

definition 1. The resolution and sharpness of an image, or the extent to which an image is brought into sharp relief. 2. The degree to which a communication system reproduces sound images or messages.

deflagration A sudden or rapid burning, as opposed to a detonation or explosion.

deflashing Removing fins or protrusions from the parting line of a die casting, forging, or molded plastics part.

deflecting force In a direct-acting recording instrument, the force produced at the marking device, for any position of the scale, by its positioning mechanism acting in response to the electrical quantity being measured.

deflection *1.* Movement of a pointer away from its zero or null position. *2.* Elastic movement of a structural member under load. *3.* Shape change or change in diameter of a tubular member without fracturing the material.

deflection temperature under load (DTUL) The temperature at which a beam deflects a specific amount under load.

deflectometer An instrument for determining minute elastic movements that occur when a structure is loaded.

deflector A device for changing direction of a stream of air or of a mixture of pulverized fuel and air.

deflocculation The processing of a powder suspension where the particles in suspension are separated from each other by electrical charges on the particle surface.

defocus To cause a beam of electrons, light, x-rays, or other type of radiation to depart from accurate focus at a specific point in space, ordinarily the surface of a workpiece or test object.

deformation A stress-induced change of form or shape.

deformation under load The dimensional change of material that occurs from the initial load deformation to continued deformation that occurs in a specific time.

degas To remove dissolved, entrained, or adsorbed gas from a solid or liquid.

degasification Removal of gases from samples of steam taken for purity tests. Removal of CO_2 from water as in the ion exchange method of softening.

degasifier *1.* An element or compound added to molten metal to remove dissolved gases. *2.* A process or type of vessel that removes dissolved gases from molten metal.

degassing The deliberate removal of gas from a material, usually by application of heat under high vacuum. *See also* bakeout.

degradation A gradual deterioration in performance as a function of time.

degradation factor A factor that, when multiplied by the predicted mean time between failures, yields a reasonable estimate of the operational mean time between failures.

degradation failure Gradual shift of an attribute or operating characteristic to a point where the device no longer can fulfill its intended purpose.

degrease To remove oil and grease from adherend surfaces.

degreasing An industrial process for removing grease, oil, or other fatty substances from the surfaces of metal parts, usually by exposing the parts to condensing vapors of a polyhalogenated hydrocarbon solvent.

degree *1.* A unit of measure of a temperature scale. *2.* A unit of angular measure equal to 1/360 of a circle. *3.* Relative intensity. *4.* One of a series of stages or steps.

degree of polymerization The number of structural units in a polymer molecule.

degree of saturation The ratio of the weight of water vapor related to a pound of dry air saturated at the same temperature.

degree rise The amount of increase in temperature caused by the flow of electrical current through a wire.

degrees of freedom *1.* The number of independent variables (such as temperature, pressure, or concentration within the phases present) that may be altered at will without causing a phase change in an alloy system at equilibrium. *2.* The number of such variables that must be fixed arbitrarily to completely define the system.

dehumidification Reducing the moisture content of air, which increases its cooling power.

deicing Using heat, chemicals, or mechanical rupture to remove ice deposits, especially those that form on motor vehicles and aircraft at low temperatures or high altitudes.

deionize *1.* To remove ions from. *2.* To restore gas that has become ionized to its former condition.

deionized water Water from which the charged species (Cl^-, Ca^{2+}, etc.) has been removed.

delamination Separation of a material into layers, especially a material such as a bonded laminate or metal in sheet form.

delayed fracture Failure whereby fracture occurs in time even though the load on the component is less than would ordinarily cause fracture. Such failure

can be caused by hydrogen embrittlement.

delay-interval timer A timing device that is electrically reset to delay energization or deenergization of a circuit.

delay line A transmission medium that delays a signal passing through it by a known amount of time; typically used in timing events.

delay modulation A method of data encoding for serial data transmission or recording.

delimiter A character that separates, terminates, or organizes elements of a character string, statement, or program.

deliquescence The absorption of water vapor by a crystalline structure until the point that the structure dissolves into a solution.

delta ferrite *See* ferrite.

delta limits The difference between initial and final readings, usually associated with the difference between zero time readings on a life test and the final readings. Determines how much parameters shift during the test.

delta network A set of three circuit branches connected in series, end to end, to form a mesh having three nodes.

delube The removal of a lubricant from a powder metal compact by burnout or chemical means.

demand meter Any of several types of instruments used to determine the amount of electricity used over a fixed period of time.

demodulation The process of retrieving intelligence (data) from a modulated carrier wave. The reverse of modulation.

demodulator Electronic device that operates on an input of a modulated carrier to recover the modulating wave as an output.

dendrite A crystal with a tree-like pattern resulting from cast metals cooling slowly through solidification.

dendritic coring Segregation in solid solution alloys during rapid solidification.

dendritic porosity Tiny shrinkage cavities formed within dendrite arms.

dendritic powder Particles with a tree-like appearance.

denier A numbering system for yarn or filament in which the denier number is equal to the gram weight of 9000 meters of the material.

densitometer *1.* Instrument for measuring the density of specific gravity of liquids, gases, or solids. *2.* An instrument for determining optical density of photographic or radiographic film by measuring the intensity of transmitted or reflected light.

density *1.* The weight of a substance for a specified volume at a definite temperature, for example, grams per cubic centimeter at 20°C. *2.* Closeness of texture or consistency. *3.* Degree of opacity, often referred to as optical density.

density, absolute *1.* The ratio of mass of a solid material to an equal mass of water. *2.* The mass per unit volume of a solid material shown in grams per cubic centimeter.

density bottle *See* specific gravity bottle.

density correction Any correction made to an instrument reading to compensate for the deviation of density from a fixed reference value; it may be applied because the fluid being measured is not at standard temperature and pressure, because ambient temperature affects density of the

fluid in a fluid-filled instrument, or because of other similar effects.

density, dry The mass per unit volume of a sintered compact that is not impregnated.

density, mass The mass of any substance per unit volume.

density ratio The percentage of density of a powder metal compact compared to the density of metal of the same composition.

density recovery efficiency The ratio of the charge density increase achieved from cooling the charge air, to the density decrease due to the temperature rise in the process of compressing the charge air.

density recovery ratio The ratio of the charge air density at the engine intake manifold to the air density at conditions of ambient temperature and boost pressure.

density, wet The mass per unit volume of a sintered compact impregnated with a nonmetallic material.

deoxidizer In metal making, a substance that can be added to molten metal to remove oxygen.

deoxidizing *1.* Removing oxygen from molten metal. *2.* Removing oxide films from material surfaces.

dependability *1.* A measure of the degree to which an item is operable and capable of performing its required function at any (random) time during a specified mission profile, given item availability at the start of the mission. (Item state during a mission includes the combined effects of the mission-related system R&M parameters, but exclude nonmission time. *See* availability). *2.* The quality of service that the network is capable of providing.

dependent variables Variables considered as a function of other variables, the latter being called independent.

depleted layer *1.* A zone between p-type and n-type semiconductor materials where free electrons and holes have combined, reducing the number of charge carriers. *2.* In steels, a specific layer that exhibits a decreased quantity of a stated chemical element.

depletion Removal of one element in an alloy, typically from the surface or a grain-boundary region.

depolarizers Optical components that scramble the polarization of light passing through them, effectively turning a polarized beam into an unpolarized beam.

depolymerization Separation of a more complex molecule into two or more simpler molecules chemically similar to and having the same empirical composition as the original.

deposit *1.* Any substance intentionally laid down on a surface by chemical, electrical, electrochemical, mechanical, vacuum, or vapor-transfer methods. *2.* Solid or semisolid material accumulated by corrosion or sedimentation on the interior of a tube or pipe. *3.* A foreign substance, from the environment, adhering to a surface of a material.

deposit attack Corrosion occurring under or around a discontinuous deposit on a metallic surface.

deposited metal In a weldment, filler metal added to the joint during welding.

deposit gage Any instrument used for assessing atmospheric quality by measuring the amount of particulate matter that settles out on a specific area during a defined period of time.

deposition Applying a material to another material by means of electrical, chemical or vapor methods.

deposition rate *1.* The amount of filler metal deposited per unit time by a specific welding procedure, usually expressed in pounds per hour. *2.* The rate at which a coating material is deposited on a surface, usually expressed as weight per unit area per unit time, or as thickness per unit time.

deposition sequence The order in which increments of a weld deposit are laid down.

depression A depressed part or piece; a hollow or low place on a surface.

depth dose Measurement and recording of the radiation penetration from the surface of a material.

depth gage An instrument or micrometer device capable of measuring distance below a reference surface to the nearest 0.001 inch.

depth of engagement The radial contact distance between mating threads.

depth of field The depth of an object space that is in focus.

depth of fusion The distance from the original surface that the molten zone extends into the base metal during welding.

depth of penetration The distance a radiation probe penetrates below the surface of a specimen.

depth of thread The radial distance from crest to root of a screw thread.

derandomizer The circuit that removes the effect of data randomizing, thereby recovering data that had been randomized for tape storage.

derating The intentional reduction of stress/strength ratio in the application of an item, usually for the purpose of reducing the occurrence of stress-related failures.

derivative *1.* Mathematically, the reciprocal of rate. 2. A control action that will cause the output signal to change according to the rate at which input signal variations occur during a certain time interval.

derivative action A type of control-system action in which a predetermined relation exists between the position of the final control element and the derivative of the controlled variable with respect to time.

derivative control Change in the output that is proportional to the rate of change of the input. Also called rate control. *See* control action, derivative (D).

descaling Removing adherent deposits from a metal surface, such as thick oxide from hot rolled or forged steel, or inorganic compounds from the interior of boiler tubes. Such removal can be done mechanically or chemically.

desiccant A substance with a propensity to absorb water, used to dry material.

design allowables Anticipated strengths of materials defined by a test program and/or statistical calculations.

design error A mistake in the design process resulting from incorrect methods or incorrect application of methods or knowledge.

design evaluation tests Tests performed to evaluate the design under environmental conditions and to verify compatibility of interfaces, adequacy of tools and test equipment, etc., for the established maintenance concept.

design for testability A design process or characteristic thereof such that deliberate effort is expended to assure that a product may be thoroughly tested with minimum effort, and that high confidence may be ascribed to test results.

design load *1.* A specified maximum load that a structural member or part should withstand without failing. 2. The load for which a steam-generating unit is designed, considered the maximum load to be carried.

design pressure The maximum allowable working pressure permitted under the rules of the ASME Construction Code.

design stress The maximum permissible load per unit area that a given structure can withstand in service, including all allowances for such things as unexpected or impact loads, corrosion, dimensional variations during fabrication, and possible underestimation of service loading.

design support tests Tests performed to determine the need for parts, materials, and component evaluation or qualification to satisfy characteristic stability, interchangeability, failure rate, tolerances, design margins, and other reliability design criteria.

design thickness The sum of thickness required to support service loads. This method of specifying material thickness is used particularly when designing boilers, chemical process equipment, and metal structures that will be exposed to atmospheric environments, soils, or seawater.

design verification tests Tests performed to verify the functional adequacy of the design.

desizing Eliminating undesirable surface materials from a material before further processing.

desorption Removing adsorbed material.

destannification The loss of tin from the surface of bronze alloys.

destaticization A treatment for plastic materials that reduces the effect of static electricity on the surface.

destructive testing Any method of determining a material property, functional attribute, or operational characteristic which renders the test object unsuitable for further use or severely impairs its intended service life.

detectability The quality of a measured variable in a specific environment that is determined by relative freedom from interfering energy or other characteristics of the same general nature as the measured variable.

detection limit The smallest concentration of a substance that can be measured.

detector *1.* Sensor, or an instrument employing a sensor, to detect the presence of something in the surrounding environment. *2.* A device that detects light, generating an electrical signal that can be measured or otherwise processed. *See* transducer.

detergent A natural material or synthetic substance having the soaplike quality of being able to emulsify oil and remove soil from a surface.

deterioration A decline in the quality of a device, mechanism, or structure over time due to environmental effects, corrosion, wear, or gradual changes in material properties.

determination The quantity or concentration of a substance in a test sample. *See also* measure.

detonation The extremely rapid chemical decomposition (explosion) of a material.

detritus Debris material resulting from friction and wear.

deuterium A heavy isotope of hydrogen having one proton and one neutron in the nucleus. Used for hydrogen 2.

deuterium fluorides Fluorides of deuterium, a heavy isotope of hydrogen.

deuterium oxide *See* heavy water.

deuteron detector A type of specialized radiation detector used in some nuclear reactors to detect the concentration of deuterium nuclei present.

deuterons The nuclei of deuterium atoms.

developed blank *See* finished blank.

developer A fine powder usually mixed with a penetrant to highlight defects in a material.

deviation *1.* The difference between control or set point, and a value of process variable. *2.* In quality control, any departure of a quality characteristic from its specified value. *3.* A statistical quantity that gives a measure of the random error that can be expected in numerous independent measurements of the same value under the same conditions of measurement.

deviation alarm An alarm that is set whenever the deviation exceeds the preset limits.

device *1.* A component or assembly designed to perform a specific function by harnessing mechanical, electrical, magnetic, thermal, or chemical energy. *2.* Any piece of machinery or computer hardware that can perform a specific task. *3.* A component in a control system, for example, the primary element, transmitter, controller, recorder, or final control element.

devitrification The formation of unwanted crystals in glass during processing.

devitrify To deprive of vitreous quality. To make glass or vitreous rock opaque and crystalline.

dewatering Removing water from solid or semisolid material, for instance, by centrifuging, filtering, settling, or evaporation.

dewetting *1.* Generally, loss of surface attraction between a solid and a liquid. *2.* Specifically, flow of solder away from a soldered joint upon reheating.

dew point Temperature at which water vapor beings to condense.

dew-point recorder An instrument that determines dew-point temperature.

dew-point temperature The temperature at which condensation of moisture from the vapor phase begins.

dezincification *1.* A corrosion phenomenon resulting in the parting of zinc from copper-zinc alloys. *2.* The loss of zinc vapor from brass compacts during sintering. *3.* Corrosion occurring in some brass alloys when exposed to sea water.

DF *See* deuterium fluorides.

DF lasers Gas lasers in which the active material is deuterium fluoride.

DGA *See* differential gravimetric analysis.

D-glass Strong glass fiber used in laminates with a high boron content.

diadic polyamide A polyamide derived from diamine and dicarboxylic acid.

diagnosis The careful investigation of symptoms and facts leading to the determination of the cause of a problem.

diagnostic *1.* Pertaining to the detection and isolation of a malfunction or mistake. *2.* Program or other system feature designed to help identify malfunctions in the system. An aid to debugging.

diagnostic accuracy The percentage of failures correctly diagnosed, based on the possible failure population

diagnostic alarm An alarm that is set whenever the diagnostic program reports a malfunction.

diagnostic capability All capabilities associated with the detection and isolation of faults, including built-in test, automatic test systems, and manual test.

diagnostic flow chart A test-oriented logical description of branching routines sued in a test sequence to describe the steps taken to diagnose a failure successfully.

diagnostic function test (DFT) A program to test overall system reliability.

diagnostic message An error message in a programming routine to help the programmer identify the error.

diagnostic programs *1.* A troubleshooting aid for locating hardware malfunctions in a system. *2.* A program to aid in locating coding errors in newly developed programs. *3.* Computer programs that isolate equipment malfunctions or programming errors.

diagnostic routine A routine used to locate a malfunction in a computer, or to aid in locating mistakes in a computer program.

diagnostic(s) Information concerning known failure modes and their characteristics which can be used in troubleshooting and failure analysis to help pinpoint the cause of a failure and aid in defining suitable corrective measures.

diagnostic test The running of a machine program or routine for the purpose of discovering a failure or a potential failure of a machine element, and to determine its location or its potential location.

dial indicator *1.* Any meter or gage with a graduated circular face and a pivoted pointer to indicate the reading. *2.* A type of measuring gage used to determine fine linear measurements, such as radial or lateral runout of a rotating member.

diallyl ester An ester produced by the reaction of dibasic acid and allyl alcohol.

diallyl isophthalate (DAIP) A high-strength, thermosetting resin used as a molding compound and laminating coating.

diallyl phthalate (DAP) Similar to DAIP resins, but with increased resistance to alkaline chemicals.

diamagnetic material A substance whose specific permeability is less than 1.00 and is therefore weakly repelled by a magnetic field.

diametrical strength The load required to crush a cylindrical compact in a direction perpendicular to the axis.

diamond The pure crystalline form of carbon.

diamond-pyramid hardness A material hardness determined by indenting a specimen with a diamond-pyramid indenter having a 136° angle between opposite faces then calculating a hardness number by dividing the indenting load by the pyramidal area of the impression. Also known as Vickers hardness.

diamond wheel A grinding wheel for cutting very hard materials which uses synthetic diamond dust as the bonded abrasive material.

diaphragm *1.* A thin, flexible disc that is supported around the edges and whose center is allowed to move in a direction perpendicular to the plane of the disc.

2. A partition of metal or other material placed in a header, duct, or pipe to separate portions thereof.

diathermous Describes a material capable of issuing thermal radiation.

dichroic filter A filter that selectively transmits some wavelengths of light and reflects others.

dichromate treatment A technique for producing a corrosion-resistant conversion coating on magnesium parts by boiling them in a sodium dichromate solution.

didymium A mixture of rare earth elements that is free of cerium.

die casting *1.* A casting process in which molten metal is forced under pressure into the cavity of a metal mold. *2.* A part made by this process.

die forging *1.* The process of forming shaped metal parts by pressure or impact between two dies. *2.* A part formed in this way.

dielectric A medium in which it is possible to maintain an electric field with little or no supply of energy from outside sources.

dielectric coating An optical coating made up of one or more layers of dielectric (nonconductive) materials. The layer structure determines what fractions of incident light at various wave lengths are transmitted and reflected.

dielectric constant *1.* The property that determines the electrostatic energy stored per unit volume for unit potential gradient. Note: This numerical value usually is given relative to a vacuum. *2.* A material characteristic expressed as the capacitance between two plates when the intervening space is filled with a given insulating material divided by the capacitance of the same plate arrangement when the space is filled with air or is evacuated.

dielectric curing The curing of a synthetic thermosetting resin by the passage of an electric charge through the resin.

dielectric heating The heating of materials by dielectric loss in a high-frequency electrostatic field.

dielectric loss A loss of energy of a dielectric that is positioned within an alternating electrical field.

dielectrics Substances that contain few or no free charges and that can support electrostatic stresses.

dielectric strength (electric strength or breakdown strength) The potential gradient at which electric failure or breakdown occurs. To obtain the true dielectric strength, the actual maximum gradient must be considered or the test piece and electrodes must be designed so that uniform gradient is obtained. The value obtained for the dielectric strength in practical tests will usually depend on the thickness of the material and on the method and conditions of test.

dielectric test 1. The high voltages impressed between a component and the frame of the alternator to check insulation characteristics. 2. Test that consists of the application of a voltage higher than the rated voltage for a specified time for the purpose of determining the adequacy against breakdown of the insulation under normal conditions.

dielectrometry Measurement of the changes in capacitance of a laminate during cure.

die scalping Improving the surface quality of bar stock, rod tubing, or wire by drawing it through a sharp-edged die to remove a thin surface layer containing minor defects.

die welding Forge welding using shaped dies.

difference limen The increment in a stimulus that is barely noticed in a specified fraction of independent observations where the same increment is imposed.

differential aeration cell An electrolytic cell, the emf of which is due to difference in air (oxygen) concentration at one electrode as compared with that at another electrode of the same material.

differential analyzer A computer (usually analog) designed and used primarily for solving many types of differential equations.

differential annealing Annealing only a portion of a metal as an aid in certain manufacturing processes.

differential coating 1. A coated product that has different coating masses and/or different coating compositions on the two surfaces of the steel substrate. 2. A coated product having a specified coating on one surface and a significantly lighter coating on the other surface, usually hot-dipped galvanized.

differential gap The smallest increment of change in a controlled variable required to cause the final control element in a two-position control system to move from one position to its alternative position.

differential gravimetric analysis (DGA) A heat analysis test used for polymer materials which measures the weight difference between a test specimen and a control sample during heating.

differential hardening Hardening only a portion of a material as an aid in certain manufacturing processes.

differential heating Heating and cooling a material such that prescribed differences of stress and properties exist within the material.

differential input The difference between the instantaneous values of two voltages, both being biased by a common mode voltage.

differential instrument Any instrument that has an output signal or indication proportional to the algebraic difference between two input signals.

differentially zinc coated A sheet (usually steel) with a zinc coating of a different thickness on one side than on the other side.

differential pressure The difference in pressure between two points of measurement.

differential-pressure gage Any of several instruments designed to measure the difference in pressure between two enclosed spaces, independent of their absolute pressures.

differential scanning calorimetry (DSC) The measurement of the amount of energy absorbed or produced as a resin is cured.

differential thermal analysis (DTA) A testing technique in which a test specimen and a control sample are heated simultaneously and the differences in their temperature is monitored to determine heat capacity, changes in structure, and chemical reactions.

differentiator A device whose output function is proportional to the derivative, i.e., the rate of change, of its input function with respect to one or more variables (usually with respect to time).

diffracted beam In x-ray crystallography, a beam of radiation composed of a large number of scattered rays mutually reinforcing one another.

diffraction *1.* Deviation of light from the paths and foci prescribed by rectilinear propagation; phenomenon responsible for bright and dark bands found within a geometrical shadow. *2.* A phenomenon associated with the scattering of waves when they encounter obstacles whose size is about the same order of magnitude as the wavelength.

diffraction contrast In electron microscopy, the contrast produced by different beam intensities diffracted from a crystalline material.

diffraction pattern The arrangement and intensity of diffracted beams.

diffuse *1.* Spread out; not concentrated. To pour in different directions. *2.* In physics, to mix by diffusion, as gases, liquids, etc.

diffuser *1.* A duct, chamber, or enclosure in which low-pressure, high-velocity flow of a fluid, usually air, is converted to high-pressure, low-velocity flow. *2.* As applied to oil or gas burners, a metal plate with openings so placed as to protect the fuel spray from high-velocity air while admitting sufficient air to promote the ignition and combustion of fuel. Sometimes termed an impeller.

diffuse transmittance The situation where transmitted beams are emitted in all directions from the transmitting body.

diffusion *1.* Conversion of gas-flow velocity into static pressure, as in the diffuser casing of a centrifugal fan. *2.* The movement of ions from a point of high concentration to low concentration. *3.* Migration of atoms, molecules, or ions spontaneously, under the driving force of compositional differences, and using only the energy of thermal excitation to cause atom movements. *See also* suffusion and perfusion.

diffusion brazing Filler material that is distributed by capillary action between heated joining surfaces.

diffusion coating Any process whereby a basis metal or alloy is either (a) coated with another metal or alloy and heated to a sufficient temperature in a suitable environment or (b) exposed to a gaseous or liquid medium containing the other metal or alloy, thus causing diffusion of the coating or of the other metal or alloy into the basis metal with resultant change in the composition of properties of its surface.

diffusion coefficient The absolute value of the ratio of the molecular flux per unit area to the concentration gradient of a gas diffusing through a gas or a porous medium.

diffusion porosity The porosity that results when one metal diffuses into another during sintering.

diffusivity A term used to describe temperature differences within a material.

digital *1.* A method of measurement using precise quantities to represent variables. *2.* Binary. *3.* A reference to the representation of data by discrete pulses, as in the presence or absence of a signal level to indicate the 1's and 0's of binary data. *4.* A type of readout in which the data are displayed as discrete, fully-informed alphanumeric characters.

digital back-up An alternate method of digital process control initiated by use of special-purpose digital logic in the event of a failure in the computer system.

digital circuits *See* digital electronics.

digital differential analyzer *1.* An incremental computer in which the principal type of computing unit is a digital integrator whose operation is similar to the operation of an integrating mechanism. *2.* A differential analyzer that uses digital representation for the analog quantities.

digital electronics The use of circuits in which there are usually only two states possible at any point. The two states can represent any of a variety of binary digits (bits) of information.

digital output Transducer output that represents the magnitude of the measurand in the form of a series of discrete quantities coded in a system of notation.

digital readout An electrically powered device that interprets a continuously variable signal and displays its amplitude, or another signal attribute, as a series of numerals or other characters that correspond to the measured value and can be read directly.

digital resolution The value of the least significant digit in a digitally coded representation.

digital signal A discrete or discontinuous signal, one whose various states are discrete intervals apart.

digital-to-analog converter (D/A or DAC) A device, or sub-system that converts binary (digital) data into continuous analog data, for example, to drive actuators of various types.

digitized signal Signal in which information is represented by a set of discrete values, in accordance with a prescribed law. Every discrete value represents a definite range of the original undigitized signal. *See also* analog-to-digital converter.

dihydroxyphenylalanine *See* dopa.

dilatant substance A material that flows under low shear stress, but whose rate of flow decreases with increasing shear stress.

dilatometer An apparatus for accurately measuring thermal expansion of materials. *See also* extensometer.

diluent A substance that reduces the viscosity and pour point of a substance.

dilution *1.* Adding solvent to a solution to lower its concentration. *2.* Melting low-alloy base metal or previously deposited weld metal into high-alloy filler metal to produce a weld deposit of intermediate composition.

dilution factor The ratio of the volume or mass of a sample after dilution to the volume or mass before dilution.

dimensional stability The ability to retain manufactured shape and size after having experienced the combination of operating stresses and temperatures.

dimer A compound formed by condensation from two monomers or molecules.

dimetcote An inorganic zinc coating composed of two materials: (1) a reactive liquid and (2) a finely divided powder, which are mixed together. The mixture reacts in place with a steel surface to form an insoluble coating.

dimorphous Describes a substance that crystallizes in different forms.

dimple A defect or imperfection resulting from foreign matter being mechanically pressed into sheet, plate, or forgings.

dimple rupture A term used in fractography to describe ductile fracture that occurs through the formation of small voids along the fracture path.

dings Accidental impact damage, similar in appearance to dimples.

diode A two-electrode electronic component containing merely an anode and a cathode.

diode laser A laser in which stimulated emission is produced at a p-n junction in a semiconductor material.

diopter A measurement of refractive power of a lens equal to the reciprocal of the focal length in meters.

DI/OU *See* data input/output unit.

dip brazing Producing a brazed joint by immersing the assembly in a bath of hot molten chemicals or hot metal.

dip coating Covering the surface of a part by immersing it in a bath containing the coating material.

diphenyl oxide resins Thermoset resins derived from diphenyl oxide with good heat and handling properties.

dipole A system composed of two, separated, equal electric or magnetic charges of opposite sign.

dipole antenna *1.* A straight radiator, usually fed in the center, and producing a maximum of radiation in the plane normal to its axis. *2.* A center-fed antenna that is approximately half as long as the wavelength of the radio waves it is primarily intended to transmit or receive.

dip soldering A process similar to dip brazing, but using a lower-melting filler metal.

dip-spin A process using a perforated basket, in which parts are placed to be dipped into an organic/inorganic finish, spun to remove excess coating, then normally placed in an oven to cure the finish.

direct access The retrieval or storage of data by a reference to its location on a volume, rather than relative to the previously retrieved or stored data.

direct access device *See* random access device.

direct current (DC) An electrical current that travels uniformly in one direction.

direct digital control (DDC) A method of control in which all control outputs are generated by the computer directly, with no other intelligence between the central computer and the process being controlled.

directional property Any mechanical or physical property of a material whose value varies with orientation of the test axis within the test specimen.

directional solidification (crystals) Controlled solidification (crystal growth) of molten metal in a casting so as to provide feed metal to the solidifying front of the casting.

direct process Any method for producing a commercial metal directly from metal ore, without an intervening step, such as roasting or smelting, which produces semirefined metal or another intermediate product.

direct quenching Quenching carburized parts directly from the carburizing temperature.

disable *1.* To remove or inhibit a computer hardware or software feature. *2.* To disallow the processing of an established interrupt until interrupts are enabled. Contrast with enable. *See* disarm.

disarm To cause an interrupt to be completely ignored. Contrast with arm. *See* disable.

disassemble To reduce an assembly to its component parts by loosening or removing threaded fasteners, pins, clips, snap rings, or other mechanical devices—in most instances, for some purpose such as cleaning, inspection, maintenance, or repair followed by reassembly.

disbond An area within bonded surfaces where adhesion failure has occurred.

discharge The reduction of an imbalanced charge at a given point (usually to zero), by a sudden flow of current through a medium (usually to ground).

discoloring In painting, a change in the absolute color of the topcoat usually due to a chemical interaction between the topcoat and the sealer.

disconnect *1.* To disengage the apparatus used in a connection and to restore it to its ready condition when not in use. *2.* To disengage the linkage between an interrupt and a designated interrupt servicing program. *See* connect.

discontinuity *1.* A break in sequence or continuity of anything. *2.* Any feature within a bulk solid that acts as a free surface; it may be a crack, lap, seam, pore, or other physical defect, or it may be a sharp boundary between the normal structure and an inclusion or other second phase.

discontinuous yielding Plastic deformation occurring at a sharp boundary between the normal structure and an inclusion.

discrepancy Any inconsistency between a requirement for a characteristic of a material or an item, and the actual characteristic of the material or item.

discrete *1.* Pertaining to distinct elements or to representation by means of distinct elements, such as characters. *2.* In data processing, data organized in specific parts. *3.* An individual bit from a selected word.

discrimination ratio A measure of the distance between two specific points of the operating characteristics curve which are used to define the acceptance sampling plan.

disengage To intentionally pull apart two normally meshing or interlocking parts, such as gears or splines, especially for the purpose of interrupting the transmission of mechanical power.

dished A distortion that appears concave on the surface of a material.

dishing A metalforming operation that forms a shallow concave surface.

disinfectant A chemical agent that destroys microorganisms, bacteria, and viruses or renders them inactive.

disintegrated Separated or decomposed into fragments; having lost its original form.

dislocation An imperfection in the crystalline displacement of atoms in a metal.

dislocation climb A lattice defect in one area that merges with many jogs, causing a complete section of dislocation climbing to the next slip plane.

dislocation locking One dislocation "locked" to another, resulting in greater yield strength.

dispatching priority A number assigned to tasks, and used to determine precedence for the use of the central processing unit in a multitask situation.

disperser *See* emulsifier.

dispersing agent *See* emulsifier.

dispersion *1.* Any process that breaks up an inhomogeneous, lumpy mixture and converts it to a smooth paste or suspension where particles of the solid component are more uniform and small in size. *2.* Breaking up globs of oil and mixing them into water to make an emulsion. *3.* Intentionally breaking up concentrations of objects of substances and scattering them over a wide area.

4. The process by which an electromagnetic signal is distorted because the various frequency components of that signal have different propagation characteristics. *5.* The relationship between refractive index and frequency (or wavelength). *6.* (linear dispersion) In wave mechanics, the rate of change of distance along a spectrum with frequency.

dispersion strength The addition of chemically stable particles of a nonmetallic phase in a metallic material to reduce dislocation at high temperatures.

displacement *1.* A vector quantity that specifies the change of position of a body or particle, usually measured from the mean position or position of rest. *2.* The volume swept out by a piston as it moves inside a cylinder from one extreme of its stroke to the other extreme. *3.* For a reciprocating engine, pump, or compressor, the volume swept out by one piston as it moves from top dead center to bottom dead center, multiplied by the number of cylinders. *4.* Forcing a fluid or granular substance to move out of a cavity or tube by forcing more of the substance in, or by means of a piston or inflatable bladder that moves or expands into the space.

displacement resonance A condition of resonance in which the external sinusoidal excitation is a force and the specified response is displacement at the point where the force is applied.

displacement-type density meter A device that measures liquid density by means of a float and balance beam, used in conjunction with a pneumatic sensing system; the float is confined within a small chamber through which the test liquid continually flows, so that density variations with time can be determined.

displacer-type meter An apparatus for detecting liquid level or determining gas density by measuring the effect of the fluid on the buoyancy of a displacer unit immersed in it.

dissipation The process of reducing the amount of electrical charge on a body; also, the loss of electric energy as heat.

dissipation factor A measure of the a-c power loss. The ratio of the loss index to its relative permittivity. The ratio of the energy dissipated to the energy stored in the dielectric per cycle. The tangent of the loss angle. Dissipation factor is proportional to the power loss per cycle (f) per potential gradient (E^2) per unit volume, as follows: dissipation factor = power loss/($E^2 \times$ f $\times$ volume $\times$ constant).

dissociation *1.* The breakdown of a substance into two or more constituents. *2.* The process by which a chemical compound breaks down into simpler constituents.

dissolved gases Gases that are "in solution" in water.

dissolved solids The solids in water that are in solution.

dissymmetry *1.* Absence of symmetry. *2.* Symmetry in opposite directions, as in left and right from centerline.

distal At the greatest distance from a central point; peripheral.

distillate *1.* The distilled product from a fractionating column. *2.* The overhead product from a distillation column. *3.* (oil and gas industry) Refers to a specific product withdrawn from the column, usually near the bottom.

distillate fuel Any of the fuel hydrocarbons obtained during the distillation of petroleum which have boiling points higher than that of gasoline.

distillation *1.* A unit operation used to separate a mixture into its individual chemical components. *2.* Vaporization of a substance with subsequent recovery of the vapor by condensation. *3.* Often used in less precise sense to refer to vaporization of volatile constituents of a fuel without subsequence condensation.

distortion *1.* A change in the shape of a material resulting from applied stress or the release of residual stress, usually during heat treatment. Change in size is more predictable and a separate consideration. *2.* In a composite laminate, the change in position of reinforcement fibers. *3.* A lens defect that causes the images of straight lines to appear geometrically other than straight lines.

distortion descriptor A non-dimensional numerical representation of the measured inlet pressure distribution.

distortion extent The circumferential arc size of a distorted region.

distortion index A quality parameter of the optical data channel.

distortion intensity The amplitude of a distortion pattern.

distortion meter An instrument that visually indicates the harmonic content of an audio-frequency signal.

distortion sensitivity Loss in compressor surge pressure ratio per unit of numerical distortion descriptor.

disturbance suppression Action that reduces or eliminates electrical disturbance.

disturbance voltage; interference voltage Voltage produced between two points on separate conductors by an electromagnetic disturbance, measured under specified conditions.

divalent metals Metals with two valence electrons per atom.

divergence *1.* A motion with progressively increasing amplitude; an instability. *2.* The expansion or spreading out of a vector field; also, a precise measure thereof. *3.* The spreading out of a laser beam with distance, measured as an angle.

divergence loss The portion of energy in a radiated beam that is lost due to non-parallel transmission, or spreading.

DL *See* dead load stress condition.

DMC *See* dough molding compound.

doctor blade A straight piece of material used to spread resin, as in application of a thin film of resin for use in hot melt prepregging or for use as an adhesive film.

doctor roll A roller mechanism that is revolving at a different surface speed, or in an opposite direction, resulting in a wiping action for regulating the adhesive supplied to the spreader roll.

document *1.* A medium and the data recorded on it for human use, for example, a report sheet or a book. *2.* By extension, any record that has permanence and that can be read by man or machine.

documentation *1.* The creating, collecting, organizing, storing, citing, and disseminating of documents, or the information recorded in documents. *2.* A collection of documents or information on a given subject. *3.* Often used in specific reference to computer program explanation.

Dodge-Romig tables A set of standard tables with known statistical characteristics that are used in lot-tolerance and AQL (acceptable quality level) acceptance sampling.

dolomite A common rock-forming rhombohedral material consisting of calcium, magnesium, and carbonates. Dolomite is used for refractory products.

domain *1.* Copolymers in which chemically different portions of a polymer chain separate, creating two or more amorphous phases. *2.* Ferromagnetic materials where magnetic zones within the crystal are parallel to one another.

dominant wavelength The wavelength of monochromatic light that matches a given color when combined in suitable proportions with a standard reference light.

dopa An intermediate organic compound produced by oxidation of tyrosine by tyramine; also, an intermediate product in the synthesis of both epinephrine and melanin.

dopant A substance added to a polymer to induce a change in physical properties.

dope A cellulose ester lacquer used as an adhesive or coating.

doped germanium A type of detector in which impurities are added to germanium to make the material respond to infrared radiation at wavelengths much longer than those detectable by pure germanium.

doping *1.* Adding a small amount of a substance to a material or mixture to achieve a special effect. *2.* Coating a mold or mandrel to prevent a molded part from sticking to it. *See also* additive.

Doppler effect The change in frequency with which energy reaches a receiver when the receiver and the energy source are in motion relative to each other.

Doppler shift A phenomenon that causes electromagnetic or compression waves emanating from an object to have a longer wavelength if the object moves away from an observer than would be the case if the object were stationary with respect to the observer, and to have a shorter wavelength if the object moves toward the observer.

dose The amount of radiation received at a specific location per unit area or unit volume, or the amount received by the whole body.

dose meter Any of several instruments for directly indicating radiation dose.

dose rate Radiation dose per unit time.

dose-rate meter Any of several instruments for directly indicating radiation dose rate.

dosimeter An instrument that measures the dose of nuclear radiation received in a given time.

double aging Employment of two different aging treatments to control the type of precipitate formed from a supersaturated alloy matrix in order to obtain the desired properties. The first aging treatment, sometimes referred to as intermediate or stabilizing, is usually carried out at a higher temperature than the second.

double amplitude displacement (DA) The peak-to-peak amplitude of an elastomer specimen measured in the direction of the applied vibrational force. Two times the single peak value in either the plus or minus direction may not be equivalent to the peak-to-peak value.

double amplitude (peak-to-peak amplitude) Peak-to-peak amplitude of displacement at a point in a vibrating system is the sum of the extreme values of displacement in both directions from the equilibrium position.

double base powder (propellant) A powder or propellant containing nitrocellulose and another principle explosive ingredient, usually nitroglycerin.

double base propellants Solid rocket propellants using two unstable compounds, such as nitrocellulose and nitroglycerin. The unstable compounds used in a double base propellant do not require a separate oxidizer.

double buffering The concept of using two buffer registers between a computer's central processing unit and an input/output device in which one buffer is being refreshed while the other is being used for the input (or output) operation thereby resulting in increased performance and reduced flicker between screen updates.

double-density A type of computer diskette that has twice the storage capacity of a single-density diskette.

double groove weld A weldment in which the joint is beveled or grooved from both sides to prepare the joint for welding.

double layer The interface between an electrode and the electrolyte where charge separation takes place. The simplest model is represented by a parallel plate condenser. In general, the electrode will be positively charged with respect to the solution.

double precision *1.* Pertaining to the use of two computer words to represent a number. *2.* In floating-point arithmetic, the use of additional bytes or words representing the number, in order to double the number of bits in the mantissa.

doubler *1.* Localized area of extra layers of reinforcement, usually to provide stiffness or strength for fastening or other abrupt load transfers. *2.* An extra piece of facing attached to strengthen or stiffen the panel or to distribute the load more widely into the core.

double sampling A type of sampling inspection in which the lot can be accepted or rejected based on results from a single sample, or the decision can be deferred until the results from a second sample are known.

double tempering A treatment in which quench hardened steel is given two complete tempering cycles at substantially the same temperature for the purpose of assuring completion of the tempering reaction and promoting stability of the resulting microstructure.

double-welded joint A weldment in which the joint is welded from both sides.

dough molding compound (DMC) A dough-like mixture of fibers, polymers, and filler material used for molding components with random orientation of fibers.

downhand welding *See* flat-position welding.

dowtherm A constant-boiling mixture of phenyl oxide and diphenyl oxide used in high-temperature heat transfer systems.

draft Also spelled draught. *1.* The side taper on molds and dies which makes it easier to remove finished parts from the cavity. *2.* The depth to which a boat or other vessel is submerged in a body of water; the value varies with vessel

weight and water density. *3.* Drawing a product in a die. *4.* The small, positive pressure that propels exhaust gas out of a furnace and up the stack. *5.* The difference between atmospheric pressure and some lower pressure existing in the furnace or gas passages of a steam generating unit. *6.* A preliminary document.

drag *1.* Imperfections on the surface of components when withdrawn from a die. *2.* A retarding force acting upon the direction of motion of the body. Drag is a component of the total fluid forces acting on the body. *3.* The bottom part of the flask for a casting mold. *See also* cope. *4.* Resistance of a vehicle body to motion through the air due to total force acting parallel to and opposite to the direction of motion. *5.* In data processing, the movement of an object on a screen by using a mouse.

draw *1.* To form cup-shaped parts from sheet metal. *2.* To reduce the size of wire or bar stock by pulling it through a die. *3.* To remove a pattern from a sand-mold cavity. *4.* A fissure or pocket in a casting caused by inadequate feeding of molten metal during solidification. *5.* A colloquial shop term meaning temper.

drawability The ability of the sheet steel to be formed or drawn into the intended end product without fracturing.

draw bead *1.* A bead or offset used for controlling metal flow during sheet-metal forming. *2.* A contoured rib or projection on a draw ring or holddown to control metal flow in deep drawing.

drawing The shaping of a flat blank into the desired contour by causing the metal to flow over a draw ring and around a punch. The flow of metal is restrained by sufficient blank holder pressure to prevent buckling.

drawing back *1.* A shop term for tempering. *2.* Reheating hardened steel below the critical temperature to reduce its hardness.

drawing compound A substance applied to minimize metal to metal contact between the sheet metal, bar, or wire and the die. Proper application of the proper lubricant can improve flow characteristics of the metal and prevent scoring, galling, and pickup on the dies or part.

drawing lubricant *See* drawing compound.

draw mark Any surface flaw or blemish that occurs during drawing, including scoring, galling, pickup, or die lines.

droplet erosion Surface damage from liquid droplets striking a surface, causing small cavities which may result in fatigue failure.

drop tests The release of a component, for example a landing gear assembly, from a specified height and orientation, onto a steel plate, to determine if it inadvertently functions or becomes inoperable as a result of the gravitational impact.

dross *1.* A scum that forms on the surface of molten metals largely because of oxidation but sometimes because of the rising of impurities to the surface. *2.* The scum that develops in a molten galvanizing pot.

drought *See* draft.

dry To change the physical state of an adhesive on an adherent by the loss of solvent constituents by evaporation or absorption.

dry ash Refuse in the solid state, usually in granular or dust form.

dry assay Determining the amount of a metal or compound in an alloy, ore, or metallurgical residue by means that do not involve the use of liquid to separate or analyze for constituents.

dry basis A method of expressing moisture content whereby the amount of moisture present is calculated as a percentage of the weight of bone-dry material; used extensively in the textile industry.

dry blast cleaning Using a dry abrasive medium such as grit, sand, or shot to clean metal surfaces by driving it against the surface with a blast of air or by centrifugal force.

dry blend Molding compound mixed to produce a dry, free-flowing material.

dry-bulb temperature The temperature of the air indicated by thermometer, not affected by the water vapor content of the air.

dry corrosion Atmospheric corrosion taking place at temperatures above the dew point.

dryer A device used to remove water or water vapor from a refrigerant or other fluid.

dry fiber area Area of fiber not totally encapsulated by resin.

dry fuel rocket A rocket that uses a mixture of fast-burning powder.

dry gas Gas containing no water vapor.

dry-gas loss The loss representing the difference between the heat content of the dry exhaust gases and their heat content at the temperature of ambient air.

drying oil One of the many natural, usually vegetable oils—the glyceryl esters of unsaturated fatty acids—that harden in air by oxidation to a resinous skin.

drying oven A closed chamber for driving moisture from surfaces or bulk materials by heating them at relatively low temperatures.

drying temperature The temperature to which an adhesive on an adherent or in any assembly, or the assembly itself, is subjected to dry the adhesive.

drying time *1.* The time required to dry a component with absorbed moisture to a specified, measurable level. *2.* The time specified for drying a component, which is considered sufficient to achieve an adequate condition for bonding, when measurement methods are not available.

dry laminate A laminate with insufficient resin to cause bonding with the reinforcement fibers.

dry lay-up Construction of a laminate by the layering of preimpregnated reinforcement in a female mold or on a male mold, usually followed by bag molding or autoclave molding.

dry spot Of a laminate, an area of incomplete surface film on laminated plastics.

dry steam Steam containing no moisture. Commercially dry steam containing not more than one half of one percent moisture.

dry strength The strength of an adhesive joint measured immediately after drying.

d$_s$ *See* surface diameter.

DSC *See* differential scanning calorimetry.

DTA *See* differential thermal analysis.

DTUL *See* deflection temperature under load.

dual beam analyzer *See* split-beam ultraviolet analyzer.

duality principle Principle whereby for any theorem in electric circuit analysis there is a dual theorem in which quantities are replaced with dual quantities. Examples are current and voltage or impedance and admittance.

duality theorem Theorem stating that if either of the two dual linear program-ming problems has a solution, then so does the other.

duct An enclosed fluid-flow passage, which may be any size up to several feet in cross section. The term is most often applied to passages for ventilating air, and to intakes and exhausts for engines, boilers, and furnaces.

ductile Describes materials that elongate more than 5% before fracture (for example, steel and aluminum). These materials distort by yielding (plastic deformation) prior to fracture.

ductile crack propagation Slow crack growth accompanied by visible plastic deformation.

ductile fracture The tearing of metal combined with relatively severe plastic deformation.

ductile iron The term preferred in the United States for cast iron containing spheroidal nodules of graphite in the as-cast condition. Also known as nodular cast iron, nodular iron, or spherulitic-graphite cast iron.

ductile nitriding *See* nitriding.

ductility The measure of how much a material an be deformed before fracture. This property is typically measured by elongation and reduction of area in a tensile test.

dulling In painting, a reduction in the level of gloss of the topcoat, usually due to an interaction between the topcoat and the sealer beneath.

Dumet wire Wire made of Fe-42Ni, covered with a layer of copper, used to replace expensive platinum as the seal-in wire in incandescent lamps and vacuum tubes; the copper coating prevents gassing at the seal.

dummy *1.* A device constructed physically to resemble another device, but without the operating characteristics. *2.* A cathode, usually corrugated to give varying current densities, that is plated at low current densities to preferentially remove impurities from an electroplating solution. *3.* A substitute cathode used during adjustment of the operating conditions in electroplating. *4.* An artificial address, instruction, or record of computer information inserted solely to fulfill prescribed conditions.

dummy block In extrusion, a thick, unattached disc placed between the ram and the billet to prevent overheating of the ram.

duodecimal number A number, consisting of successive characters, representing a sum, in which the individual quantity represented by each character is based on a radix of twelve. The characters used are 0, 1, 2, 3, 4, 5, 6, 7, 8, 9, T (for ten,) and E (for eleven). *See* number system.

duplex film An adhesive film containing two different adhesives separated by a scrim cloth which are manufactured into one film.

durability *1.* The ability of an item of material to resist cracking, corrosion, thermal shock and degradation, delamination, wear, and the effects of foreign object damage for a specified period of time. *2.* A measure of the resistance of a material to wear and physico-chemical change. *3.* A measure of useful life (a special case of reliability).

durability and damage tolerance analysis An analysis performed to determine the growth of flaws, cracks, and other damage in a structure versus time.

duralumin A common name for aluminum-copper high-strength alloys.

durometer An instrument that measures the hardness of rubber by the penetration (without puncturing) of an indentor point into the surface of rubber.

durometer hardness An arbitrary numerical value that indicates the resistance to penetration of the indentor point into the rubber surface. Value may be taken immediately or after a very short specified time.

dust Particles of gas-borne solid matter larger than one micron in diameter.

dust counter A photoelectric instrument that measures the number and size of dust particles in a known volume of air. Also known as a Kern counter.

dust loading The amount of dust in a gas, usually expressed in grains per cubic feet or pound per thousand pounds of gas.

duty cycle *1.* For a device that operates repeatedly, but not continuously, the time intervals involved in starting, running, and stopping, plus any idling or warm-up time. *2.* For a device that operates intermittently, the ratio of working time to total time, usually expressed as a percent. Also known as duty factor. *3.* The percent of total operating time during which current flows in an electric resistance welding machine.

duty factor *See* duty cycle.

dye penetrant A low-viscosity liquid containing a dye and used in nondestructive examination to detect surface discontinuities such as cracks and laps in both magnetic and nonmagnetic materials.

dynamic Moving or with high velocity. Contrast with static.

dynamic calibration A calibration procedure in which the quantity of liquid is measured while liquid is flowing into or out of the measuring vessel.

dynamic compensation A technique used in control to compensate for dynamic response differences to different input streams to a process. A combination of lead and lag algorithms will handle most situations.

dynamic creep Creep that results under variable loads and temperatures.

dynamic impedance (stiffness) The impedance associated with the output deflections of an active closed-loop actuation system caused by externally applied dynamic forces (usually sinusoidal) over a specified frequency range. Dynamic impedance includes the load mass or inertia, load friction, system stiffness, and any other load-related compliance effects. Since it is a complex quantity, the term dynamic impedance is preferred to the term dynamic stiffness.

dynamic information Information that changes during the time of relevant task performance.

dynamic load The portion of a service load that varies with time, and cannot be characterized as a series of different, unvarying (static) loads applied and removed successively.

dynamic mechanical measurement The measurement of the modulus or damping of a material under load or displacement related to temperature, frequency, or time.

dynamic model A model in which the variables are functions of time. Contrast with static model and steady-state model.

dynamic modulus The ratio of stress to strain when under vibration.

dynamic pressure The increase in pressure above the static pressure that results from complete transformation of the kinetic energy of the fluid into potential energy.

dynamic programming In operations research, a procedure for optimization of a multi-stage problem whereby a number of decisions are available at each stage of the process.

dynamic rated load The maximum load that can be lifted under specified dynamic conditions, without exceeding allowable strength limits.

dynamics The branch of mechanics that treats the motion of bodies (kinematics) and the action of forces that produce or change the motion (kinetics).

dynamic sensitivity In leak testing, the minimum leak rate that a particular device is capable of detecting.

dynamic stability The characteristics of a body, that causes it, when disturbed from steady motion, to damp oscillations set up and gradually return to the original state.

dynamic stiffness The stiffness associated with the output deflections of an active actuation system caused by externally applied dynamic forces, usually sinusoidal, over a specified frequency range.

dynamic stop A loop stop consisting of a single jump instruction.

dynamic subroutine A subroutine that involves parameters, such as decimal point position or item size, from which a relatively coded subroutine is derived.

dynamic test A test of anchorages through the application of forces at onset rates that result from a barrier impact or equivalent sled simulation.

dynamic unbalance A condition in rotating equipment whereby the axle of rotation does not exactly coincide with one of the principal axes of inertia for the mechanism; it produces additional forces and vibrations which, if severe, can lead to failure or malfunction.

dynamic variable Process variables that can change from moment to moment due to unspecified or unknown sources.

dynamometer *1.* An energy-absorbing device designed to allow measured dynamic operation of a vehicle's drive train while the vehicle remains stationary. *2.* An electrical instrument in which current, voltage, or power is measured by determining the force between a fixed coil and a moving coil.

E

e The base of natural logarithms.

E *See* modulus of elasticity.

earing (scalloping) *See* scalloping.

EBCDIC *See* Extended Binary Coded Decimal Interchange Code.

ebullition The act of boiling or bubbling.

EC *See* ethyl cellulose.

E-coat *See* electrocoat.

eccentric Describes any rotating mechanism whose center of rotation does not coincide with the geometric center of the rotating member.

eccentricity A measure of the center of a conductor's location with respect to the circular cross section of the insulation. Expressed as a percentage of center displacement of one circle within another.

echo(es) *1.* Waves that have been reflected or otherwise returned with sufficient magnitude and delay to be detected as a wave distinct from that directly transmitted. *2.* (MS-DOS) A command that prints the text of the commands on the screen as they are executed.

ECM *See* electrochemical machining.

E coat/ELPO Common names for the electrocoating process.

econometrics The application of mathematics and statistical techniques to the testing and quantifying of economic theories and the solution of economic problems.

E$_{cp}$ *See* critical pitting potential.

ECTFE *See* ethylene chlorotrifluoroethylene.

eddy A whirlpool of fluid. *See also* vortices.

eddy currents (general) Currents that exist as a result of voltages induced in the body of a conducting mass by a variation of magnetic flux. Note: The variation of magnetic flux is the result of a varying magnetic field or of a relative motion of the mass with respect to the magnetic field.

eddy-current tachometer A device for measuring rotational speed which has been used extensively in automotive speedometers.

eddy-current test A nondestructive test in which changes in eddy-current flowing through a test specimen are measured to detect potential faults.

eddy-sonic Describes a process in which sonic or ultrasonic energy is produced in a test part by a coil on or near the surface of the test part. The coil is used to produce eddy currents in the test part. Vibrations in the test part result from the interaction of the magnetic field from the eddy currents in the test part with the magnetic field of the coil.

eddy-sonic bondtester A dry-coupled bondtester operating on the eddy-sonic principle.

edge bleed Removal of volatiles and excess resin through the edge of the laminate, as in matched die molding of a laminate.

edge break (side strain) *See* edge strain.

edge close-outs Members placed around the panel sides to protect a sandwich from damage or to attach the panel to a support or other panel.

edge delamination A separation of the detail parts along an edge after the assembly has been cured.

edge stability A measure of strength in a green compact.

edge strain Strain lines located from 1 to 12 inches from the edges of steel strip or sheet metal.

EDM *See* electrical discharge machining.

EDS *See* energy-dispersive spectroscopy.

eductor *1.* A device that withdraws a fluid by aspiration and mixes it with another fluid. *2.* Using water, steam, or air to induce the flow of other fluids from a vessel. *See* injector.

EELS *See* electron energy loss spectroscopy.

effective case depth The perpendicular distance from the surface of a hardened case to the furthest point where a specified level of hardness is maintained. The hardness criterion is 50 HRC normally. Effective case depth should always be determined on the part itself, or on samples or specimens having a heat-treated condition representative of the part under consideration.

effective decarburization Any measurable loss of carbon content that results in mechanical properties, for example, hardness, below the minimum acceptable specifications for hardened material.

effectiveness The probability that a material will operate successfully when required.

effective value The root-mean-square value of a cyclically varying quantity; it is determined by finding the average of the squares of the values throughout one cycle and taking the square root of the average.

efficiency *1.* The ratio of output to input. The efficiency of a steam-generating unit is the ratio of the heat absorbed by water and steam to the heat in the fuel fired. *2.* In manufacturing, the average output of a process or production line expressed as a percent of its expected output under ideal conditions. *3.* The ratio of useful energy supplied by a dynamic system to the energy supplied to it over a given period of time.

effluent Liquid waste discharged from an industrial processing facility or waste-treatment plant.

E finish A designation indicating the material is to be used for an exposed part requiring a good painted surface.

E-glass An electric glass with a calcium aluminoborosilicate composition used as a reinforcement fiber in laminates requiring electrical resistance.

eigenvalue A numerical value, derived from mathematical modeling and equations of motion, which describes the frequency, amplitude, and damping properties of a mode of motion.

ejection *1.* Physical removal of an object from a specific site—such as removal of a cast or molded product from a die cavity—by hand, compressed air or mechanical means. *2.* Emergency expulsion of a passenger compartment from an aircraft or spacecraft. *3.* Withdrawal of fluid from a chamber by the action of a jet pump or eductor.

ejector A device that utilizes the kinetic energy in a jet of water or other fluid to remove a fluid or fluent material from tanks or hoppers.

elastic after effect Recovery of a test specimen to its original dimension after a load is removed, measured by time.

elastic chamber The portion of a pressure-measuring system that is filled with the medium whose pressure is being measured, and that expands and collapses elastically with changes in pressure; examples include a Bourdon tube, bellows, flat or corrugated diaphragm, spring-loaded piston, or a combination of two or more single elements, which may be the same or different types.

elastic collision A collision between two or more bodies in which the internal energy of the participating bodies remains constant, and in which the kinetic energy of translation for the combination of bodies is conserved.

elastic deformation The changes in dimensions of items, caused by stress, provided a return to original dimensions occurs when stress is removed.

elasticity The property of a material that causes it to return to its original shape after deformation not exceeding its elastic limit.

elastic limit The maximum stress that a material can withstand without permanent strain.

elastic modulus See modulus of elasticity.

elastic ratio The yield point divided by the tensile strength.

elastic recovery *1.* The fraction of a given deformation that behaves elastically. A perfectly elastic material has an elastic recovery of 1; a perfectly plastic material has an elastic recovery of 0.

elastic scattering A collision between two particles, or between a particle and a photon, in which total kinetic energy and momentum are conserved.

elastomer *1.* An elastic rubberlike substance, such as natural or synthetic rubber. *2.* Macromolecular material that returns rapidly to approximately the initial dimensions and shape after substantial deformation by a weak stress and release of the stress.

Elber equation In fatigue crack propagation studies, expresses the effective stress range ratio as $U = 0.5 + 0.4R$, where R is the stress ratio.

electrical apparatus All items applied as a whole or in part for the utilization of electrical energy. These include, among others, items for generation, transmission, distribution storage measurement, control, and consumption of electrical energy and items for telecommunications.

electrical conductivity A material characteristic indicative of the relative ease with which electrons flow through the material; usual units are %IACS (International Annealed Copper Standard), which relates the conductivity to that of annealed pure copper. Electrical conductivity is the reciprocal of electrical resistivity.

electrical discharge machining A machining method in which stock is removed by melting and vaporization under the action of rapid, repetitive spark discharges between a shaped electrode and the workpiece through a dielectric fluid flowing in the intervening space.

electrical engineering A branch of engineering that deals with practical applications of electricity, especially the generation, transmission, and utilization of electric power by means of current flow in conductors.

electrical overstress (EOS) The electrical stressing of electronic components beyond specifications.

electrical resistivity A material characteristic indicative of its relative resistance to the flow of electrons. Electrical resistivity is the reciprocal of electrical conductivity.

electrical steel Low carbon steel that contains 0.5 to 5% Si or other material; produced specifically to have enhanced electromagnetic properties suitable for making the cores of transformers, alternators, motors, and other iron-core electric machines. Contrast with electric steel.

electric contact Either of two opposing, electrically conductive buttons or other shapes which allow current to flow in a circuit when they touch each other.

electric discharge The flowing of electricity through a gas, resulting in the emission of radiation that is characteristic of the gas and the intensity of the current.

electric field A condition within a medium or evacuated space which imposes forces on stationary or moving electrified bodies in direct relation to their electric charges.

electric field strength The magnitude of an electric field vector.

electric furnace steel *See* electric steel.

electric instrument An indicating device for measuring electrical attributes of a system or circuit. Contrast with electric meter.

electric meter A recording or totalizing instrument that measures the amount of electric power generated or used as a function of time. Contrast with electric instrument.

electric potential In electrostatics, the work done in moving unit positive charge from infinity to the point whose potential is being specified.

electric steel Any steel melted in an electric furnace, which allows close control of composition. Also known as electric-furnace steel. Contrast with electrical steel.

electric strength The property of a dielectric which opposes a disruptive discharge.

electric stroboscope A device that uses an electric oscillator or similar element to produce precisely timed pulses of light; oscillator frequency can be controlled over a wide range so that the device can be used to determine the frequency of a mechanical oscillation.

electric tachometer An electrically powered instrument for determining rotational speed, usually in rpm.

electric telemeter An apparatus for remotely detecting and measuring a quantity—including the detector intermediate means, transmitter, receiver, and indicating device—in which the transmitted signal is conducted electrically to the remote indicating or recording station.

electroacoustic transducer A type of transducer in which the input signal is an electric wave and the output a sound wave, or vice versa.

electrochemical A chemical reaction that is driven by a difference in electrode potential from one site to another on the same or different parts.

electrochemical cleaning Removing soil by the chemical action induced by passing an electric current through an electrolyte. Also known as electrolytic cleaning.

electrochemical coating A coating formed on the surface of a part due to chemical action induced by passing an electric current through an electrolyte.

electrochemical corrosion Corrosion of a metal due to chemical action induced by electric current flowing in an electrolyte. Also known as electrolytic corrosion.

electrochemical equivalent The weight of an element or group of elements oxidized or reduced at 100% efficiency by the passage of a unit quantity of electricity. Usually expressed as grams per coulomb.

electrochemical machining A machining method in which stock is removed by electrolytic dissolution under the action of a flow of electric current between a tool cathode and the workpiece, through an electrolyte flowing in the intervening space. Abbreviated ECM. Also known as electrochemical milling or electrolytic machining.

electrochemical milling *See* electrochemical machining.

electrochemical potential, electrochemical tension The partial derivative of the total electrochemical free energy of a constituent with respect to the number of moles of this constituent, where all factors are kept constant.

electrochromism A phenomenon whereby a select number of solid materials will change color when an electric field is applied.

electrocoat (E-coat, ELPO) A coating for metals deposited by the application of high voltages between an anode and a cathode in an electrolyte.

electrocoating A method of coating a metal requiring the application of high voltages between an anode and a cathode in an electrolyte. Based on the method of deposition, two types of electropaint have been developed: anodic and cathodic.

electrode *1.* Terminal at which electricity passes from one medium into another. The positive electrode is called the anode; the negative electrode is called the cathode. *2.* An electrically conductive member that emits or collects electrons or ions, or that controls movement of electrons or ions in the interelectrode space by means of an electric field.

electrode characteristic The relation between electrode voltage and electrode amperage for a given electrode in a system, with the voltages of all other electrodes in the system held constant.

electrode force The force that tends to compress the electrodes against the workpiece in electric-resistance spot, seam or projection welding. Also known as welding force.

electrodeposition Any electrolytic process that results in deposition of a metal from a solution of its ions. Electrodeposition includes processes such as electroplating and electroforming. Also known as electrolytic deposition.

electrode potential Voltage or potential of an electrode with respect to a reference half cell.

electrode voltage The electric potential difference between a given electrode and the system cathode or a specific point on the cathode. The latter is especially applicable when the cathode is a long wire or filament.

electrodynamics The science dealing with the forces and energy transformations of electric currents and the magnetic fields associated with them.

electroemissive machining *See* electrical discharge machining.

electroepitaxy Crystal growth process achieved by passing an electric current through a substrate solution.

electroforming Shaping a component by electrodeposition of a thick metal plate on a conductive pattern; the part may be used as formed, or it may be sprayed on the back with molten metal or other material to increase its strength.

electrogalvanizing Electroplating of zinc or zinc alloys upon iron or steel.

electrograph *1.* A tracing produced on prepared sensitized paper or other material by passing an electric current or electric spark through the paper. *2.* A plot or graph produced by an electrically controlled stylus or pen.

electroless deposition Controlled autocatalytic reduction method of depositing coatings.

electroless plating Deposition of a metal from a solution of its ions by chemical reduction, induced when the basis metal is immersed in the solution, without the use of impressed electric current.

electroluminescence Emission of light caused by an application of electric fields to solids or gases.

electrolysis The chemical change in an electrolyte resulting from the passage of electricity.

electrolyte *1.* A chemical substance or mixture, usually liquid, containing ions that migrate in an electric field. *2.* The ionically conductive alkaline solution used in the nickel cadmium cell.

electrolytic cell *See* galvanic cell.

electrolytic cleaning The process of degreasing or descaling a metal by making it an electrode in a suitable bath. *See also* electrochemical cleaning.

electrolytic corrosion *See* electrochemical corrosion.

electrolytic deposition *See* electrodeposition.

electrolytic etching Engraving a pattern on a metal surface by electrolytic dissolution.

electrolytic machining *See* electrochemical machining.

electrolytic pickling Removal of scale and surface deposits by electrolytic action in a chemically active solution.

electrolytic powder Metal powder that is produced directly or indirectly by electrodeposition.

electromagnet Any magnet assembly whose magnetic field strength is determined by the magnitude of an electric current passing through some portion of the assembly.

electromagnetic acceleration The use of perpendicular components of electric and magnetic fields to accelerate a current carrier.

electromagnetic compatibility (EMC) The ability of an equipment or system to function satisfactorily in its electromagnetic environment without introducing intolerable electromagnetic disturbance to anything in that environment.

electromagnetic disturbance Any electromagnetic phenomenon that may degrade the performance of a device, equipment, or system, or adversely affect living or inert matter.

electromagnetic instrument Any instrument in which the indicating means or recording means is positioned by mechanical motion controlled by the strength of an induced electromagnetic field.

electromagnetic interference (EMI) Degradation of the performance of equipment, transmission channel, or system caused by an electromagnetic disturbance.

electromagnetic lens An electromagnet designed to produce a specific magnetic field necessary for the focusing and deflection of electrons in electro-optical instruments.

electromagnetic radiation Energy propagated through space or through material media in the form of an advancing disturbance in electric and magnetic fields existing in space or in media.

electromagnetic waves The radiant energy produced by the oscillation of electric charge.

electrometallurgy The technology associated with recovery and processing of metals using electrolytic and electrical methods.

electrometer Instrument for measuring differences in electric potential.

electrometric titration A technique in which the location of the end-point of a titration requires some measurement of an electrical quantity.

electromotive force (EMF) The force that causes current to flow when there is a difference of potential between two points.

electromotive force series (EMF series) An orderly listing of elements according to their standard electrode potentials. (Hydrogen electrode is a reference point and is given the value of zero.)

electron *1.* One of the natural elementary constituents of matter. It carries a negative electric charge of one electronic unit. *2.* An elementary subatomic particle having a rest mass of 9.107×10^{-28} g and a negative charge of 4.802×10^{-10} statcoulomb. Also known as negatron.

electron bands Energy potential for the free electrons in a material.

electron beam A narrow, focused ray of electrons which streams from a cathode or emitter and which can be used to cut, machine, melt, heat treat, or weld metals.

electron beam welding Welding technique in which a high-energy, finely focused electron beam is used to fuse materials.

electron device Any device whose operation depends on conduction by the flow of electrons through a vacuum, gas-filled space, or semiconductor material.

electron diffraction Producing diffraction patterns from crystalline materials by the use of electron beams.

electron emission Ejection of free electrons from the surface of an electrode into the adjacent space.

electron energy loss spectroscopy (EELS) A spectrographic test in which the electron microscope analyzes the transmitted energy from the specimen.

electron gun An electron-tube subassembly that generates a beam of electrons, and may additionally accelerate control, focus, or deflect the beam.

electronic Any device or system in which electrons flow through a vacuum, gas, or semiconductor. Normally operative in frequency ranges above 1,000 Hz (cycles per second).

electronic engineering A branch of engineering that deals chiefly with the design, fabrication, and operation of electron-tube or transistorized equipment, which is used to generate, transmit, analyze, and control radio-frequency electromagnetic waves or similar electrical signals.

electronic equipment Equipment in which electricity is conducted principally by electrons moving through a vacuum, gas, or semiconductor.

electronic heating Producing heat by the use of radio-frequency current generated and controlled by an electron-tube oscillator or similar power source. Also known as high-frequency heating or radio-frequency heating.

electronic photometer *See* photoelectric photometer.

electronics The branch of physics that treats of the emission, transmission, behavior, and effects of electrons.

electronic switch A circuit element that causes a start and stop action or a switching action electronically, usually at high speeds.

electronic transition A transition in which an electron in an atom or molecule moves from one energy level to another.

electron metallography Using an electron microscope to study the structure of metals and alloys.

electronmicroprobe analysis A technique for determining concentration and distribution of chemical elements over a microscopic area of a specimen by bombarding the specimen with high-energy electrons in an evacuated chamber and performing x-ray fluorescent analysis of secondary x-radiation emitted by the specimen.

electron microscopy The interpretive application of an electron microscope for the magnification of materials that cannot be properly seen with an optical microscope.

electron probe x-ray microanalysis (EMPA) An analytical chemistry test in which a beam of electrons is used to stimulate an x-ray spectrum characteristic of the elements in a sample.

electron runaway (plasma physics) High acceleration of electrons in a collisional plasma caused by a suddenly applied electric field.

electron spectroscopy The study and interpretation of atomic, molecular, and solid-state structure based on x-ray-induced electron emission from substances.

electron spin resonance spectroscopy (ESR) Similar to magnetic resonance, except that the class studied is an unpaired electron.

electron tube Any device whose operation depends on conduction by the flow of electrons through a vacuum or gas-filled space within a gastight envelope.

electron volt Abbreviated eV (preferred) or EV. A unit of energy equal to the work done in accelerating one electron through an electric potential difference of one volt.

electro-optic effect A change in the refractive index of a material under the influence of an electric field.

electropainting Electrodeposition of a thin layer of paint on metal parts which are made anodic. Also known as electrophoretic painting.

electrophoretic display A passive display whose operating principle is the movement of particles in a fluid under the influence of an electric field.

electrophoretic painting *See* electro-painting.

electroplated coatings Coatings applied in a low-temperature continuous process whereby negatively charged steel sheet is passed between positively charged anodes. Metallic ions in an electrolyte bath are reduced and plated on the surface of the steel sheet forming the coating.

electroplating Electrodeposition of a thin layer of metal on a surface of a part that is in contact with a solution, or electrolyte, containing ions of the deposited metal.

electropolishing Smoothing and polishing a metal surface by closely controlled electrochemical action similar to electrochemical machining or electrolytic pickling.

electroscope An instrument for detecting an electric charge by observing the effects of mechanical force exerted between two or more electrically charged bodies.

electroslag remelting (ESR) A melting technique whereby a previously cast electrode is remelted under a special chemistry slag through electrical resistance heating through the slag and an air gap between the electrode and the slag blanket. The ingot is cast into a water-cooled copper mold.

electrospark machining *See* electrical discharge machining.

electrostatic *1.* Of, pertaining to, or produced by static electric charges. *2.* The scientific study of static electricity.

electrostatic bonding Use of the particle-attracting property of electrostatic charges to bond particles of one charge to those of the opposite charge.

electrostatic charge Electric charge on the surface of a body. Causes a surrounding electrostatic field.

electrostatic deposition A common method of coating metal shapes with polymers.

electrostatic discharge A large electrical potential moving from one surface or substance to another.

electrostatic force The force that electrically charged bodies exert upon one another.

electrostatic instrument Any instrument whose operation depends on forces of electrostatic attraction or repulsion between charged bodies.

electrostatic lens A set of electrodes arranged so that their composite electric field acts to focus a beam of electrons or other charged particles.

electrostatic memory A memory device that retains information by means of electrostatic charge, usually involving a special type of cathode-ray tube and its associated circuits.

electrostatic painting A spray painting process in which the paint particles are charged by spraying them through a grid of wires which is held at a d-c potential of about 100 kV; the parts being painted are connected to the opposite terminal of the high-voltage circuit so that they attract the charged paint particles.

electrostatic potential The electrostatic potential at any point is the potential difference between the point and an agreed-upon reference point, usually the point at infinity.

electrostatic precipitation Fine particles in a gas vapor are attracted to an electrically charged base material.

electrostatic voltmeter An instrument for measuring electrical potential by means of electrostatic forces between elements in the instrument.

electrostriction The phenomenon whereby some dielectric materials experience an elastic strain when subjected to an electric field, this strain being independent of polarity of the field.

electrostriction transducer A device that consists of a crystalline material that produces elastic strain when subjected to an electric field, or that produces an electric field when strained elastically. Also known as a piezoelectric transducer.

electrothermal process Any process that produces heat by means of an electric current—especially when temperatures higher than those obtained by burning a fuel are required.

element *1.* In data processing, one of the items in an array. *2.* A substance that cannot be decomposed by chemical means into simpler substances.

elemental error The bias and/or precision error associated with a single error source.

element end load proof test A nondestructive end load test in which the filter must remain free of visible damage and must not leak after being subjected to a specified end load for a specified time.

element melt time The time elapsed from the moment a fusing current begins to flow to the moment the current sharply drops in value and arcing commences.

elevation error A type of error in temperature-measuring or pressure-measuring systems that incorporate capillary tubes partly filled with liquid.

elinvar An iron-nickel-chromium alloy that also contains varying amounts of manganese and tungsten and that has low thermal expansion and almost invariable modulus of elasticity.

ellipse Plane curve constituting the locus of all points, the sum of whose distances from two fixed points called focuses or foci is constant; an elongated circle.

ellipsoid Surface whose plane sections (cross sections) are all ellipses or circles, or the solid enclosed by such a surface.

ellipsometer An optical instrument that measures the constants of elliptically polarized light. It is most often used in thin-film measurements.

elongation The increase in length of a specimen due to a tensile force expressed as a percentage of the original specimen length; an expression of ductility.

elongated alpha The hexagonal crystal phase appearing as stringer-like arrays, considerably larger in appearance than the primary alpha. Commonly exhibits an aspect ratio of 3:1 or higher.

elongation at break Dimension change recorded at the moment of specimen rupture. *See* elongation.

Elphal process Coating steel with aluminum by electrophoretic deposition.

ELPO *See* electrocoat.

elutriation Separation of fine, light particles from coarser, heavier particles by passing a slow stream of fluid upward through a mixture so that the finer particles are carried along with it.

embossing Raising a design in relief against a surface.

embrittlement Severe loss of ductility of a material due to a chemical or physical change.

embrittlement cracking A form of metal failure that occurs in steam boilers at riveted joints and at tube ends, the cracking being predominantly inter-crystalline.

EMC (electromagnetic compatibility) *1.* The ability of electronic equipment to operate in its intended environment without suffering or causing unacceptable degradation of performance as a result of unintentional electromagnetic radiation or response. *2. See* electromagnetic compatibility.

emery An abrasive material composed of pulverized, impure corundum; used in various forms including cloth or paper.

EMF *See* electromotive force.

EMF series *See* electromotive force series.

EMI *See* electromagnetic interference.

emission Electromagnetic energy propagated from a source by radiation or conduction.

emission characteristic The relation between rate of electron emission and some controlling factor—temperature, voltage, or current of a filament or heater, for instance.

emission spectrograph An instrument that identifies the presence of elements by burning the sample in an air plasma and analyzing the resulting optical spectrum.

emission spectrometer An instrument that measures the length and relative intensities of electromagnetic radiation to determine the percent concentration of elements in a sample.

emission spectroscopy The science treating the unique, characteristic radiation emission spectrum of each element.

emissivity A material characteristic determined as the ratio of radiant-energy emission rate due solely to temperature for an opaque, polished surface of a material divided by the emission rate for an equal area of a blackbody at the same temperature.

emittance The ability of a surface to emit radiant energy. It is expressed as the ratio of the radiant energy emitted per unit time, per unit area, by an opaque material to that by a blackbody at the same temperature.

EMPA *See* electron probe x-ray microanalysis.

EMS *See* engine monitoring system.

emulate To imitate one system with another such that the imitating system accepts the same data, executes the same programs, and achieves the same results as the imitated system.

emulator *1.* A device or program that emulates, usually done by microprogramming the imitating system. Contrast with simulate. *2.* A computer that behaves very much like another computer by means of suitable hardware and software.

emulsification time The total time that an emulsifier is permitted to combine with the penetrant prior to removal by water.

emulsifier A substance that can be mixed with two immiscible liquids to form an emulsion. Also known as a disperser or dispersing agent.

emulsion The dispersion of one liquid into another liquid, usually by use of an emulsifier.

emulsion corrosion test A test used to evaluate the rust preventative property of grease-water emulsions when coated steel objects are exposed to the severely corrosive atmosphere of the salt spray.

enable *1.* To restore a computer system to ordinary operating conditions. *2.* To "arm" a software or hardware element to receive and respond to a stimulus. *3.* To allow the processing of an established interrupt. *4.* To remove a blocking device, i.e., switch, to permit operation. Contrast with disable. *See* arm.

enamel *1.* A type of oil paint that contains a finely ground resin and that dries to a harder, smoother, glossier finish than other types of paint. *2.* Any relatively glossy coating, but especially a vitreous coating on metal or ceramic obtained by covering it with a slurry of glass frit and firing the object in a kiln to fuse the coating. Also known as glaze or porcelain enamel.

encapsulation *1.* Sealing an item in a polymer material. *2.* In microstructural language, to have one phase completely surround another

encipher *See* encode.

encode *1.* To apply a code, frequently one consisting of binary numbers, to represent individual characters or groups of characters in a message. Synonymous with encipher. *2.* To represent computer data in digital form. *3.* To substitute letters, numbers, or characters for other numbers, letters, or characters, usually to intentionally hide the meaning of the message except to certain individuals who know the enciphering scheme.

encoder *1.* A device capable of translating from one method of expression to another method of expression. *2.* A device that transforms a linear or rotary displacement into a proportional digital code. *3.* A hardware device that converts analog data into digital representations.

encrustation The buildup of slag, corrosion products, biological organisms such as barnacles, or other solids on a structure or exposed surface.

encryption Converting data into codes that cannot be read without a key or password.

end device The last device in a chain of devices that performs a measurement function, that is, the one performing final conversion of a measured value into an indication, record, or control-system input signal.

end mark (exit mark) A slight corrugation caused by the entry or exit rolls of a roller leveling unit.

endothermic reaction A reaction that occurs with the absorption of heat.

end point In titration, an experimentally determined point close to the equivalence point, which is used as the signal to terminate titration.

endpoint control The exact balancing of process inputs required to satisfy the stoichiometric demands of the process.

endpoint linearity The linearity of an object taken between the end points of calibration.

end-quench hardenability test (Jominy test) A test for determining the hardenability of steel in which a cylindrical test specimen is heated above the upper critical temperature and then quenched with a water spout at one end. The resulting hardness is measured at intervals from the quenched end.

end scale value The value of an actuating electrical quantity that corresponds to the high end of the indicating or recording scale on a given instrument.

end-to-end data system Comprehensive data system that demonstrates the processing of sensor data to the user, thus reducing data fragmentation.

endurance limit The maximum stress that a material can withstand without breaking for an infinitely large number of cycles.

endurance ratio The ratio of endurance limit to the tensile strength of a material.

endurance testing The process of subjecting material to stress levels within design limits until failure occurs or until the desired life has been demonstrated.

energetic particles Charged particles having energies equaling or exceeding a hundred MeV.

energize To apply rated voltage to a circuit or device in order to activate it.

energy *1.* Any quantity with dimension that can be represented as mass $\times$ length$^2 \div$ time^2. *2.* The capacity of a body to do work or its equivalent.

energy balance The balance relating the energy in and energy out of a column.

energy beam An intense ray of electromagnetic radiation, such as a laser beam, or of nuclear particles, such as electrons, that can be used to test materials or to process them by cutting, drilling, forming, welding, or heat treating.

energy density Light energy per unit area, expressed in joules per square meter—equivalent to the radiometric term "irradiance." *See also* flux density.

energy-dispersive spectroscopy (EDS) X-ray analysis that measures the energy levels of characteristic x-rays emitted from a sample.

energy gap (solid state) A range of forbidden energies in the band theory of solids.

energy level Any one of different values of energy that a particle, atom, or molecule may adopt under conditions where the possible values are restricted by quantizing conditions.

engineering plastics A general term for plastics with mechanical, chemical, or thermal properties that make them suitable for use as construction materials.

engineering, reliability The science of including factors in the basic design which will assure the required degree of reliability.

engineering units Terms of data measurement, such as degrees Celsius, pounds, grams, and so on.

engine monitoring system (EMS) A complete system approach to define engine, engine component, and sub-system health status through the use of sensor inputs, data collection, data processing, data analysis, and the human decision process.

Engler viscosity A standard time-based viscosity scale used primarily in Europe.

Enhanced Performance Architecture (EPA) An extension to MAP (manufacturing automation protocol) which provides for low-delay communication between nodes on a single segment. *See* MINI-MAP.

enhancement, serial data A method whereby a continuous string of logical ONEs or ZEROs is modified to introduce bit transitions to enable bit synchronization for recording purposes; also preserves bandwidth.

enthalpy The heat content of a substance.

entity An active element within an OSI (open system interconnection) layer (for example, Token Bus MAC [media access control] is an entity in the Layer 2).

entrainment The conveying of particles of water or solids from boiler water by steam.

entropy A measure of the extent to which the energy of a system is unavailable.

envelope *1.* Generally, the boundaries of an enclosed system or mechanism. *2.* Specifically, the glass or metal housing of an electron tube, or the glass enclosure of an incandescent lamp. Also known as a bulb.

environment *1.* The aggregate, at a given moment, of all external conditions and influences to which the data channel is subjected. *2.* The surroundings or conditions (physical, chemical, mechanical) in which a material exists. *3.* The aggregate of all external conditions and influences affecting the life and development of the product.

environmental contaminant Contaminant(s) present in the immediate surroundings, introduced into a fluid system or component.

environmental damage/deterioration Physical deterioration of an item's strength or resistance to failure as a result of interaction with climate or environment.

environmental engineering A branch of engineering that deals with the technology related to control of the surroundings in which humans live, especially the control or mitigation of contamination or degradation of natural resources such as air quality and water purity.

environmentally sealed The provision or characteristic of a device that enables it to be protected against the entry of moisture, fluids, and foreign, particulate contaminants which could otherwise affect the performance of the device.

environmental simulation tests Exposure of a system, subassemblies, or components to varying temperature, humidity, vibration, and shock tests to simulate the vehicle environment.

environmental stress cracking (ESC) The propensity of plastic materials to crack or craze in specific environments.

environmental stress screening A series of tests conducted under environmental stresses to disclose weak parts and workmanship defects for correction.

environmental test Any laboratory test conducted under conditions that simulate the expected operating environment in order to determine the effect of the environment on component operation or service life.

EOS *See* electrical overstress.

E$_p$ *See* critical pitting potential.

EPA *See* Enhanced Performance Architecture.

epichlorhydrin polymers An elastomeric polymer similar to synthetic rubber with good resistance to oils.

epitaxy Electrodeposits where the orientation of the crystals in the deposit are related to the crystals in a base metal.

epoxy Resin formed by the reaction of bisphenol and epichlorhydrin.

epoxy adhesive An adhesive made of epoxy resin.

epoxy matrix composite High-strength composition consisting of epoxy resin and a reinforcing matrix of filaments or fibers of glass metal, or other materials.

epoxy plastics A thermoset polymer with one or more epoxide groups which is used in composites and adhesives.

E$_{pp}$ *See* critical pitting potential.

EPROM Acronym for erasable programmable read only memory. A PROM is a memory chip that can be programmed once but cannot be reprogrammed. An EPROM is a special type of PROM that can be erased by exposing the chip to ultraviolet light. Then it can be reprogrammed.

EPS *See* polystyrene foam.

equalization The use of feedback to achieve close coincidence between the outputs of two or more elements or channels in a fault-tolerant control system.

equalizer *1.* A device that connects parts of a boiler to equalize pressures. *2.* The electronic circuit in a tape reproducer whose gain across the spectrum of interest compensates for the unequal gain characteristic of the record/reproduce heads, thereby providing "equalized" gain across the band.

equations of state Equations relating temperature, pressure, and volume of a system in thermodynamic equilibrium.

equiaxed Crystals or grains in a microstructure where all axes are equal.

equiaxed structure In alpha-beta titanium alloys, commonly refers to a microstructure in which most of the alpha phase appears spheroidal, primarily in the transverse direction. Generally, a structure in which the grains have approximately the same dimensions in all directions.

equi-cohesive temperature The temperature where the strength of grain boundaries and crystal grains is equal.

equilibrium diagram A graphical representation of the temperature, pressure, and composition limits of phase fields in an alloy system as they exist under conditions of complete equilibrium. In metal systems, pressure is usually considered constant.

equilibrium flow Gas flow in which energy is constant along streamlines and the composition of the gas at any point is not time-dependent.

equilibrium state Any set of conditions that results in perfect stability—mechanical forces that completely balance each other and do not produce acceleration, or a reversible chemical reaction in which there is no net increase or decrease in the concentration of reactants or reaction products, for instance.

equivalence point Point on the titration curve where the acid ion concentration equals the base ion concentration.

equivalent binary digits The number of binary digits required to express a number in another base to the same precision, for example, approximately 3 1/3 binary digits are required to express in binary form each digit of a decimal number.

equivalent evaporation Evaporation expressed in pounds of water evaporated from a temperature of 212°F to dry saturated steam at 212°F.

equivalent network A network that can perform the functions of another network under certain conditions; the two networks may be of different forms— one mechanical and one electrical, for instance.

erg The unit of energy in the CGS system is the erg. One erg = 10^{-7} joules.

ergonomics The science of designing machines and work environments to suit the needs of people. *See also* human factors engineering.

Erichsen test A cupping test for determining the suitability of metal sheet for use in a deep drawing operation.

erosion *1.* Destruction of materials by the abrasive action of moving fluids, usually accelerated by the presence of solid particles. *2.* The wearing away of refractory or of metal parts by gas-borne dust particles.

erosion corrosion A corrosion reaction accelerated by the relative movement of the corrosive fluid and the metal surface.

err Colloquial term for the more proper term—error. Used frequently in computer programming. *See also* error.

error *1.* Any discrepancy between a computed, observed, or measured quantity and the true, specified, or theoretically correct value or condition. 2. The algebraic difference between the indicated value and the true value of the measurand. *3.* In a single automatic control loop, the setpoint minus the controlled variable measurement.

error band The band of maximum deviations of output values from a specified reference line or curve due to those causes attributable to the device.

error burst In data transmission, a sequence of signals containing one or more errors, but counted as only one unit in accordance with some specific criterion or measure.

error checking Data quality assurance usually attempted by calculating some property of the data block before transmission.

error correcting code A code in which each acceptable expression conforms to specific rules of construction that also define one or more equivalent nonacceptable expressions, so that if certain errors occur in an acceptable expression, the result will be one of its equivalents, and thus the error can be corrected.

error detecting code A code in which each expression conforms to specific rules of construction, so that if certain errors occur in an expression, the resulting expression will not conform to the rules of construction, and thus the presence of the errors is detected. Synonymous with self-checking code.

error detection routine A routine used to detect whether or not an error has occurred, usually without special provision to find or indicate its location.

error message An audible or visual indication of a software or hardware malfunction, or a non-acceptable data entry attempt.

error range *1.* The range of all possible values of the error of a particular quantity. *2.* The difference between the highest and the lowest of these values.

error ratio The ratio of the number of data units in error to the total number of data units.

error signal The output of a comparing element.

error, span *See* span error.

ESC *See* environmental stress cracking.

ESR *1.* Electroslag remelting. *2. See* electron spin resonance spectroscopy.

established reliability A quantitative maximum failure rate demonstrated under controlled test conditions specified in a military specification and usually expressed as percent failures per thousand hours of test.

ester The product of an alcohol and acid reaction.

estimate A predicted value of performance.

etalons Two adjustable parallel mirrors mounted so that either one may serve as one of the mirrors in a Michelson interferometer; used to measure distance in terms of wavelengths of spectral lines.

etch cleaning Removing soil by electrolytic or chemical action that also removes some of the underlying metal.

etch cracks Shallow cracks in the surface of hardened steel due to hydrogen embrittlement that sometimes occurs when the metal comes in contact with an acidic environment.

etched medium A filter medium having passages produced by chemical or electrolytic removal of unwanted material.

etching *1.* Controlled corrosion of a metal surface to reveal its metallurgical structure. *2.* Controlled corrosion of a metal part to create a design; the design may consist of alternating raised and depressed areas, or it may consist of alternating polished and roughened areas, depending on the conditions and corrodent used. *See also* macroetching.

ETFE *See* ethylene tetrafluoroethylene.

ethanol or ethyl alcohol An alcohol of formula C_2H_5OH.

ethene An unsaturated hydrocarbon used in the production of polyethylene and vinyl compounds.

ethyl cellulose (EC) A tough polymer with low flammability and water absorption.

ethylene chlorotrifluoroethylene (ECTFE) A copolymer with outstanding creep and tensile properties, plus wear, radiation, and fire resistance.

ethylene glycol A reactant used in the production of some polyesters, and a major ingredient in antifreeze.

ethylene-propylene copolymer A copolymer with good impact strength and resistance to stress cracking and fatigue.

ethylene propylene elastomer A thermoplastic copolymer with high resistance to oxidation.

ethylene tetrafluoroethylene (ETFE) A copolymer with high impact strength and stiffness.

ethylene vinyl acetate (EVA) A copolymer with good tear resistance and flexibility.

ETPC *See* copper, ETP.

E_u *See* uniform elongation.

eutectic *1.* A process by which a liquid solution undergoes isothermal decomposition to form two homogeneous solids— one richer in solute than the original liquid, and one leaner. *2.* The composition of the liquid that undergoes eutectic decomposition and possesses the lowest coherent melting point of any composition in the range where the liquid remains single-phase. *3.* The solid resulting from eutectic decomposition, which consists of an intimate mixture of two phases.

eutectic alloys Multi-element alloys with a low melting point, such as solder.

eutectic composites Composite materials with a metal matrix of a mixture of solids including eutectoids.

eutectic mixture A mixture of two or more substances in such a ratio that it has the lowest melting point of any combination.

eutectoid A decomposition process having the same general characteristics as a eutectic, but taking place entirely within the solid state.

eV or EV *See* electron volt.

EVA *See* ethylene vinyl acetate.

evaluation A broad term used to encompass measurement, prediction, and demonstration.

evaluation kit A small microcomputer system used for learning the instruction set of a given microcomputer.

evaporation The physical process by which a liquid or solid is transformed into the gaseous state; the opposite of condensation.

evaporation gage *See* atmometer.

evaporation rate The number of weight units of water evaporated in a unit of time.

evaporative cooling *1.* Lowering the temperature of a mass of liquid by evaporating part of it, using the latent heat of vaporization to dissipate a significant amount of heat. *2.* Cooling ambient air by evaporating water into it. *3. See* vaporization cooling.

evaporimeter *See* atmometer.

event *1.* The occurrence of some programmed action within a process which can affect another process. *2.* In TELEVENT, an occurrence recognized by telemetry hardware such as frame synchronization, buffer complete, start, halt, and the like.

event oriented Pertaining to a physical occurrence.

exactness *See* precision.

examination An element of inspection consisting of investigation, without the use of special laboratory appliances or procedures, of materials and services to determine conformance to those specified requirements that can be determined by such investigations.

exception reporting An information system that reports on situations only when actual results differ from planned results. When results occur within a normal range, they are not reported.

excessive intergranular corrosion A degree of intergranular corrosion that is greater than what is usually experienced with test samples of the same alloy, heat treated to the same applicable specification.

excimer laser A laser in which the active medium is an "excimer" molecule—a diatomic molecule that can exist only in its excited state.

excimers Molecules characterized by repulsive or very weakly bound ground electronic states.

excitation *1.* Addition of energy to a nuclear, atomic, or molecular system transferring it to another energy state. *2.* Voltage supplied by a signal conditioner to certain types of physical measurement transducers (bridges, for example).

excitation index The ratio of the intensities of two spectral lines of an element with different excitation energies.

excited state *See* excitation.

EXEC *See* execute.

execute (EXEC) *1.* To interpret machine instructions or higher-level statements and perform the indicated operations on the operands specified. *2.* In computer terminology, to run a program.

execution The act of performing programmed actions.

execution time (regarding computers) *1.* The time required to execute a program. *2.* The period during which a program is being executed. *3.* The time at which execution of a program is initiated. *4.* The period of time required for a particular machine instruction. *See also* instruction time.

executive The controlling program or set of routines in an operating system. The executive coordinates all activities in the system including I/O supervision, resource allocation, program execution, and operator communication.

executive commands In TELEVENT, the several commands that establish modes of operation, such as SETUP, END, CONNECT, and the like.

executive mode A central processor mode characterized by a lack of memory protection and relocation by the normal execution of all defined instruction codes.

executive program A program that controls the execution of all other programs in the computer based on established hardware and software priorities and real-time or demand requirements.

executive software The portion of the operational software that controls on-line, response-critical events, responding to urgent situations as specified by the application program. This software is also known as the real-time executive.

executive system An integrated collection of service routines for supervising the sequencing of programs by a computer.

exfoliation Spalling on the surface of a compact as a result of trapped air or faulty pressing.

exfoliation corrosion Localized subsurface corrosion in zones parallel to the surface which result in thin layers of uncorroded metal to be elevated by the formation of corrosion products.

exhaust *1.* Discharge of working fluid or gas from an engine cylinder or from turbine vanes after it has expanded to perform work on the piston or rotor. *2.* The fluid or gas discharged. *3.* A duct for conducting waste gases, fumes, or odors from an enclosed space.

exhaust sampling The technique of obtaining an accurate sample of exhaust gas for analysis. Sampling may be by bag, continuous, or proportional method.

exit The time or place at which the control sequence ends or transfers out of a par-

ticular computer program or subroutine.

exit mark *See* end mark.

exotherm The liberation or evolution of heat during the curing of a plastic product or during any chemical reaction.

exothermic reaction A reaction that occurs with the evolution of heat.

expandable plastic Plastics that can be made into cellular foam.

expanded metal A form of coarse screening made by lancing sheet metal in alternating rows of short slits, each offset from the adjacent rows, then stretching the sheet in a direction transverse to the rows of slits so that each slit expands to give a roughly diamond-shaped opening.

expanded plastic A light, spongy plastics material made by introducing air or gas into solidifying plastic to make it foamy. Also known as foamed plastic or plastic foam.

expanded rubber Cellular rubbers having closed cells made from a solid rubber compound.

expansion factor Correction for the change in density between two pressure-measurement stations in a constricted flow.

experimental environment Surrounding conditions that may affect an experiment in known or unknown ways.

experimental specimen Objects or materials to which treatments are applied.

expert system A computer program that uses stored data to reach conclusions, unlike a database, which presents data unchanged. *See also* artificial intelligence.

explosion Combustion that proceeds so rapidly that a high pressure is generated suddenly.

explosion welding A solid-state process for creating a metallurgical bond by driving one piece of metal rapidly against another with the force of a controlled explosive detonation.

explosive cladding Producing a bimetallic material by explosion welding a thin layer of one metal on a substrate.

exponent In floating-point representation, one of a pair of numerals representing a number that indicates the power to which the base is raised. Synonymous with characteristic.

exponential case The reliability characteristics of those products known to exhibit a constant failure rate.

exponential model In reliability engineering, a model based on the assumption that times t between successive failures are described by the exponential distribution.

exponential notation A way of expressing very large or small numbers in data processing.

exponentiation A mathematical operation that denotes increases in the base number by a previously selected factor.

exposed material Used in lenses or optical devices exposed to direct sunlight as installed on the vehicle.

expression *1.* A combination of operands and operators that can be evaluated to a distinct result by a computing system. *2.* Any symbol representing a variable or a group of symbols representing a group of variables, possibly combined through the use of symbols representing operators in accordance with a set of definitions and rules. *3.* In computer programming, a set of symbols that can have a specific value.

Extended Binary Coded Decimal Interchange Code (EBCDIC) An 8-bit code that represents an extension of a 6-bit "BCD" code, which has been widely used in computers of the first and second generations. EBCDIC can represent up to 256 distinct characters and is the principal code used in many of the current computers.

extender Fillers and inert materials added to polymers to reduce voids and crazing without changing the physical properties of the material.

extensibility The ability of a material to elongate under load, expressed as a percent of the initial length. See elongation.

extensional shear coupling The property of some laminates whereby they show shear strains when under load.

extension bending coupling The property of some laminates whereby they show bending curvatures when under load.

extensometer *1.* An apparatus for studying seismic displacements by measuring the change in distance between two reference points that are separated by 20 to 30 meters or more. *2.* An instrument for measuring minute elastic and plastic strains in small objects under stress, especially the strains prior to fracture in standard tensile-test specimens.

external multiplexors Scanivalves, switching temperature indicators, and other devices that permit input of several signals on one computer input channel.

extract instruction An instruction that requests the formation of a new expression from selected parts of given expressions.

extra hard temper A level of hardness and strength in nonferrous alloys and some ferrous alloys corresponding approximately to a cold-worked state one-third of the way from full hard to extra spring temper.

extra spring temper A level of hardness and strength for nonferrous alloys and some ferrous alloys corresponding to a cold-worked state above full hard beyond which hardness and strength cannot be measurably increased by further cold work.

extremum values In statistics, the upper or lower bound of the random variable which is not expected to be exceeded by a specified percentage of the population within a given confidence interval.

extrudate The product resulting from an extrusion process.

extrusion *1.* A process for forming elongated metal or plastic shapes of simple to moderately complex cross section by forcing ductile, semisoft solid material through a die orifice. *2.* A length of product made by this process.

extrusion billet A slug of metal, usually heated into the forging temperature range, which is forced through a die by a ram in an extrusion process.

extrusion blow molding Extruding a melted material into a mold then pushing it against the mold cavity to form a shape.

extrusion coating Extruding a melted material onto a base material then pressing it for adhesion.

extrusion pressing *See* cold extrusion.

exudation The action whereby the low-melting elements of a compact are forced to the surface during sintering.

F

F *1.* Symbol for farad. *2.* Symbol for Fahrenheit. *3.* Chemical symbol for fluoride.

fabric In plastics engineering, organic fibers often spun into yarns then woven into fabric which is used as reinforcement material in composites. In metals engineering, meshes of steel used for reinforced concrete.

fabrication *1.* A general term for parts manufacture, especially structural or mechanical parts. *2.* Assembly of components into a completed structure.

fabric warp face The side of a fabric where most yarns are parallel to the selvage.

Fabry-Perot cavity A pair of highly reflecting mirrors, whose separation can be adjusted to select light of particular wavelengths. When used as a laser resonator, this type of cavity can narrow the range of wavelengths emitted by the laser.

face *1.* An exposed structural surface. *2.* In a weldment, the exposed surface of the fusion zone.

face centered cubic (FCC) A metal crystalline structure with cubic symmetry, such as aluminum and copper.

face dimpling Buckling of a compressive facing into a honeycomb cell.

facesheet *See* facings.

facet The plane surface of a crystal or fracture surface.

face *See* facing.

face wrinkling Buckling of the compressive facing into or away from the core.

facings *1.* Skins and doublers in any lay-up. *2.* The outermost layer or composite component of a sandwich construction, generally thin and of high density, which resists most of the edgewise loads and flatwise bending moments. Synonymous with face, skin, and facesheet.

factor One of the variables, qualitative or quantitative, being studied in an experiment.

factor, acceleration The ratio between the times necessary to obtain a stated proportion of failures for two different sets of stress conditions involving the same failure modes and/or mechanisms.

fadeometer An instrument to determine the amount of fading in a material.

Fahrenheit A temperature scale in which the freezing point of pure water occurs at 32°F and the span between freezing point and boiling point of pure water at standard pressure is defined to be 180 scale divisions (180 degrees).

failure *1.* Describes the condition when a material or component ceases to fulfill the design specified responses essential to the successful operation as a subunit of a system. A part may fail from tearing, cracking, rupture, embrittlement, softening, heat or chemical degradation, creep, or a combination thereof. *2.* Sudden and complete loss of the engine or steering power source output. *3.* A condition of a machine in which some part, component, or system does not function in the manner intended.

failure analysis The identification of the failure, the failure mechanism, and the cause. Often includes physical dissection and extensive laboratory study.

failure, catastrophic A sudden change in the operating characteristics of an item resulting in a complete loss of useful performance.

failure coverage The ratio of failures detected to failure population, expressed as a percentage.

failure, degradation A failure that occurs as a result of a gradual or partial change in the operating characteristics of an item.

failure, detectable A failure that can be detected with 100% test detection efficiency.

failure effect The consequences a failure has on the operation, function, or status of an item.

failure, incipient A degradation failure that is just beginning to exist.

failure, induced A failure caused by a physical condition external to the failed item.

failure, inherent A failure basically caused by a physical condition or phenomenon internal to the failed item.

failure, latent A malfunction that occurs as a result of a previous exposure to a condition that did not exist in an immediately detectable failure.

failure mechanism The mechanical, chemical, or other process that results in a failure.

failure mode The effect or manner by which a failure is observed. Generally describes the way the failure occurs.

failure, modes, and effects analysis (FMEA) A systematic, organized procedure for evaluating potential failures in an operating system.

failure, modes, effects and criticality analysis (FMECA) An analysis of possible modes of failure, their causes and effects, their criticalities, and expected frequencies of occurrence.

failure, nonrelevant A failure not applicable to the computation or the reliability.

failure, primary A failure whose occurrence is not caused by other failures.

failure, random A failure whose occurrence is not predictable in an absolute sense, but is predictable in a probabilistic sense.

failure rate *1.* The conditional probability that an item will fail just after time t, given the item has not failed up to time t. *2.* The number of failures of an item per unit measure of (life cycles, time, miles, events, etc.) as applicable for the time. The symbol lambda is used to represent failure rate.

failure, relevant A failure attributable to a deficiency of design, manufacture, or materials of the failed device, applicable to the computation of reliability.

failure, secondary A failure caused directly or indirectly by the failure of another item.

failure, wearout A failure whose time of occurrence is governed by rapidly increasing failure rate.

fairing or feathering *1.* A shape that produces a smooth transition from one angular direction to another; or the act of producing this smooth contour. *2.* A stationary member or structure whose primary function is to produce a smooth contour.

falling weight test An impact test to determine toughness of materials.

fall time The time required for the output voltage of a digital circuit to change from a logical high level (1) to a logical low level (0).

false add To form a partial sum, that is, to add without carries.

false brinelling Fretting between the rolling elements and races of ball or roller bearings.

false indication In nondestructive testing, a result that may be misinterpreted as a fault.

family The complete series of compatible materials from one manufacturer designed to perform a specific process of penetrant inspection.

farad (F) Unit of electrical capacitance; the capacitance of a capacitor that, when charged with one coulomb, gives a difference of potential of one volt.

faraday cage An apparatus for measuring electric charge. Consists of two conducting enclosures: an inner and an outer, insulated from each other. The grounded outer enclosure shields the inner from external fields; the object to be measured is placed within the inner enclosure.

Faraday rotation A rotation of the plane of polarization of light caused by the application of a magnetic field to the material transmitting the light.

Faraday rotator A device that relies on the Faraday effect to rotate the plane of polarization of a beam of light passing through it.

Faraday's laws of electrolysis *1.* Law stating that the amount of substance deposited or dissolved in an electrolytic cell is directly proportional to the amount of electricity passing through it. *2.* Law stating that the quantity of different substances deposited or dissolved by the same amount of electricity is directly proportional to their chemical equivalents.

far field Distant from the source of light. This qualification is often used in measuring beam quality, to indicate that the measurement is made far enough away from the laser that local aberrations in the vicinity of the laser have been averaged out.

far-infrared laser Generically, can be taken to mean any laser emitting in the far infrared, a vaguely defined region of wavelengths from around 10 micrometers to 1 millimeter.

fascia Flexible material commonly used as a bumper cover (may extend below bumper).

fast break In magnetic particle testing of ferromagnetic materials, interrupting the current in the magnetizing coil to induce eddy currents and strong magnetization as the magnetizing field collapses.

fast cure polyesters Alkyd resin materials.

fastener *1.* Any of several types of devices used to hold parts firmly together in an assembly; some fasteners hold parts firmly in position, but allow free or limited relative rotation. *2.* A device for holding a door, gate, or similar structural member closed.

Fast-Fourier Transform A type of frequency analysis on data that can be done by computer using special software, or by an array processor, or by a special-purpose, hardware device.

fatigue The process of progressive localized permanent structural changes occurring in a material or component subject to conditions that produce fluctuating stresses and strains at some point or points, which may culminate in loss of load-bearing ability, cracks, or complete fracture after a sufficient number of fluctuations.

fatigue crack growth rate The rate by which a crack grows under fatigue loading, expressed as crack growth per load cycle.

fatigue ductility Plastic deformation that occurs before fracture.

fatigue ductility coefficient (ε'_f) The "true" strain required to cause failure in one reversal. It is taken as the intercept of the log ($\Delta\varepsilon_p/2$) versus log ($2N_f$) = 1.

fatigue ductility exponent (c) The power to which the life in reversals must be raised to be proportional to the "true" strain amplitude. It is taken as the slope of the log ($\Delta\varepsilon_p/2$) versus log ($2N_f$) plot.

fatigue life The number of cycles of stress or related strain of a specified character that a given specimen sustains before failure of a specified nature occurs.

fatigue limit The level of applied stress at which an infinite number of reversals can be endured without the material failing.

fatigue notch factor The ratio of the fatigue strength of an unnotched specimen to the fatigue strength of a notched specimen of the same material and condition.

fatigue notch sensitivity An estimate of the effect of a notch or hole on the fatigue properties of a material; it is expressed as $q = (K_f - 1)/K_t - 1)$, where q is the fatigue notch sensitivity, K_f is the fatigue notch factor, and K_t is the stress concentration factor for a specimen of the material containing a notch of a specific size and shape.

fatigue ratio The ratio of fatigue strength to tensile strength.

fatigue strength The stress to which a material can be subjected for a specified number of fatigue cycles without failure.

fatigue strength coefficient (σ'_f) The "true" stress required to cause failure in one reversal. It is taken at the intercept of the log $\Delta\sigma/2$ versus log ($2N_f$) plot at $2N_f = 1$.

fatigue strength exponent (b) The power to which life in reversals must be raised to be proportional to "true" stress amplitude. It is taken as the slope of the log $\Delta\sigma/2$ versus log ($2N_f$) plot.

fatigue striations Parallel lines seen in a fractograph of failure surfaces.

fatigue test Tests to determine the fatigue characteristics of a material by subjecting a test specimen to cyclic loads.

fatigue test, combined pressure pulse and vibration A more sophisticated test method on which the laboratory test simulates simultaneously the service condition of impulse and fatigue.

fault *1.* An attribute that adversely affects the reliability of a device. *2.* A defect of any component upon which the intrinsic safety of a circuit depends. *3.* The failure of any part of a computer system.

fault tolerance Ability of a system to survive a certain number of failures with at least downgraded performance.

fault tree analysis (FTA) A method of reliability analysis in which a logical block diagram is used to indicate contributing lower level events.

faying surface Either of two surfaces in contact with each other in a welded, fastened, or bonded joint, or in one about to be welded, fastened, or bonded.

FCAW *See* flux-cored arc welding.

FCC *See* face centered cubic.

F-center laser A solid-state laser in which optical pumping by light from a visible-wavelength laser produces tunable near-infrared emission from defects—called "color centers" or "F centers"—in certain crystals.

Fe Chemical symbol for iron.

feasibility study Any evaluation of the worth of a proposed project based on specific criteria.

feathering *See* fairing.

feedback *1.* Process signal used in control as a measure of response to control action. *2.* The part of a closed-loop system that automatically brings back information about the condition under control. *3.* The part of a closed loop system that provides information about a given condition for comparison with the desired condition.

feedback loop The components and processes involved in correcting or controlling a system by using part of the output as input.

feed-forward control action Control action in which information concerning one or more external conditions that can disturb the controlled variable is converted into corrective action to minimize deviations of the controlled variable. Feed-forward control is usually combined with other types of control to anticipate and minimize deviations

of the controlled variable. *See also* open loop.

feedhead A reservoir of molten metal that extends above a casting to supply additional molten metal and compensate for solidification shrinkage. Also known as riser or sinkhead.

felsite A light-colored, fine-grained igneous rock composed chiefly of quartz or feldspar.

felt A nonwoven material made up of fibers that are meshed by mechanical or chemical action.

Fenimore-Martin test A test used to determine the combustion characteristics of a polymer.

FEP *See* fluorinated ethylene propylene.

Feret's diameter (of a particle) *See* particle diameter (Feret's diameter).

Fermat's principle Also called the principle of least time. Principle holding that a ray of light traveling from one point to another, including reflections and refractions that may be suffered, follows that path that requires the least time.

Fermi-Dirac statistics The statistics of an assembly of identical half-integer spin particles; such particles have wave functions antisymmetrical with respect to particle interchange and satisfy the Pauli exclusion principle.

ferric percentage Actual ferric iron in slag, expressed as the percentage of total iron calculated as ferric iron.

ferrite *1.* A body-centered cubic crystalline phase of iron-base alloys. One of the main constituents of steel; it is iron with a very small amount of dissolved carbon and possibly alloys. It is magnetic and very soft. *2.* Chemical compound of iron oxide and other metallic oxides combined with ceramic material.

These compounds have ferromagnetic properties, but are poor conductors of electricity. Hence, they are useful where ordinary ferromagnetic materials (which are good electrical conductors) would cause too great a loss of electrical energy.

ferritic nitrocarburizing A process applied to ferrous materials which involves the diffusion of nitrogen and carbon into the ferrite phase and the formation of thin white layer of epsilon carbonitrides.

ferritizing anneal A treatment given to as-cast gray or ductile (nodular) iron to produce an essentially ferritic matrix. For the term to be meaningful, the final microstructure desired or the time-temperature cycle used must be specified.

ferroalloy An alloy, usually a binary alloy, of iron and another chemical element which contains enough of the second element for the alloy to be suitable for introduction into molten steel to produce alloy steel, or in the case of ferrosilicon or ferroaluminum, to produce controlled deoxidation during the melting process.

ferrodynamic instrument An electrodynamic instrument in which the presence of ferromagnetic material (such as an iron core for an electromagnetic coil) enhances the forces ordinarily developed in the instrument.

ferroelectric materials Dielectric materials.

ferrograph An instrument that measures the size of wear particles in lubrication oils.

ferrography Wear analysis conducted by withdrawing lubricating oil from an oil reservoir and using a ferrograph analyzer to determine the size distribution of wear particles.

ferromagnetic material Material whose relative permeability is greater than unity and depends on the magnetizing force. A ferromagnetic material usually has relatively high values of relative permeability and exhibits hysteresis. Examples are iron, cobalt, and nickel.

ferrometer An instrument for measuring magnetic permeability and hysteresis in iron, steel, and other ferromagnetic materials.

ferrous alloy Any alloy containing at least 50% of the element iron by weight.

FFT *See* fast Fourier transform.

fiber *1.* A particle whose length is greater than 100 μm and at least 10 times its width. *2.* The characteristic of wrought metal that indicates directionality, and can be revealed by etching or fractography. *3.* The pattern of preferred orientation in a polycrystalline metal after directional plastic deformation such as by rolling or wiredrawing. *4.* A filament or filamentary fragment of natural or synthetic materials used to make thread, rope, matting, or fabric. *5.* In stress analysis, a theoretical element representing a filamentary section of solid material aligned with the direction of stress.

fiberboard A broad general term for fibrous structures produced on any of the several types of fiber forming machines. The primary composition of these boards is normally refined cellulosic or matted wood fibers which may or may not be supplemented by the use of synthetic materials or chemical additives. The manufacture of fiberboards normally involves the formation of a wet web of suspended fibers, which is subsequently pressed, dried, and often calendered or laminated to develop desired end use properties.

fiber composite Structural material consisting of combinations of metals or alloys or plastics reinforced with one or more types of fibers.

fiber content The percent volume of a fiber in a composite.

fiber count The number of fibers per unit of ply in a specific portion of a composite.

fiber direction The alignment of a fiber on a longitudinal axis related to the reference axis.

fiberglass A continuous reinforcement filament drawn from molten glass.

fiberglass reinforcement Material used to reinforce a resin matrix using continuous or discontinuous glass fibers.

fiber metal A material composed of metal fibers that have been pressed or sintered together, and that may also have been impregnated with resin, molten metal, or other material that subsequently hardened.

fiber metallurgy The science of producing solid materials that include fibers or chopped filaments.

fiber migration Carry-over of fibers from filter or separator media material into the effluent.

fiber optics A medium that uses light conducted through glass or plastic fibers for data transmission, optical measurements, or optical observations.

fiber pattern The fibers that can be seen on the surface of a composite or laminate.

fiber-reinforced ceramics Ceramic materials reinforced with organic or metallic fibers.

fiber-reinforced plastic (FRP) A very general term referring to any composite that is reinforced with fibers.

fiber show Strands of fibers showing above the surface of a composite.

fiber stress Stress in a small area on a component where the stress is not uniform.

fiber wash Fanning out of fibers caused by the movement of molten resin during cure.

fibrillation The creation of fiber from a film.

fibrous composite A material consisting of natural, synthetic, or metallic fibers embedded in a matrix, usually a matrix of molded plastics material or hardenable resin.

fibrous fracture A type of fracture surface appearance characterized by a smooth, dull gray surface.

fibrous structure *1.* In fractography, a ropy fracture-surface appearance, which is generally synonymous with silky or ductile fracture. *2.* In forgings, a characteristic macrostructure indicative of metal flow during the forging process, which is revealed as a ropy appearance on a fracture surface or as a laminar appearance on a macroetched section. *3.* In wrought iron, a microscopic structure consisting of elongated slag fibers embedded in a matrix of ferrite.

fictive temperature The temperature at which a specific type of glass would become thermodynamically stable.

FID *See* flame ionization detector.

fidelity The degree to which a system, subsystem, or component accurately reproduces the essential characteristics of an input signal in its output signal. *See also* accuracy.

field emission microscopy An imaging technique using induced electron emission from an unheated metal specimen with the electrons captured on film to show variations in emission over the surface.

field ionization Ionizing atoms and molecules by a strong electric field on the surface of a metal.

field ion microscopy An imaging technique using ionized atoms from a metal specimen where the atom patterns are projected on a screen.

field of view The area or solid angle that can be viewed through or scanned by an optical instrument.

field strength For any physical field, the flux density, intensity, or gradient of the field at the point in question.

field test A test performed in the actual environment in which the product will be used.

filament A very fine single strand of metal wire, extruded plastic, or other material.

filamentary shrinkage Small cavities due to shrinkage which produce a radiographic image with a lacy appearance.

filament winding Fabricating a composite structure by winding a continuous fiber reinforcement on a rotating core under tension.

filament-wound structure A composite structure made by fabricating one or more structural elements by filament winding, then curing them and assembling them.

file hardness A hardness determined through the use of a file of standardized hardness on the assumption that a material that cannot be cut with the file is as hard as, or harder than, the file. Files covering a range of hardnesses may be employed.

file transfer access and management (FTAM) One of the application protocols specified by Manufacturing Automation Protocol (MAP) and Technical Office Protocol (TOP).

filiform corrosion Corrosion that occurs under a coating appearing as threadlike filaments.

fill Yarn at right angles to the lengthwise yarns in a woven fabric.

filled composite A plastics material made of short-strand fibers or a granular solid mixed into thermoplastic or thermosetting resin prior to molding.

filler *1.* An inert material added to paper, resin, elastomers, and other materials to modify their properties or improve quality in end products. *2.* A material used to fill holes, cracks, pores, and other surface defects before applying a decorative coating such as paint. *3.* A metal or alloy deposited in a joint during welding, brazing, or soldering; usually referred to as filler metal.

fillet *1.* A concave transition surface between two surfaces that meet at an angle. A larger radius will promote less likelihood of stress concentration. *2.* A molding or corner piece placed at the junction of two perpendicular surfaces to lessen the likelihood of cracking.

fillet weld A roughly triangular weld that joins two members along the intersection of two surfaces that are approximately perpendicular to each other.

film *1.* A flat, continuous sheet of thermoplastic resin or similar material that is extremely thin in relation to its width and length. *2.* A very thin coating, deposit, or reaction product that completely covers the surface of a solid.

film adhesive A thermosetting resin produced as a thin film with or without some carrier.

filmogen The material or binder in paint that imparts continuity to the coating.

film strength *1.* Generally, the resistance of a film to disruption. *2.* In lubricants, a measure of the ability to maintain an unbroken film over surfaces under varying conditions of load and speed.

film thickness In a dynamic seal, the distance separating the two surfaces that form the primary seal.

filter *1.* In electronic, acoustic, and optical equipment, a device that allows signals of certain frequencies to pass, while rejecting signals having frequencies in another range. *2.* A machine word that specifies which parts of another machine word are to be operated upon, thus the criterion for an external command. Synonymous with mask. *3.* A porous material or structural element designed to allow fluids to pass through it while collecting and retaining solids of a certain particle size or larger.

filter aid An inert powdery or granular material such as diatomaceous earth, fly ash, or sand which is added to a liquid that is about to be filtered in order to form a porous bed on the filter surface, thereby increasing the rate and effectiveness of the filtering process.

filter element burst pressure test A test procedure that determines the least pressure drop across a filter element, arising from inside to outside flow, that causes structural failure or medium rupture in a filter element.

filter element collapse pressure test A test procedure that determines the least pressure drop across a filter element arising from outside to inside flow that causes structural failure or medium rupture in a filter element.

filter element end load failure test A test procedure that determines the least axial force applied to the end of a filter element that causes seal failure or failure due to permanent deformation of the element.

filter element media flow fatigue test A method of test to determine the ability of filter element media to withstand the flexing caused by cyclic differential pressure.

filter life test A type of filter capacity test in which a clogging contaminant is added to the influent of a filter, under specified test conditions, to produce a given rise in pressure drop across the filter or until a specified reduction of flow is reached. Filter life may be expressed as test time required to reach terminal conditions, at a specified contaminant addition rate.

filter mechanical tests Laboratory testing of filters to prove their mechanical integrity. Filters are tested to determine vibration fatigue life, pressure impulse fatigue life, and the hydrostatic burst pressure strength of the housing material and construction. The tests simulate conditions of engine vibrations and engine lubrication system cyclic pressure pulsation and maximum pressure surges. The test condition values should correspond with those actually measured on the intended engine applicable when such information is available.

filter medium The portion of a filter or filtration system that actually performs the function of separating out the solid material.

filter (or element) pressure differential The drop in pressure due to flow across a filter or element at any time. The term may be qualified by adding one of the words "initial," "final," or "mean."

filter rated flow The maximum flow rate, in liters/hour, of a fluid of specified viscosity, for which a filter is designed. The standard fluid viscosity for flow rating of lubricating oil filters is 20 cst (100 S.U.S.). Suitability of the filter for use at flows and/or viscosities beyond the rated flow is left to the discretion of the user.

filter rated operating pressure The maximum steady-state pressure at which a filter is designed to operate without structural damage to the filter housing.

filtration ratio The ratio of the number of particles greater than a given size in the influent to the number of particles of the same size in the filter effluent.

filtration ratio test A form of particle retention rating test in which the ratio of the number of particles larger than a specified size per unit volume in the influent, to the number of particles larger than the same size per unit volume in the effluent fluid is determined.

fin *1.* Fixed or adjustable airfoil or vane attached longitudinally to an aircraft, rocket, or a similar body to provide a stabilizing effect. *2.* A thin, flat, or curved projecting plate, typically used to stabilize a structure surrounded by flowing fluid or to provide an extended surface to improve convective or radiative heat transfer. *3.* A defect consisting of a very thin projection of excess material at a corner, edge, or hole in a cast, forged, molded, or upset part,

which must be removed before the part can be used.

final annealing—nonferrous or ferrous An imprecise term used to denote the anneal used to prepare a material for shipment to the user.

final density The end density of a sintered material.

fine grinding *1.* Mechanical reduction of a powdery material to a final size of at least 100 mesh, usually in a ball mill or similar grinding apparatus. *2.* In metallography or abrasive finishing, producing a surface finish of fine scratches by use of an abrasive having a particle size of 320 grit or smaller.

fineness Purity of gold or silver expressed in parts per thousand; for instance, gold having a fineness of 999.8 has only 0.02%, or 200 parts per million, of impurities by weight.

fines *1.* In a granular substance having mixed particle sizes, those particles smaller than the average particle size. *2.* Fine granular material that passes through a standard screen on which the coarser particles in the mixture are retained. *3.* In a powdered metal, the portion consisting of particles smaller than a specified particle size.

finish *1.* A chemical or other substance applied to the surface of virtually any solid material to protect it, alter its appearance, or modify its physical properties. *2.* The degree of reflectivity of a lustrous material, especially metal. *3.* Generally, the surface quality, condition, or appearance of a metal or plastic part.

finished blank (developed blank) A blank that requires little or no trimming after being formed.

finished steel Steel that is ready for manufacturing purposes.

finishing temperature In a rolling or forging operation, the metal temperature during the last reduction and sizing step, or the temperature at which hot working is completed.

fire assay Determining the metal content of an ore or other substance through the use of techniques involving high temperatures.

fireproof Resistant to combustion or to damage by fire under all but the most severe conditions.

fire-resistant Resistant to combustion and to heat of standard intensity for a specified time without catching fire or failing structurally.

fire retardant *1.* Treated by coating or impregnation so that a combustible material—wood, paper, or textile, for instance—catches fire less readily and burns more slowly than untreated material. *2.* The substance used to coat or impregnate a combustible material to reduce its tendency to burn.

fire scale Copper oxide in silver-copper alloys that remains after annealing.

firmware Programs or instructions that are permanently stored in hardware memory devices (usually read-only memories) which control hardware at a primitive level.

first-order system A system definable by a first-order differential equation.

first order transition A change in a material. Related to melting or crystallization.

fisheye(s) *1.* Circular areas devoid of topcoat caused by the wet paint "drawing" in on itself due to surface contaminants such as oil or silicones or a general incompatibility of paint and sealer. *2.* An area on a fracture surface having

a characteristic white crystalline appearance, usually caused by internal hydrogen cracking. *3.* A small, globular mass in a blended material such as plastic or glass which is not completely homogeneous with the surrounding material. *4. See* flake.

fishmouthing *See* alligatoring.

fishscale A scaly effect on porcelain enamel caused by loss of adhesion between the enamel and the substrate.

fishtail Excess metal at the trailing end of an extrusion or a rolled billet or bar, which is generally cropped and either discarded or recycled into a melting operation.

fission The splitting of an atomic nucleus into two more-or-less equal fragments.

fissionable material Material containing nuclides capable of undergoing fission only by fast neutrons with energy greater than 1 MeV, for example thorium-232 and uranium-238.

fissure A small, cracklike surface discontinuity, often one whose sides are slightly opened or displaced with respect to each other.

fit The closeness of mating parts in an assembly, as determined by their respective dimensions and tolerances.

fixed carbon The carbonaceous residue less the ash remaining in the test container after the volatile matter has been driven off in making the proximate analysis of a solid fuel.

fixed point *1.* A reproducible standard value, usually derived from a physical property of a pure substance, which can be used to standardize a measurement or check an instrument calibration. *2.* Pertaining to a numeration system in which the position of the radix point is fixed with respect to one end of the numerals, according to some convention. *See* fixed-point arithmetic.

fixed-point arithmetic *1.* A method of calculation in which operations take place in an invariant manner, and in which the computer does not consider the location of the radix point. *2.* A type of arithmetic in which the operands and results of all arithmetic operations must be properly scaled so as to have a magnitude between certain fixed values.

fixed-point data In data processing, the representation of information by means of the set of positive and negative integers. Fixed-point data is faster than floating point data and requires fewer circuits to implement.

fixed point notation In data processing, numbers that are expressed by a set of digits with the decimal point in the correct position.

fixed-point part In a floating-point representation, the numeral of a pair of numerals representing a number, that is, the fixed-point factor by which the power is multiplied. Synonymous with mantissa.

fixed-program computer A computer in which the sequences of instructions are permanently stored or wired in, and perform automatically and are not subject to change either by the computer or the programmer except by rewiring or changing the storage input. *See* wired program computer.

fL *See* footlambert.

flag *1.* A bit of information attached to a character or word to indicate the boundary of a field. *2.* An indicator frequently used to tell some later part of a program that some condition occurred earlier. *3.* An indicator used to identify the members of several sets which

are intermixed. *4.* A storage bit whose location is usually reserved to indicate the occurrence or nonoccurrence of some condition.

flag register An 8-bit register in which each bit acts as a flag.

flake *1.* Dry, unplasticized cellulosic plastics base material. *2.* Plastics material in chip form used as feed in a molding operation. *3.* An internal crack associated with the presence of hydrogen such as may be formed in steel during cooling from high temperature. Also known as fisheye, shattercrack, or snowflake. *4.* Metal powder in the form of fish-scale particles. Also known as flaked powder.

flaked powder *See* flake.

flame A luminous body of burning gas or vapor.

flame annealing Annealing in which the heat is applied directly by a flame.

flame cutting Using an oxyfuel-gas flame and an auxiliary oxygen jet to sever thick metal sections or blanks.

flame hardening A surface hardening process in which only the surface layer of a suitable workpiece is heated by a suitably intense flame to above the upper transformation temperature and immediately quenched.

flame ionization detector (FID) An analytical instrument used for determining the carbon concentration of hydrocarbons in a gas sample.

flame resistance The tendency of the material, when burning, to self-extinguish once the ignition source is removed.

flame retardants Certain chemicals that are used to reduce or eliminate the tendency of a resin to burn.

flame retarded resin A resin compounded with certain chemicals to reduce or eliminate its tendency to burn.

flame shield The metal shield adjacent to the case insulation which prevents erosion of the insulation and objectionable insulation pyrolysis products from entering the gas stream.

flame spraying *1.* Applying a plastic coating on a surface by projecting finely powdered plastic material mixed with suitable fluxes through a cone of flame toward the target surface. *2.* Thermal spraying by feeding an alloy or ceramic coating material into an oxyfuel-gas flame.

flame treating Making inert thermoplastics parts receptive to inks, lacquers, paints, or adhesives by bathing them in open flames to promote surface oxidation.

flammability Susceptibility to combustion.

flammable A characteristic of a material that is demonstrated by the tendency of the material to ignite and burn when an ignition source is brought sufficiently close.

flammable liquid A liquid, usually a liquid hydrocarbon, that gives off combustible vapors.

flare test A test for tubing, involving tapered expansion over a cone, to determine the amount of flare that can be achieved before rupture.

flaring Increasing the diameter at the end of a pipe or tube to form a conical section.

flash *1.* In plastics molding, elastomer molding, or metal die casting, a portion of the molded material that overflows the cavity at the mold parting line. *2.* A fin of material attached to a molded, cast, or die forged part.

flash coating A very thin coating of paint or plating applied to provide limited corrosion protection or to improve the adhesion of subsequent coatings.

flashed glass *See* cased glass.

flash groove *See* cutoff.

flashing A rapid change in fluid state, from liquid to gaseous. In a dynamic seal, this can occur when frictional energy is added to the fluid as the latter passes between the primary sealing faces, or when fluid pressure is reduced below the fluid's vapor pressure because of a pressure drop across the sealing faces.

flash line A raised line on the surface of a molded or die cast part which corresponds to the parting line between mold faces.

flash plating Electrodeposition of a very thin film of metal, usually just barely enough to completely cover the surface.

flash point Temperature at which a fluid gives off sufficient vapor to cause it to ignite.

flash radiography A radiography technique using high intensity radiation for a very short exposure time.

flash-resistant Not susceptible to burning violently or rapidly when ignited.

flash welding A resistance welding process commonly applied to wide, thin members, irregularly shaped parts, and tube-to-tube joints, in which the faying surfaces are brought into close proximity, electric current is passed between them to partly melt the surfaces by combined arcing and resistance heating, and the surfaces are then upset forged together to complete the bond.

flat die forging Shaping metal by pressing or hammering it between simple flat or regularly contoured dies.

flat lay *1.* The property of nonwarping in laminating adhesives. *2.* An adhesive material with good noncurling and nondistension characteristics.

flat-position welding Welding from above the work, with the face of the weld in the horizontal plane. Also known as downhand welding.

flattening Straightening metal sheet or plate by passing it through a set of staggered and opposing rollers that bend the sheet slightly to flatten it without reducing its thickness.

flattening test A test that evaluates the ductility, formability, and weld quality of metal tubing by flattening it between parallel plates to a specified height.

flatting agent A chemical additive that promotes a nonglossy, matte finish in paints and varnishes.

flat wire Rectangular flat metal that is narrower than strip metal.

flaw A discontinuity or other physical attribute in a material that exceeds acceptable limits; the term flaw is nonspecific, and more specific terms such as defect, discontinuity, or imperfection are often preferred.

flexibility, cold *See* cold flexibility.

flexibilizer An additive that makes a material tougher (more flexible).

flexible *1.* Describes a material that has continued structural integrity and resistance to rupture during bending. *2.* Describes a cellular organic polymeric material that will not rupture when a specimen 200 × 25 × 25 mm (8 × 1 × 1 in) is bent around a 25 mm (1 in.) diameter mandrel at a uniform rate of 1 lap in 5 s at a temperature between 18 and 29°C (65 and 85°F).

flexible cellular rubber A cellular organic polymeric material that will not rupture when a specimen 200 × 25 × 25 mm (8 × 1 × 1 in) is bent around a 25 mm (1 in) diameter mandrel at a uniform rate of 1 lap in 5 s at a temperature between 18 and 29°C (65 and 85°F). In the case of latex foam rubbers, these cells are open and interconnecting.

flexible foam Foams with a high tensile to compression strength ratio and good recovery rate.

flexible line tests Tests that measure burst strength, estimate rates of deterioration in service, and determine flexure fatigue and fluid compatibility.

flexible manufacturing systems (FMS) A manufacturing system under computer control, with automatic material handling; primarily designed for batch manufacturing.

flexivity Temperature rate of flexure for a bimetal strip of given dimensions and material composition.

flex life The length of time to failure which indicates the relative ability of a material to withstand dynamic bending or flexing under specific test conditions.

flex roll A movable jump roll designed to push up against the sheet as it passes through the roller lever. The roll can be adjusted to produce varying amounts of deflection of the sheet up to the diameter of the roll.

flex rolling Passing sheets through a flex roll unit to minimize yield point elongation so as to reduce the tendency for stretcher strains to appear in forming.

flex, test A laboratory method used to evaluate the resistance of a material to repeated bonding.

flexural modulus The ratio of load on a reinforced fiber polymer test specimen to the strain on the outermost fibers.

flexural strength The maximum load that can be borne by fibers in a reinforced fiber polymer.

flinching In quality control inspection, failure of an inspector to call a borderline defect a defect.

flip-flop *1.* A bistable device, i.e., a device capable of assuming two stable states. *2.* A control device for opening or closing gates, i.e., a toggle.

floating point *1.* An arithmetic notation in which the decimal point can be manipulated; values are sign, magnitude, and exponent ($+0.833 \times 10^2$). *2.* A form of number representation in which quantities are represented by a bounded number (mantissa) and a scale factor (characteristic or exponent) consisting of a power of the number base.

floating point arithmetic A method of calculation that automatically accounts for the location of the radix point. This usually is accomplished by handling the number as a signed mantissa times the radix raised to an integral exponent.

floating point base In floating-point representation, the fixed positive integer that is the understood base of the power. Synonymous with floating-point radix.

floating point notation In data processing, numbers that are expressed as a fraction coupled with an integer exponent of the base.

floating point radix *See* floating point base.

floating-point routine A set of subroutines that causes a computer to execute floating-point arithmetic. These routines may be used to simulate floating-point operations on a computer with no built-in floating-point hardware.

floppers Lines or ridges that are transverse to the direction of rolling and generally confined to the section midway between the edges of the coil as rolled. They are somewhat irregular and tend toward a flat arc shape.

flospinning Forming cylindrical, conical, or curvilinear parts from light plate by power spinning the metal over a rotating mandrel.

flotation A process for separating particulate matter in which differences in surface chemical properties are used to make one group of particles float on water while other particles do not.

flow *1.* The order of events in the computer solution to a problem. *2.* The movement of material in any direction. *See also* block flow.

flowability A general term describing the ability of a slurry, plasticized material, or semisolid to behave like a fluid.

flow brazing A brazing process in which the joint is heated by pouring hot, molten nonferrous filler metal over the assembled parts until brazing temperature is attained.

flow chart *1.* A chart to represent, for a problem, the flow of data, procedures, growth, equipment, methods, documents, machine instructions, etc. *2.* A graphical representation of a sequence of operations by using symbols to represent the operations.

flow coat To apply a coating by pouring liquid over an object and allowing the excess to drain off.

flow diagram *See* flow chart.

flow line(s) *1.* The connecting line or arrow between symbols on a flow chart. *2.* A mark on a molded plastic part where two flow fronts met during molding. Also known as a weld mark. *3.* In mechanical metallurgy, a path followed by minute volumes of metal during forming. *4.* A series of sharp parallel kinks or creases occurring in the arc when sheet steel is roll formed into a cylindrical shape.

flow marks An uneven surface on a molded component caused by incomplete flow of resin in the acid.

flow rate The quantity of fluid that moves through a pipe or channel within a given period of time.

flow soldering *See* wave soldering.

flow stress The stress at the beginning of plastic deformation.

fluid A gas or liquid, both of which have the property of undergoing continuous deformation when subjected to any finite shear stress as long as the shear stress is maintained.

fluidity The degree to which a substance flows freely. The ability of a liquid to fill a mold cavity.

fluidized bed A heat treatment medium. A dynamic mixture of a gas and/or vapor and minute solid particles of such a size that the mixture resembles a fluid in motion.

fluid side That side of the seal that in normal use faces toward the fluid being sealed.

fluid temperature A temperature of 120 and 180°F (49 and 82°C) for standard tests. The test cylinder should be stroked prior to test runs to eliminate air from the circuit and to allow the cylinder components to heat up to the test temperature.

fluorescence *1.* Emission of electromagnetic radiation from a surface upon absorption of energy from other electromagnetic or particulate radiation. *2.* The electromagnetic radiation produced by the above process. *3.* Characteristic x-rays produced due to absorption of higher-energy x-rays.

fluorescence spectroscopy The study of materials by means of the light they emit when irradiated by other light. Many materials emit visible light after they have been illuminated by ultraviolet light. The intensity and wavelengths of the emitted light can be used to identify the material and its concentration.

fluorescent The property of emitting visible light due to the absorption of radiation of a shorter wavelength which may be outside the visible spectrum.

fluorescent magnetic particle testing Magnetic particles coated with fluorescent material to increase the visibility of results.

fluorescent penetrant test Coating a surface with a fluorescent liquid, wiping the surface clean, and then inspecting for flaws under ultraviolet light.

fluoride A binary compound of fluorine with another element.

fluorinated ethylene propylene A copolymer with high impact strength, good electrical properties, and resistance to chemical attack.

fluorine A pale yellow, poisonous, highly corrosive gas element used in a wide variety of important industrial compounds.

fluorocarbon A compound containing fluorine and carbon (including other elements).

fluoroelastomer A saturated polymer in which hydrogen atoms have been replaced with fluorine. Fluoroelastomers are characterized by excellent chemical and heat resistance.

fluorometer An instrument for measuring the fluorescent radiation emitted by a material when excited by monochromatic incident radiation.

fluorometric analysis Chemical analysis in which the fluorescent intensity of a substance is measured to differentiate between elements.

fluoroplastics A family of plastics resins based on fluorine substitution of hydrogen atoms in certain hydrocarbon molecules. *See also* fluoropolymers.

fluoropolymer(s) A family of polymers based on fluorine replacement of hydrogen atoms in hydrocarbon molecules. Compounds are characterized by chemical inertness, thermal stability, and low coefficient of friction.

fluoroscopy X-ray examination similar to radiography, but in which the image is produced on a fluorescent screen instead of on radiographic film.

fluorosilicone elastomers Silicone synthetic rubbers with high chemical resistance.

fluted core A woven reinforcement material between two skins of a sandwich construction.

flutter *1.* In tape recorders, the higher-frequency variations in record and/or reproduce speed which cause time base errors in the record/reproduce process. *2.* Irregular alternating motion of a control surface, often due to turbulence in a fluid flowing past it. *3.* Repeated speed variation in computer processing.

flux *1.* (magnetic) The sum of all the lines of force in a magnetic field crossing a unit area per unit time. *2.* In metal refining, a substance added to the melt to remove undesirable substances such as sand, dirt, or ash, and sometimes to absorb undesirable elements or compounds such as sulfur in steelmaking or iron oxide in copper refining. *3.* In

welding, brazing, and soldering, a substance preplaced in the joint or fed into the molten zone to prevent formation of oxides or other undesirable compounds, or to dissolve them and make it easy to remove them.

flux-cored arc welding Abbreviated FCAW. A form of electric arc welding in which the electrode is a continuous tubular wire of filler metal whose central cavity contains welding flux.

flux density Flux per unit area perpendicular to the direction of the flux.

fluxmeter An instrument for measuring the intensity of magnetic flux.

flux pinning In superconductors, the interaction between the magnetic and the metallurgical microstructures. Flux pinning controls the critical current density in a given superconducting material.

FM *See* modulation.

FME *See* failure modes and effects analysis.

FMECA *See* failure, modes, effects and criticality analysis.

FMS *See* flexible manufacturing system.

F number The ratio of the principle length of a lens to its diameter.

foam Material with a core of foamed polymers which may or may not be reinforced with another material.

foamed plastics *1.* Resins in sponge form, flexible or rigid, with cells closed or interconnected and density over a range from that of the solid parent resin to 0.030 g/cm^3. *2. See* expanded plastic.

foaming *1.* Any of various methods of introducing air or gas into a liquid or solid material to produce a foam. *2.* The continuous formation of bubbles that have sufficiently high surface tension to remain as bubbles beyond the disengaging surface.

foaming agent Chemicals added to plastics and rubbers which generate inert gases on heating, causing the resin to assume a cellular structure.

foaming film adhesive An adhesive film used to join honeycomb core in bonded assemblies.

foam porosity Numerous, small openings in the foam structure; coarse and fine classifications are indicative of the size of the openings and are based on air pressure drop tests.

focal length The distance from the focal point of a lens or lens system to a reference plane at the lens location, measured along the focal axis of the lens system.

focal plane device Radiation sensitive device positioned at the focal area of electromagnetic detectors.

focal point *1.* The location on the opposite side of a lens or lens system where rays of light from a distant object meet at a point. Also known as a focus. *2.* The point in space where a beam of electromagnetic energy (such as light, x-rays, or laser energy) or of particles (such as electrons) has its greatest concentration of energy.

focal spot The area of the target in an x-ray tube where the stream of electrons from the cathode strikes the target.

focus *See* focal point.

focusing magnet An assembly containing one or more permanent magnets or electromagnets which is used to focus an electron beam.

fog *1.* A suspension of very small water droplets in the air. *2.* A defect in developed radiographic, photographic, or spectrographic emulsions consisting of uniform blackening due to unintentional exposure to low-intensity light or penetrating radiation.

fogged metal A metal surface whose luster has been greatly reduced by the creation of a film of oxide or other reaction products.

fog quenching Rapidly cooling an item by subjecting it to a fine mist, usually of water.

foil Very thin metal sheet, usually less than 0.006 inch (0.15 millimeter) thick.

foil strain gage A type of metallic strain gage, usually made in the form of a back-and-forth grid by photoetching a precise pattern on foil made of a special alloy having high resistivity and low temperature coefficient of resistivity.

foot A fundamental unit of length in the British and U.S. Customary systems of measurement equal to 12 inches.

footcandle A measure of the intensity or level of illumination. One footcandle (fc) is the intensity of illumination at a point on a surface one foot from a uniform point source of one standard candle.

footlambert (fL) A unit of brightness equal to the uniform brightness of a perfectly diffusing surface emitting or reflecting light at the rate of one lumen per square foot.

foot-pound A force of one pound applied to a lever one foot long.

force The cause of the acceleration of material bodies measured by the rate of change of momentum produced on a free body.

forced vibration Vibration during which variable forces outside the system determine the period of the vibration.

force factor *1.* The complex ratio of the force required to block the mechanical system of an electromechanical transducer to the corresponding current in the electrical system. *2.* The complex ratio of open-circuit voltage in the electrical system of an electromechanical transducer to corresponding velocity in the mechanical system.

foreground/background A control system that uses two computers, one performing the control functions and the other used for data logging, off-line evaluation of performance, financial operations, and so on. Either computer is able to perform the control functions.

foreground/background processing A computer system organized so that primary tasks dominate computer processing time when required, and secondary tasks fill the remaining time.

foreground program A time-dependent program initiated via request, whose urgency preempts operation of a background program. Contrast with background program.

forehand welding Welding in which the palm of the welder's torch or electrode hand faces the direction of weld travel. Contrast with backhand welding.

foreign material The presence of an object or material that comes from some source external to the part or system.

forging *1.* Using compressive force to plastically deform and shape metal; it is usually done hot, in contained, closed dies or between flat dies. *2.* A shaped part made by impact, compression, or rolling; if by rolling, the part is usually referred to as a roll forging.

forging billet *See* forging stock.

forging cracks Rupture of a metal that develops during the forging operation due to overstressing the metal by forging at too low a temperature.

forging press A press, usually vertical, used to operate dies to deform metal plastically. Mechanical presses are used for smaller closed die forging; hydraulic, or steam hydraulic presses are used for flat die forgings and larger closed die forgings.

forging range In hot forging, the optimum (or at least acceptable) temperature range for shaping the metal.

forging stock A piece of semifinished metal used to make a forging. Also known as forging billet. If it is a single piece to make a single forging, it can be referred to as a mult or a multiple.

formability The degree to which a metal can be shaped through plastic deformation.

formed-in-place gasket materials Liquids of varying consistencies which can be applied to one of the mating joint surfaces before assembly. When parts are mated, the FIPG material is capable of flowing into voids, gaps, scratch marks, etc. and cures to form a durable seal.

forming *1.* The shaping of sheet metal by bending, uniaxial stretching, biaxial stretching, compression, or a combination thereof. *2.* Applying pressure to shape a material by plastic deformation without intentionally altering its thickness.

forming die A die for producing a contoured shape in sheet metal or other material.

formula A set of parameters that distinguish the products defined by procedures. The formula may include types and quantities of ingredients, along with information such as the magnitude of process variables. It may effect procedures.

formula translating system (FORTRAN) A procedure-oriented language for solution of arithmetic and logical programs.

FORTRAN *See* formula translating system. *See also* high-level language(s).

forward power The power supplied by the output of an amplifier (or generator) traveling towards the load.

foundry A commercial enterprise, plant, or portion of a factory where metal or glass is melted and cast.

four-ball tester An apparatus for determining lubrication efficiency by driving one ball against three stationary balls clamped together in a cup filled with the test lubricant. Such an apparatus can also be designed to test for rolling contact fatigue properties of the metal.

Fourier analysis The representation of physical or mathematical data by the use of the Fourier series of Fourier integral.

Fourier optics *1.* Optical components used in making Fourier transforms and other types of optical processing operations. *2.* A prism or grating monochromator that essentially performs a Fourier transform on the light incident upon the entrance slit.

Fourier transform infrared spectrometry (FT-IR) Infrared spectrometry in which data are produced as an interferogram, then Fourier transformed to indicate physical or mathematical data.

FP fiber A ceramic fiber used for composites that must withstand high temperatures.

fraction *1.* In classification of powdered or granular solids, the proportion of the sample (by weight) that lies between two stated particle sizes. *2.* In chemical distillation, the proportion of a solution of two liquids consisting of a specific chemical substance.

fractional distillation A thermal process whereby a mixture of liquids that boil at different temperatures is heated at a series of increasing temperatures, and the distillates boiled off at each temperature are collected separately.

fraction defective In quality control, the average number of units of product containing one or more defects for each 100 units of product in a given lot.

fractography The study of fracture surfaces, especially for the purpose of determining the causes of failure and relating these causes to macrostructural and microstructural characteristics of parts and materials.

fracture The partial rupture on the surface of a material or a complete separation of a material under load.

fracture mechanics A quantitative analysis for evaluating structural reliability in terms of applied stress, crack length, and specimen geometry.

fracture stress The stress on a minimum cross-section of a material at the beginning of rupture.

fracture test A method for determining grain size, case depth, or material quality by breaking a test specimen and examining the fracture surface for certain characteristic features.

fracture toughness A measure of the resistance of a metal or polymer that has small cracks or flaws to, under load, progress to rupture.

fragmentation In data processing, the effect whereby often used files that are growing in size become non-contiguous when stored on a soft or hard disk.

frame *1.* The image in a computer display terminal. *2.* In time-division multiplexing, one complete commutator revolution that includes a single synchronizing signal or code.

framing error An error resulting from transmitting or receiving data at the wrong speed. The character of data will appear to have an incorrect number of bits.

Frank-Condon principle Principle whereby the transition from one energy state to another is so fast that the nuclei of the atoms are considered stationary during transition.

fraying The unraveling of a material.

free air Air present as a dispersed phase in a fluid.

free carbon The portion of carbon in steel or cast iron that is present as graphite or temper carbon.

free-electron laser Multifrequency laser utilizing optical radiation amplification by a beam of free electrons passing through a vacuum in a transverse periodic magnetic field.

free electrons Electrons that are not bound to an atom.

free ferrite Ferrite that is formed from the decomposition of hypoeutectoid austenite without the formation of cementite. This term is used when speaking of a phase internal to the steel. It is not used in reference to surface decarburization.

free machining Pertains to the machining characteristics of an alloy to which an ingredient has been introduced to promote small broken chips, better surface finish, and longer tool life while lowering powering consumption.

free radicals Atoms or groups of atoms broken away from stable compounds by application of external energy, and although containing unpaired electrons, remaining free for transitory or longer periods.

free water *1.* Water present in a fluid which may separate as a result of the difference in densities. *2.* The amount of water released when a wet solid is dried to its equilibrium moisture content.

freeze To hold the contents of a register (time, for example) until they have been transferred to another device.

freezing point The temperature at which equilibrium is attained between liquid and solid phases of a pure substance. This term also is applied to compounds and alloys that undergo isothermal liquid-solid phase transformation.

Frenkel defect A defect in a crystal where an expected lattice site is empty, and the atom that normally would occupy that site is located at a different position on the lattice.

frequencies The harmonics of a periodic variable.

frequency (f) *1.* The number of complete cycles, whole periods, of forced vibrations per unit of time caused and maintained by a periodic excitation, usually sinusoidal. *2.* Rate of signal oscillation in Hertz.

frequency band The continuum between two specified limiting frequencies. *See also* frequency.

frequency meter An instrument for determining the frequency of a cyclic signal, such as an alternating current or radio wave.

frequency modulation *1.* A type of electronic circuit that produces an output signal whose frequency has been modified by one or more input signals. *See also* modulated wave. *2.* In telemetry, modulation of the frequency of an oscillator to indicate data magnitude.

frequency of vibration The number of periods occurring in unit time.

frequency output An output in the form of frequency which varies as a function of the applied measurand (for example, angular speed or flow rate).

frequency ratio The ratio of the exciting frequency to the natural frequency.

frequency response *1.* A measure of how the gain or loss of a circuit device or system varies with the frequencies applied to it. Also, the portion of the frequency spectrum that can be sensed by a device within specified limits of error. *2.* The change with frequency of the output/measurand amplitude ratio (and of the phase difference between output and measurand), for a sinusoidally varying measurand applied to a transducer within a stated range of measurand frequencies. *3.* The response of a component, instrument, or control system to input signals at varying frequencies.

Fresnel lens(es) *1.* Thin lenses constructed with stepped setbacks so as to have the optical properties of much thicker lenses. *2.* A lens in which the surface is composed of a number of concentric lens sections with the same focal length desired for the larger lens.

Fresnel reflector Device characterized by a set of mirrors with varying orientation, arranged so as to have the optical properties of a smooth reflector, for example, a parabolic reflector.

fretting A form of wear that occurs between closely fitting surfaces subjected to cyclic relative motion of very small amplitude; it is usually accompanied by corrosion, especially of the very fine wear debris.

fretting corrosion Deterioration at the interface between two contacting surfaces under load, accelerated by relative motion between them of sufficient amplitude to produce slip.

friable Capable of being easily crumbled, pulverized, or otherwise reduced to powder.

friction The resistance to relative motion between two bodies in contact.

friction damper A device in which the reacting force or torque is roughly constant with velocity, and results from Coulomb friction.

friction gouges/scratches A series of relatively short scratches, variable in form or severity.

frigorimeter A thermometer for measuring low temperatures.

fringe multiplication The duplicating effect of a family of curves superimposed on another family of curves so that the curves intersect at angles less than 45 degrees. A new family of curves appears which pass through intersections of the original curves.

fringes Zones of variable light magnitude used to determine surface roughness or texture.

frit Fusible ceramic mixture used to make ceramic glazes and porcelain enamels.

frit seal A hermetic seal for enclosing integrated circuits and other electronic components, which is made by fusing a mixture of metallic powder and glass binder.

from-to tester A type of electronic test equipment for checking continuity between two points in a circuit.

front-end processor *1.* A device that receives computer data from other input devices, organizes such data as specified, and then transmits this data to another computer for processing. *2.* The computer equipment used to receive plant signals, including analog-to-digital converters and the associated controls.

frothing Production of a layer of relatively stable bubbles at an air-liquid interface.

Froude number The nondimensional ratio of the inertial force to the force of gravity for a given fluid flow; the reciprocal of the Reech number.

FRP *See* fiber-reinforced plastic.

FTA *See* fault tree analysis.

FTAM *See* file transfer access and management.

FT-IR *See* Fourier transform infrared spectrometry.

fuel cell Device that converts chemical energy directly into electrical energy.

fuel, test A fuel for use in a given test and having specific chemical and physical properties required for that test.

fugitive binder A substance added to a metal powder which strengthens the bond between particles during compaction and decomposes during sintering.

full adder A computer logic device that accepts two addends and a carry input, and produces a sum and a carry output.

full annealing An imprecise term that implies heating to a suitable temperature followed by controlled cooling to produce a condition of minimum strength and hardness.

full center A fullness in the center portion of the sheet or strip.

Fuller's earth A highly absorbent, claylike material formerly used to remove grease from woolen cloth, but now used principally as a filter medium.

full hard temper A level of hardness and strength for nonferrous alloys and some ferrous alloys corresponding to a cold worked state beyond which the material can no longer be formed by bending.

full scale *1.* The maximum undistorted signal level for each instrument. *2.* For an amplifier, the maximum output signal level. Input full scale can change with amplifier gain. *3.* For a tape recorder, the maximum signal amplitude defined under "maximum signal amplitude." *4.* For an indicating instrument, the input voltage for maximum indication. *5.* The maximum usable linear range of an instrument.

full-scale value *1.* The largest value of a measured quantity that can be indicated on an instrument scale. *2.* For an instrument whose zero is between the ends of the scale, the sum of the absolute values of the measured quantity corresponding to the two ends in the scale.

function *1.* A specific purpose of an entity, or its characteristic action. *2.* In communications, a machine action such as a carriage return or line feed. *3.* A closed subroutine that returns a value to the calling routine upon conclusion. *4.* The operation called for in a computer software instruction.

functional design The specification of the working relations between the parts of a system in terms of their characteristic actions.

functional failure A failure whereby a device does not perform its intended function when the input or controls are correct.

function keys Special keys on a computer keyboard which instruct the computer to perform a specific operation.

function table *1.* The two or more sets of information so arranged that an entry in one set selects one or more entries in the remaining sets. *2.* A dictionary. *3.* A device constructed of hardware, or a subroutine, which can either decode multiple inputs into a single output or encode a single output into multiple outputs. *4.* A tabulation of the values of a function for a set of values of the variable.

furan *1.* Resin formed from reactions involving furfuryl alcohol itself or in combination with other constituents. *2.* Organic heterocyclic compound containing diunsaturated rings of four carbon atoms and one oxygen atom; also known as furfuran or tetrol.

furfuran *See* furan.

furfuryl resins Polymerized furfuryl alcohol provides a binder with good chemical and heat resistance.

furnace An apparatus for liberating heat and using it to produce a physical or chemical change in a solid or liquid mass.

furnace brazing A mass production brazing process in which the filler metal is preplaced on the joint, then the entire assembly is heated to brazing temperature in a furnace. Usually, a protective furnace atmosphere is required, and wetting of the joint surfaces is accomplished without using a brazing flux.

fuse *1.* A replaceable circuit protecting device depending on the melting of a conductor for circuit interruption. *2.* An igniting device used to communicate fire or other initiation stimuli.

fused fiber optics A number of separate fibers that are melted together to form a rigid, fused bundle to transmit light. Fused fiber optics may be used for transmitting images or simply illumination; they are not necessarily coherent bundles of fibers.

fused silica Term usually applied to synthetic fused silica, formed by the chemical combination of silicon and oxygen to produce a high-purity silica.

fused slag Slag that has coalesced into a homogeneous solid mass by fusing.

fusibility Property of slag whereby it fuses and coalesces into a homogeneous mass.

fusible alloy An alloy with a very low melting point, in some instances approaching 150°F (65°C), usually based on Bi, Cd, Sn, or Pb.

fusion The combining of atoms and consequent release of energy. A change of state from solid to liquid; melting.

fusion welding Any welding process that involves melting of a portion of the base metal.

fusion zone In a weldment, the area of base metal melted, as determined on a cross section through the weld.

fuzz Broken reinforcement filaments resulting from abrasive contact with a harder material.

G

g *See* gram.

G The basic unit of acceleration. 1G = 32.2 ft/s². *See also* specific gravity.

Ga Chemical symbol for gallium.

gadolinium alloy Mixture of gadolinium, a rare earth metal, with other metals.

gage Also spelled gauge. *1.* The thickness of metal sheet, or the diameter of rod or wire. *2.* A device for determining dimensions such as thickness or length. *3.* A visual inspection aid that helps an inspector to reliably determine whether size or contour of a formed, stamped, or machined part meets tolerances.

gage block A rectangular stainless steel block having two flat parallel surfaces, with flatness and parallelism guaranteed within a few millionths of an inch.

gage factor *See* strain sensitivity.

gage glass A glass or plastic tube for measuring liquid level in a tank or pressure vessel, usually by direct sight.

gage length In materials testing, the original length (prior to application of stresses of temperature) of a specimen over which measurements of strain, thermal expansion, or other properties are taken. In tensile testing of a round specimen, the gage length is typically four times the diameter.

gage pressure *1.* Pressure measured relative to ambient pressure. *2.* The difference between the local absolute pressure of the system and the atmospheric pressure at the place of the measurement. *3.* Static pressure as indicated on a gage.

gain *1.* Any increase in power when a signal is transmitted from one point to another. Usually expressed in decibels. *2.* The relative degree of amplification in an electronic circuit. *3.* The ratio of the change in output to the change in input that caused the change. *4.* In a controller, the reciprocal of proportional band. For example, if the proportional band is set at 25%, the controller gain is .25. Proportional band can be expressed as a dimensionless number (gain) or as a percent.

gal *See* gallon.

Gal A unit of acceleration equal to 1 cm/s². The term "milligal" is frequently used because it is about 0.001 times the earth's gravity.

galling Localized adhesive welding with subsequent spalling and roughening of rubbing metal surfaces as a result of excessive friction and metal-to-metal contact at high spots.

gallium arsenide A semiconductor material used in electroluminescent instruments.

gallon A unit of capacity (volume) usually referring to liquid measure in the

British or U.S. Customary system of units. The capacity defined by the British (Imperial) gallon equals 1.20095 U.S. gallons; one U.S. gallon equals four quarts or 3.785×10^{-3} m^3.

galvanic cell A cell where a chemical reaction produces electrical energy.

galvanic corrosion Corrosion associated with the current resulting from the electrical coupling of dissimilar electrodes in an electrolyte.

galvanic series A series of metals organized according to their electrode potentials in a specific environment.

galvanizing Coating a metal with zinc, using any of several processes, the most common being hot dipping and electroplating.

galvanneal To create a zinc alloy coating on iron or steel by keeping the zinc molten after galvanizing to allow the zinc to alloy with the base metal.

galvannealed sheets Galvannealed sheets are hot dipped zinc-coated sheets that have been processed to produce a zinc-iron alloy coating. This product has a dull finish and is suitable for painting after cleaning. The alloy produced lacks ductility, and powdering of the coating can occur during forming.

galvanometer An instrument for measuring small electric currents using electromagnetic or electrodynamic forces to create mechanical motion, such as changing the position of a suspended moving coil.

galvanometer recorder A sensitive moving-coil instrument having a small mirror mounted on the coil; a small signal voltage applied to the coil causes a light beam reflected from the mirror to move along the length of a slit, producing a trace on a light-sensitive recording medium that moves transverse to the slit at constant speed.

galvanostatic Refers to a constant-current technique of applying current to a specimen in an electrolyte. Also called intentiostatic.

gamma A measure of the contrast properties of a photographic or radiographic emulsion which equals the slope of the straight-line portion of its H-D curve.

gamma counter An instrument for detecting gamma radiation—either by measuring integrated intensity over a period of time or by detecting each photon separately.

gamma loss peak The third peak in the dampening curve in dynamic measurement.

gamma radiation *See* gamma ray.

gamma ray *1.* Quantum of electromagnetic radiation emitted by nuclei, each such photon being emitted as the result of a quantum transition between two energy levels of the nucleus. Gamma rays have energies usually between 10 thousand electron volts and 10 million electron volts, with correspondingly short wavelengths and high frequencies. *2.* A term sometimes used to describe any high-energy electromagnetic radiation, such as x-rays exceeding about 1 MeV or photons of annihilation radiation.

gamma-ray spectrometer *1.* An instrument for deriving the physical constants of materials by using induced gamma radiation as the emission source. *2.* An instrument for measuring the energy distribution in a beam of gamma rays.

gamma-ray spectroscopy A technique for deriving the physical constants of a material by measuring the energy of gamma radiation emissions.

gap *1.* An interval of space or time that is used as an automatic sentinel to indicate the end of a word, record, or file of data on a tape; for example, a word gap at the end of a word, a record or item gap at the end of a group of words, or a file gap at the end of a group of records or items. *2.* The absence of information for a specified length of time, or space on a recording medium, as contrasted with marks and sentinels that indicate the presence of specific information to achieve a similar purpose. *3.* The space between the reading or recording head and the recording medium, such as tape, drum, or disk. Related to gap, head. *4.* In a weldment, the space between members, prior to welding, at the point of closest approach for opposing faces. *5.* In filament winding, the separation between fibers.

gap-filling adhesives Adhesives with very low shrinkage when drying.

gap gage A "go-no/go" gage used to determine if specific manufacturing tolerances have been met.

gap scanning In ultrasonic examination, projecting the sound beam through a short column of fluid produced by pumping couplant through a nozzle in the ultrasonic search unit.

garnets Groups of minerals that are silicates of cubic crystalline form.

gas analysis The determination of the constituents of a gaseous mixture.

gas carburizing A surface hardening process in which steel or an alloy of suitable alternative composition is exposed at elevated temperature to a gaseous atmosphere with a high carbon poten-

tial; hardening of the resulting carbon-rich surface layer is accomplished by quenching the part from the carburizing temperature or by reheating and quenching.

gas chromatogram The recorder output versus time of a detector signal from a gas chromatograph, which shows deflections to indicate, for example, the presence of individual hydrocarbons.

gas chromatography A separation technique in which a sample in the gaseous state is carried by a flowing gas (carrier gas) through a tube (column) containing stationary material. The stationary material performs the separation by means of its differential affinity for the components of the sample.

gas classification Separation of submesh-size particles by using a gas stream flowing counter to the fall of the particles.

gas composition tests The quantitative and qualitative measurement of effluent gas from an inflator.

gas constant The pressure of a gas times its volume divided by its temperature.

gas cyaniding A misnomer. *See* carbonitriding.

gaseous cavitation *See* cavitation flow.

gaseous reduction *1.* Producing a metal by the reaction of a metallic compound with a reducing gas. *2.* Producing metallic particles by the conversion of a metal with a reducing gas.

gas etching Removing material from a semiconductor material by reacting it with a gas to form a volatile compound.

gas generant A solid phase material composed of both oxidizer/reducer and fuel elements which rapidly burns to produce the gas that inflates the airbag.

gas holes (porosity) Holes in castings or weldments caused by gas escaping from the hot metal during solidification.

gasket A sealing member, usually made by stamping from a sheet of cork, rubber, metal, or impregnated synthetic material; it is then clamped between two essentially flat surfaces to prevent pressurized fluid from leaking through the crevice.

gas mass spectrometry A technique for the quantitative analysis of gas mixtures.

gas metal-arc welding (GMAW) A form of electric arc welding in which the electrode is a continuous filler metal wire and in which the welding arc is shielded by supplying a gas such as argon, helium, or CO_2 through a nozzle in the torch or welding head; the term GMAW includes the methods known as MIG welding.

gasohol Synthetic fuel consisting of a mixture of gasoline and grain alcohol (ethanol).

gasometer A piece of apparatus typically used in analytical chemistry to hold and measure the quantity of gas evolved in a reaction; similar equipment is used in some industrial applications.

gas path analysis Mathematical process of determining overall engine performance, individual module performances, and sensor performances from any specific set of engine-related measurements.

gas permeability The transfer rate of a gas through a specific material.

gas plasma display A data display screen used on some laptop computers. Characters are easier to read than those on liquid crystal display screens, but the unit is more expensive.

gas plating The plating of a thin layer of a material on another material by gas vapor deposition.

gas-shielded arc welding An all-inclusive term for any arc welding process that utilizes a gas stream to prevent direct contact between the ambient atmosphere and the welding arc and weld puddle. The gas is typically argon, helium, or CO_2.

gassing *1.* Absorption of gas by a material. *2.* Formation of gas pockets in a material. *3.* Evolution of gas during a process, for example, evolution of hydrogen at the cathode during electroplating, gas evolution from a metal during melting or solidification, or desorption of gas from internal surfaces during evacuation of a vacuum system; the last is sometimes referred to as outgassing.

gas-solid interaction Effect of the impingement of gases (particles) on solid surfaces in various environments.

gas specific gravity balance A weighing device consisting of a tall gas column with a floating bottom; a pointer mechanically linked to the floating bottom indicates density or specific gravity directly, depending on scale calibration.

gas thermometer A temperature transducer that converts temperature to pressure of gas in a closed system. The relation between temperature and pressure is based on the gas laws at constant volume.

gas tungsten-arc welding (GTAW) A form of metal-arc welding in which the electrode is a nonconsumable pointed tungsten rod; shielding is provided by a stream of inert gas, usually helium or argon; filler metal wire may or may not be fed into the weld puddle; and pressure may or may not be applied to the joint. The term GTAW includes the method known as Heliarc or TIG welding.

gate *1.* A device or element that, depending upon one or more specified inputs, has the ability to permit or inhibit the passage of a signal. *2.* In electronic computers, a device having one output channel and one or more input channels, such that the output channel state is completely determined by the contemporaneous input channel states, except during switching transients; a combinational logic element having at least one input channel; an AND gate; an OR gate. *3.* In a field effect transistor, the electrode that is analogous to the base of a transistor or the grid of a vacuum tube. *4.* A movable barrier. *5.* A device such as a valve or door which controls the rate at which materials are admitted to a conduit, pipe, or conveyor. *6.* A device for positioning film in a movie camera, printer, or projector. *7.* The passage in a casting mold that connects the sprue to the mold cavity. Also known as the in-gate.

gauge *See* gage.

gauss The CGS unit of magnetic flux density or magnetic induction; the SI unit, the tesla, is preferred.

Gaussian beam A laser beam in which the intensity has its peak at the center of the beam, then drops off gradually toward the edges. The intensity profile measured across the center of the beam is a classical Gaussian curve.

Gaussian noise *See* random noise.

gaussmeter A magnetometer for measuring only the intensity, not the direction, of a magnetic field; its scale is graduated in gauss or kilogauss. *See* magnetometer.

gauze *1.* A sheer, loosely woven textile fabric; one of its widest uses is for surgical dressings, but it also has some industrial uses such as for filter media. *2.* Plastic or wire cloth of fine to medium mesh size.

GB Gigabytes; billions of bytes.

Gd Chemical symbol for gadolinium.

Ge Chemical symbol for germanium.

Geiger-Mueller counter A radiation-measuring instrument whose active element is a gas-filled chamber usually consisting of a hollow cathode with a fine-wire anode along its axis. Often referred to simply as a Geiger counter.

gel The initial solid phase that develops during the polymerization of a resin.

gelation The state in a resin cure in which the material is nearly solid.

gelation time In connection with the use of synthetic thermosetting resins, the interval of time extending from the introduction of a catalyst into a liquid adhesive system until the start of gel formation.

gel coat A resin gelled on the internal surface of a plastics mold prior to filling it with a molding material; the finished part is a two-layer laminate, with the gel coat providing improved surface quality.

gel permeation chromatography (GPC) Liquid chromatography in which polymer molecules are separated by their ability or inability to penetrate material in the separation column.

gel point The level of viscosity at which a polymer liquid begins to show elastic properties.

gel time The elapsed time from initial reaction of a liquid material to when gelation occurs.

general acceptance test A test to demonstrate the degree of compliance of a device or material with purchaser's requirements.

general corrosion A form of deterioration that is distributed more or less uniformly over a surface.

general-purpose simulation system A generic class of discrete, transaction-oriented simulation languages based on a block (diagramming) approach to problem statement. Abbreviated GPSS.

generating electric field meter An instrument for measuring electric field strength in which a flat conductor is alternately exposed to the field and shielded from it; potential gradient of the field is determined by measuring the rectified current through the conductor.

generating magnetometer An instrument for measuring magnetic field strength by means of the electromotive force generated in a rotating coil immersed in the field being measured.

geodesic isotensoid The state of a constant stress in a reinforcement filament at each point in its path.

geophysical fluid General term for the liquids and gases on or in the earth (from water in all forms, to petroleum and hydrocarbons in liquid and gaseous form, and molten rock material within the earth).

getter(s) *1.* A material exposed to the interior of a vacuum system in order to reduce the concentration of residual gas by absorption or adsorption. *2.* A substance that absorbs or chemically bonds unwanted elements in a sintering atmosphere.

getter-ion pump A type of vacuum pump that produces and maintains high vacuum by continuously or intermittently depositing chemically active metal layers on the wall of the pump, where they trap and hold inert gas atoms that have been ionized by an electric discharge and drawn to the activated pump wall. Also known as sputter-ion pump.

ghost An indication that has no direct relation to reflected pulses from discontinuities in the material being tested.

ghost lines (ghost welt lines) Lines running parallel to the rolling direction which appear in a panel when it is stretched. These lines may not be evident unless the panel has been sanded or painted. (Not to be confused with leveler lines.)

GHz Gigahertz.

giga A numerical prefix for ten to the ninth power, for example, gigabyte.

gilbert The CGS unit of magnetomotive force; the SI unit, the ampere (or ampere-turn) is preferred.

gimbal Device with two mutually perpendicular and intersecting axes of rotation, thus giving free angular movement in two directions, on which engines or other objects may be mounted in gyros, supports that provide the spin axes with degrees of freedom.

gin A hoisting machine consisting of a windlass, pulleys, and ropes in a tripod frame.

Giotto mission The European Space Agency's mission to fly through the head of Halley's Comet in order to make in situ measurements of the composition and physical state as well as the structures of the head.

glass A hard, brittle, amorphous, inorganic material, often transparent or translucent, made by fusing silicates (and sometimes borates and phosphates) with certain basic oxides and cooling rapidly to prevent crystallization.

glass ceramic A polycrystalline material with better mechanical properties at elevated temperatures than glass.

glass ceramic composites Fiber reinforcements in a glass matrix.

glass cloth Woven glass fibers.

glass fiber A glass thread less than 0.001 inch (0.025 millimeter) thick; it is used in loose, matted, or woven form to make thermal, acoustical, or electrical insulation; in matted, woven, or filament-wound form to make fiber-reinforced composites; or in loose, chopped form to make glass-filled plastics parts.

glass filament Glass fiber that has been drawn to its minimum effective diameter.

glass finish Material applied to a glass reinforcement to improve the bond between the glass and a resin matrix.

glass flake Small flakes of glass used as a reinforcement in some composites.

glass former An oxide that forms a glass easily.

glassine A thin, dense, transparent, super-calendered paper made from highly refined sulfite pulp; it is used industrially as insulation between layers of iron-core transformer windings.

glass matrix composites Ceramic fiber reinforcements in a glass matrix.

glass mat thermoplastic (GMT) A tough, wear-resistant composite of polymers reinforced with a glass "mat" which can be formed into specific shapes.

glass, percent by volume The product of the specific gravity of a laminate and the percent glass by weight, divided by the specific gravity of the glass.

glass-plastic glazing material A laminate of one or more layers of glass and one or more layers of plastic in which a plastic surface of the glazing faces inward when a glazing is installed in a vehicle.

glass reinforced polymers (GRP) *See* fiberglass.

glass sand The raw material for glassmaking; it normally consists of high-quartz sand containing small amounts of the oxides of Al, Ca, Fe, and Mg.

glass stress In a glass composite, the stress determined by using the cross-sectional area of the reinforcement and load.

glass transition The change in an amorphous polymer from a rubberlike condition to a brittle condition, or vice-versa.

glass transition temperature The temperature at which an amorphous polymer changes from a hard and relatively brittle condition to a viscous or rubbery condition.

glassy alloy A metallic material having an amorphous or glassy structure. Also known as metallic glass.

glassy carbon A form of carbon with unique properties and characteristics. Formed by carbonizing phenolic resins made by reacting phenols with cellulosics, aldehydes, and ketones.

glaze A glossy, highly reflective, glasslike, inorganic fused coating. *See* enamel.

glazed surface An undesirable, hard, glossy surface developed where rubbing action (friction) breaks down the parts of the surface or the lubricant.

glazing A smooth, nonabsorptive surface fired onto a ceramic.

glitch Undesirable electronic pulses that cause processing errors.

glitter Decorative flaked powder having a particle size large enough so that the individual flakes produce a visible reflection or sparkle; used in certain decorative paints and in some compounded plastics stock.

global *1.* Describes a name that has as its scope the entire system in which it resides. *2.* A computer instruction that causes the computer to locate all occurrences of specific data. *3.* A value defined in one program module and used in others. Globals are often referred to as entry points in the module in which they are defined, and externals in the other modules that use them.

global array A set of data listings that can be referenced by other parts of the software.

global common An unnamed data area that is accessible by all programs in the system. Sometimes referred to as a blank common.

global variable Any variable available to all programs in the system. Contrast with reserved variable.

globular alpha A spheroidal form of equiaxed alpha.

gloss *1.* The shine, sheen, or luster of a dried film. *2.* Degree to which a surface simulates a perfect mirror in its capacity to reflect light.

glossimeter An instrument for measuring the "glossiness" of a surface—that is, the ratio of light reflected in a specific direction to light reflected in all directions—usually by means of a photoelectric device. Also known as a glossmeter.

glossmeter *See* glossimeter.

glow *See* luminescence.

glow discharge A discharge of electrical energy through a gas, in which the space potential near the cathode is substantially higher than the ionization potential of the gas.

glue *1.* A now general term referring to any bonding or adhesive material. *2.* Specifically, a crude, impure form of commercial gelatine which softens to a gel consistency when wetted with water and dries to form a strong adhesive layer.

gluons The carriers of the strong force that holds atomic nuclei together (holding together groups of quarks making up stable particles, which in turn are bound together in the atomic nuclei).

GMAW *See* gas metal-arc welding.

GMT *See* glass mat thermoplastic.

Golay cell An infrared detector in which the incident radiation is absorbed in a gas cell, thereby heating the gas. The temperature-induced expansion of the gas deflects a diaphragm, and a measurement of this deflection indicates the amount of incident radiation.

goniometer *1.* Specifically, an instrument used in crystallography to determine angles between crystal planes, using x-ray diffraction or other means. *2.* An instrument used to measure refractive index and other optical properties of transparent optical materials or optically scattering in materials at UV, visible or IR wavelengths.

goodness of fit A statistical term that quantifies how likely a sample was to have come from a given probability distribution.

gouges Elongated grooves or cavities, generally undesirable.

gouging Forming a groove in an object by electrically, mechanically, thermo-mechanically, or manually removing material; the process is typically used to remove shallow defects prior to repair welding.

gouraud Shading technique used for the display of objects in which varying degrees of light intensity are calculated across each polygonal surface making up the object. This technique provides a more realistic image than provided by Phong shading.

GPC *See* gel permeation chromatography.

GPSS *See* general-purpose simulation system.

G-P zone The Guinier-Preston zone.

grab sampling A method of sampling bulk materials for analysis, which consists of taking one or more small portions (usually only imprecisely measured) at random from a pile, tank, hopper, railcar, truck, or other point of accumulation.

graded index fiber (GRIN) An optical fiber in which the refractive index changes gradually between the core and cladding, in a way designed to refract light so it stays in the fiber core. Such fibers have lower dispersion and broader bandwidth than step-index fibers.

graded interfaces The study of the characteristics and properties of the region between fibers and a matrix in a composite.

gradient The rate of change of some variable with respect to another, especially a regular, uniform or stepwise rate of change.

gradient elution A technique for determining phase separations by liquid chromatography.

gradient index optics Optical systems with components whose refractive indexes vary continuously within the material used for the optical elements.

grading *1.* Analysis of particle size distribution in metal powders or ceramics. *2. See* classification.

graduation Any of the major or minor index marks on an instrument scale.

graft copolymers Copolymer in which individual units are joined together to form a branch or protuberance along the main chain.

grain *1.* The appearance or texture of wood, or a woodlike appearance or texture of another material. *2.* In paper or matted fibers, the predominant direction most fibers lie in, which corresponds to directionality imparted during manufacture. *3.* In metals and other crystalline substances, an individual crystallite in a polycrystalline mass. *See also* grain size. *4.* In crumbled or pulverized solids, a single particle too large to be called powder.

grain boundary The plane of mismatch between adjacent crystallites in a polycrystalline mass, as revealed on a polished and etched cross section through the material; it appears as a line engulfing an individual grain.

grain boundary alpha Primary or transformed alpha outlining prior beta grain boundaries. It may be continuous unless broken up by subsequent work.

grain fineness number The weighted average of grain sizes in a specific granular material.

grain flow Fibrous appearance on a sanded or polished, then etched, section through metal. Seen most typically in a sectioned forging. The appearance is caused by orientation of impurities and inhomogeneities along the direction of working during the forging process. Not to be confused with grain or grain size.

grain growth An increase in the average size of the grains in polycrystalline metal, usually as a result of heating at elevated temperature. Notes: (a) A grain is an individual crystal in a polycrystalline metal and includes twinned regions and subgrains when present. (b) Grain size is a measure of the mean diameter, area, or volume of all individual grains observed in a polycrystalline metal. In metals containing two or more phases, the grain size refers to that of the matrix unless otherwise specified.

grains (water) A unit of measure commonly used in water analysis for the measurement of impurities in water (17.1 grains = 1 part per million—ppm).

grain size The size of crystallites in a polycrystalline solid, which may be expressed as a diameter, number of grains per unit area, or standard grain size number determined by comparison with a chart such as those published by ASTM. *See also* grain, grain boundary, and grain growth.

grains per cubic feet The term for expressing dust loading in weight per unit of gas volume (7000 grains equals one pound).

gram The CGS unit of mass; it equals 0.001 kilogram, which has been adopted as the SI unit of mass.

gram molecular weight The gram weight of a compound equal to its molecular weight.

grand unified theory (GUT) A theory describing the unification of gravity with the other elementary forces in physics, i.e., the weak force, the strong force, and the electromagnetic force.

granular fracture A rough, irregular fracture surface, which can be either transcrystalline or intercrystalline, and which often indicates that fracture took place in a relatively brittle mode, even though the material involved is inherently ductile.

granular metal powder Grains of metal powder of nearly equal size in crystalline shapes.

granular polyesters Molding materials made up of alkyd resins.

granular structure Nonuniform appearance of molded or compressed material due to the presence of particles of varying composition.

granulated metal Spherical pellets formed by pouring molten metal through a screen into water.

graphic Pertaining to representational or pictorial material, usually legible to humans and applied to the printed or written form of data such as curves, alphabetic characters, and radar scope displays.

graphical display unit An electronics device that can display both text and pictorial representations.

graphic character *See* graphic.

graphic panel A master control panel that, pictorially and usually colorfully, traces the relationship of control equipment and the process operation. It permits an operator, at a glance, to check on the operation of a far-flung control system by noting dials, valves, scales, and lights.

graphite The allotropic, crystalline form of carbon.

graphite-epoxy composites Structural materials composed of epoxy resins reinforced with graphite.

graphite fiber A fiber made from a precursor by an oxidation, carbonization, and graphitization process.

graphite flake A form of graphite present in gray cast iron which appears in the microstructure as an elongated, curved inclusion.

graphite-polyimide composite Composite material utilizing graphite reinforcing fibers in a resin matrix.

graphite rosette A form of graphite present in gray cast iron which appears in the microstructure as graphite flakes extending radially outward from a center of crystallization.

graphitic carbon Free carbon present in the microstructure of steel or cast iron; it is an essential feature of most cast irons, but is almost always undesirable in steel.

graphitic corrosion Deterioration of gray cast iron in which the metallic constituents are selectively leached or converted to corrosion products leaving the graphite intact.

graphitic steel Alloy steel in which some of the carbon is present in the form of graphite.

graphitization Formation of graphite in iron or steel; it is termed primary graphitization if it forms during solidification, and secondary graphitization if it forms during subsequent heat treatment or extended service at high temperature.

graphitizing Annealing a ferrous alloy in such a way that some or all of the carbon is precipitated as graphite.

graphoepitaxy The use of artificial surface relief structures to induce crystallographic orientation in thin films.

Grashof number A nondimensional parameter used in the theory of heat transfer. The Grashof number is associated with the Reynolds number and the Prandtl number in the study of convection.

gravelometer Machine used to cause consistent intentional paint damage to samples prior to or during laboratory corrosion testing. A gravelometer propels gravel or metal shot of a particular size and shape at the surface of a painted test specimen.

gravimetric value The weight of suspended solids per unit volume of fluid.

gravitation A natural force of attraction that tends to draw bodies together.

gravity The force of gravitation, being, for any two sufficiently massive bodies, directly proportional to the product of their masses and inversely proportional to the square of the distance between them—especially the gravitational force exerted by a celestial body such as the earth.

gray Metric unit for absorbed dose.

gray code A generic name for a family of binary codes which have the property whereby a change from one number to the next sequential number can be accomplished by changing only one bit in a code for the original number. This type of code is commonly used in rotary shaft encoders to avoid ambiguous readings when moving from one position to the next. *See also* cyclic code and shaft encoder.

gray iron Cast iron in which the graphite is present as flakes instead of temper carbon nodules, as in malleable iron, or small spherulites, as in ductile iron.

grease *1.* Rendered, inedible animal fat. *2.* A semisolid to solid lubricant consisting of a thickening agent, such as metallic soap, dispersed in a fluid lubricant, such as petroleum oil.

green Unfired, uncured, or unsintered.

green compact Unfired or unsintered powder metal compact.

green rot Corrosion caused by the formation of chromium carbide at grain boundaries which oxidizes due to the lack of chromium.

greenstick fracture Fracture in which the crack does not go right through the material, but is deflected part way through, allowing the remainder of the cross section to deform elastically.

green strength The mechanical strength of a ceramic or powder metallurgy part after molding or compacting, but before firing or sintering; it represents the quality needed to maintain sharpness of contour and physical integrity during handling and mechanical operations to prepare it for firing or sintering.

greenware Unfired ceramic ware.

greige goods Any fabric before finishing, as well as any yarn or fiber before bleaching or dyeing; therefore, fabric with no finish or size.

grid *1.* A network of lines, typically forming squares, used in layout work or in creating charts and graphs. *2.* A crisscross network of conductors used for shielding or controlling a beam of electrons.

GRIN *See* graded index fiber.

grindability index A measure or rating (of a material to be ground) referring solely to the ease of removing metal by grinding. A ratio of the volume of metal removed per cubic inch of wheel wear under a set of fixed surface grinding conditions. It should not be interpreted as any indication of the resistance of steel to develop grinding checks or burn.

grinding *1.* Removing material from the surface of a workpiece using an abrasive wheel or belt. *2.* Reducing the particle size of a powder or granular solid.

grinding checks Fine thermal cracks that develop from localized overheating of the surface area being ground. Such cracks are generally at right angles to the direction of grinding, but may appear as a complete network.

grinding cracks Shallow cracks in the surface of a ground workpiece; they appear most often in relatively hard materials due to excessive grinding friction or high material sensitivity. *See* grinding checks.

grinding sensitivity Propensity of a material to surface damage when being ground. *See* grinding checks and grinding cracks.

grinding wheel A cutting tool of circular shape made of abrasive grains bonded together.

grit A particulate abrasive consisting of angular grains.

grit blasting Abrasively cleaning metal surfaces by blowing steel grit, sand, or other hard particulate against them to remove soil, rust, and scale. Also known as sandblasting.

groove *1.* A long, narrow channel or furrow in a solid surface. 2. In a weldment, a straight-sided, angled, or curved gap between joint members prior to welding which helps confine the weld puddle and ensure full joint penetration to produce a sound weld.

gross decarburization Decarburization with sufficient carbon loss to show only clearly defined ferrite grains under metallographic examination. This is sometimes called "complete decarburization."

gross porosity In weld metal or castings, large or numerous gas holes, pores, or voids that are indicative of substandard quality or poor technique.

gross vehicle weight *See* GVW.

gross vehicle weight rating *See* GVWR.

ground *1.* A (neutral) reference level for electrical potential, equivalent to the level of electrical potential of the earth's crust. *2.* A secure connection to earth which is used to reference an entire system. Usually the connection is in the form of a rod driven or buried in the soil or a series of rods connected into a grid buried in the soil.

grouping Combining two or more computer records into one block of information to conserve storage space or disk or tape. Also known as blocking.

group velocity The velocity corresponding to the rate of change of average position of a wave packet as it travels through a medium.

grouting Placing or injecting a fluid mixture of cement and water (or of cement, sand, and water) into a grout hole, crevice, seam, or joint.

growth Relating to cast iron, a permanent increase in size caused by repeated or prolonged exposure to temperatures above 900°F (480°C) due either to graphitizing of carbides or to oxidation.

GRP *See* fiberglass.

GTAW *See* gas tungsten-arc welding.

guanamine A hardener for epoxy resins.

guanosines Guanine riboside; a nucleoside composed of guanine and ribose.

guided bend test A bend test in which the specimen is bent to a predetermined shape in a jig or around a grooved mandrel.

Guinier-Preston zone (G-P zone) The first stage of precipitation in a supersaturated molten metal, usually accompanied by a change in properties.

gum A now general term referring to a substance that is sticky when wet or has "gummy" characteristics when heated.

GUT *See* grand unified theory.

gutter *1.* A drainage trough or trench, usually surrounding a raised surface. *2.* A groove around the cavity of a forging or casting die to receive excess flash.

GVW (gross vehicle weight) Value specified by the manufacturer as the maximum loaded weight of a single vehicle including all equipment, fuel, body, payload, driver, etc.

GVWR (gross vehicle weight rating) The maximum total vehicle rated capacity, measured at the tire ground surface, as rated by the chassis manufacturer.

gyratory screen A sieving machine having a series of nested screens whose mesh sizes are progressively smaller from top to bottom of the stack; the mechanism shakes the stacked screens in a nearly circular fashion, which causes fines to sift through each screen until an entire sample or batch has been classified.

gyro *See* gyroscope.

gyroscope A device consisting of a spinning wheel or disk, suspended so that its spin axis maintains a fixed angular orientation when not subjected to external torques.

gyrostat *See* gyroscope.

gyrotron *See* cyclotron.

H

H *1.* Abbreviation for henry. *2.* Chemical symbol for hydrogen.

h *See* Planck's constant.

habit The characteristic growth of crystal.

hackles Raised strips or striations on a fracture surface caused by an array of small cracks produced during a shear failure.

HAD *See* high aluminum defect.

hafnium A metal element similar to zirconium.

Haigh fatigue test An axial fatigue test capable of providing alternating or fluctuating stress cycles.

hair-line craze In a plastic reinforced material, fine cracks which in depth are less than the thickness of ply.

half-adder A logic circuit that accepts two binary input signals and produces corresponding sum and carry outputs; two half-adders and an OR gate can be combined to realize a full-adder. *See also* full adder.

half-adjust To round a number so that the least significant digit(s) determines whether or not a "one" is to be added to the digit next higher in significance that the digit(s) used as criterion for the determination.

half-and-half solder A lead-tin alloy (50Pb-50Sn) used primarily to join copper tubing and fittings.

half hard A temper of metal alloys with tensile strength about halfway between dead soft and full hard.

half-life The average time required for one-half the atoms in a sample of radioactive element to decay.

half-thickness The thickness of an absorbing medium that will depreciate the intensity of radiation beam by one-half

Hall coefficient *See* Hall effect.

Hall current *See* Hall effect.

Hall effect *1.* The development of a transverse electric potential gradient or voltage in a current-carrying conductor or semiconductor upon the application of a magnetic field. *2.* An electromotive force developed as a result of interaction when a steady-state current flows in a steady-state magnetic field; the direction of the emf is at right angles to both the direction of the current and the magnetic field vector, and the magnitude of the emf is proportional to the product of current intensity, magnetic force, and sine of the angle between current direction and magnetic field vector.

halocarbon Compound consisting of halogen atoms and carbon atoms.

Halpin-Tsai equations Equations derived from theoretical mechanical properties of reinforcement fiber stiffness.

hand lay-up The process of building various layers of reinforcement material by hand in a reinforced composite.

handling breaks Irregular breaks caused by improper handling of sheets during processing. These breaks result from the bending or sagging of the sheets while being handled.

hang up A term to describe the phenomena whereby higher-molecular-weight hydrocarbons are retained in the sample train, causing an initial low analyzer reading, followed by higher readings in subsequent tests. Excessive hang up causes errors in the analysis of the hydrocarbons in exhaust gas.

hardboard A generic term for a sheet manufactured primarily from interfelted lignocellulosic fibers (usually wood) consolidated under heat and pressure in a hot press to a density of 55–65 lb/ft^3 (880–1041 kg/m^3) (specific gravity 0.9–1.0) or greater, and to which other materials may have been added during manufacture to improve certain properties.

hard chromium Chromium plated on a base metal for wear resistance rather than decoration.

hard clad silica fibers Silica optical fibers that are coated with hard plastic material, not with the soft materials typically used in plastic clad silica.

hard conversion The process of changing a measurement from inch-pound units to nonequivalent metric units, which necessitates physical configuration changes of the item outside those permitted by established measurement tolerances. "Hard conversion" often is a concomitant of international standardization. *See also* soft conversion.

hard drawn Describes wire or tubing that is cold drawn without annealing.

hard drawn copper wire Copper wire that has been drawn to size and not annealed.

hard-drawn wire Cold-drawn metal wire of relatively high tensile strength and low ductility.

hardenability In a ferrous alloy, the property associated with chemical composition of the base metal which determines the depth and distribution of hardness that can be induced by quenching. Contrast with hardening ability.

hardener *1.* A substance added to a plastic to enhance curing action. *2.* An alloy added to a metal that results in a better composition than adding a pure metal. *3.* A substance added to film to control the measure of hardness.

hardening Increasing the hardness by suitable treatment, usually involving heating and cooling. When applicable, the following more specific terms should be used: age hardening, case hardening, precipitation hardening, quench hardening, or surface hardening.

hardening ability In a ferrous alloy, the property that describes to what maximum hardness a base metal can be heat treated. This differs from hardenability.

hard facing Depositing a wear-resistant metal on the surface of another metal.

hard lead Any of a series of lead-antimony alloys of low ductility; typically, hard lead contains 1 to 12% Sb.

hard metal A sintered metal compact with significant hardness, strength, and wear resistance.

hardness *1.* The resistance to indentation. Measured by the relative resistance of the material to an indentor point of any one of a number of standard hardness-testing instruments. *2.* A measure of the amount of calcium and magnesium salts in boiler water. Usually expressed as grains per gallon or ppm as $CaCO_3$. *See also* SHORE and REX "A" hardness.

hardness number *See* Vickers hardness number, Brinell hardness number, and Rockwell hardness number.

hardness test *See* Vicker's hardness test, Brinell hardness test, and Rockwell hardness test.

hardware *1.* In data processing, the physical equipment associated with the computer. *2.* The electrical, mechanical, and electromechanical equipment and parts associated with a computing system, as opposed to its firmware and software.

hardware priority interrupt *See* priority interrupt and software priority interrupt.

hard water Water that contains calcium or magnesium in amounts that require an excessive amount of soap to form a lather.

harmonic analysis A statistical method for determining the amplitude and period of certain harmonic or wave components in a set of data with the aid of Fourier series.

harmonic distortion *1.* The production of harmonic frequencies at the output due to the nonlinearity of a system when a sinusoidal input is applied. *2.* Distortion caused by the presence of harmonics of a desired signal.

harmonics Eigenfrequency oscillations excited in a vibrating system.

Hartley information unit In information theory, a unit of logarithmic measurement of the decision content of a set of 10 mutually exclusive events, expressed as the logarithm to the base 10; for example, the decision content of an eight-character set equals log 8, or 0.903 Hartley.

Hartree-Fock-Slater method A refined approximation method for the calculation from wave function of electron total energies, kinetic energies, etc. for chemical elements.

HAZ (metallurgy) *See* heat affected zone.

hazard A potentially unsafe condition resulting from failures, malfunctions, external events, errors, or combinations thereof.

hazardous area An area in which explosive gas/air mixtures are, or may be expected to be, present in quantities such as to require special precautions for the construction and use of electrical apparatus.

hazardous atmosphere *1.* A combustible mixture of gases and/or vapors. *2.* An explosive mixture of dust in air.

hazardous material Any substance that requires special handling to avoid endangering human life, health, or well being. Such substances include poisons; corrosives; and flammable, explosive, or radioactive chemicals.

haze *1.* The cloudy or turbid appearance of an otherwise transparent specimen caused by light scattered from within the specimen or from its surface. *2.* Diffuse scattering of reflected light in directions adjacent to the direction of specular or mirror reflection.

H-band steel A steel with prescribed limits of hardenability.

HBN *See* hexagonal boron nitride.

HCL argon laser Gas laser in which the active material is gaseous hydrogen chloride and argon.

HCL laser Gas laser in which the active material is gaseous hydrogen chloride.

HDI *See* high density inclusion.

HDLC *See* high-level data link control.

HDPE *See* high-density polyethylene.

HDT Heat deflection temperature.

He Chemical symbol for helium

header *1.* A conduit or chamber that receives fluid flow from a series of smaller conduits connected to it, or that distributes fluid flow among a series of small conduits. *2.* In data processing, data placed at the beginning of a file for identification. *3.* The portion of a batch recipe that contains information about the purpose, source, and version of the recipe, such as recipe and product identification, originator, issue date, and so on. *4.* In metalworking, a machine that produces a headed configuration, for example a fastener, from bar or wire.

head gap *1.* The space between the reading or recording head and the recording medium, such as tape, drum, or disk. *2.* The space or gap intentionally inserted into the magnetic circuit of the head in order to force or direct the recording flux into the recording medium.

heading The upsetting of wire, rod, or bar stock in dies to form parts having some of the cross-sectional area larger than the original. *See also* header.

head loss Pressure loss in terms of a length parameter, such as inches of water or millimeters of mercury.

head pressure Expression of a pressure in terms of the height of fluid. $P = ypg$, where p is fluid density and y is the fluid column height.

head-to-head A configuration in which functional polymer groups are on adjacent carbon atoms.

head-to-tail A configuration in which functional polymer groups are far apart.

head up display A method of presenting images to the pilot of an aircraft while he/she is looking forward through the windshield. These images are generated from a device out of the pilot's field of view and reflected from a transparent surface in front of the pilot.

healed over scratch A scratch that occurred in an earlier mill operation and was partially masked in subsequent rolling. It may open up during forming.

heat *1.* In metal production, material that, in the case of batch melting, was cast at the same time from the furnace and was identified with the same heat number; or, in the case of continuous melting, was poured without interruption. *2.* Energy that flows between bodies because of a difference in temperature; same as thermal energy.

heat absorbing filter A glass filter that transmits most visible light, but strongly absorbs infrared light.

heat activated adhesive An adhesive that becomes tacky through the application of heat or heat and pressure.

heat affected zone (HAZ) The portion of the base metal that was not melted during brazing, cutting, or welding, but whose microstructure and properties were altered by the heat of these processes.

heat available The thermal energy above a fixed datum that is capable of being absorbed for useful work.

heat balance An accounting of the distribution of the heat input and output.

heat buildup The rise in temperature in a part resulting from the dissipation of applied strain energy, as heat or from applied mold cure heat.

heat cleaning Placing reinforcement fibers in an elevated temperature atmosphere to remove unwanted elements in or on a reinforcement.

heat content The amount of heat per unit mass that can be released when a substance undergoes a drop in temperature, a change in state, or a chemical reaction. *See also* enthalpy.

heat-convertible resin A thermosetting resin convertible by heat into an infusible and insoluble mass.

heat deflection temperature (HDT) The temperature at which a test specimen will distort under a specific load.

heat dissipation The quantity of heat, usually expressed in British thermal units per minute (kilowatts), that a heat transfer component can dissipate under specified conditions.

heat distortion point The temperature at which a standard test bar deflects a specified amount under a stated load.

heat equation *See* thermodynamic efficiency.

heat exchanger A vessel in which heat is transferred from one medium to another.

heat-fail temperature The temperature at which a laminate ruptures under load.

heat flow *See* heat transmission.

heat of fusion The increase in enthalpy accompanying the conversion of one mole, or a unit mass, of a solid to a liquid at its melting point at constant pressure and temperature.

heat release The total quantity of thermal energy above a fixed datum introduced into a furnace by the fuel; considered to be the product of the hourly fuel rate and its high heat value, expressed in Btu per hour per cubic foot of furnace volume.

heat resistance The ability of a material to resist the deteriorating effects of elevated temperature.

heat resistant alloys Alloys developed for very high-temperature service in which relatively high stresses (tensile, thermal, vibratory, and shock) are encountered, and where oxidation resistance is frequently required.

heat-resistant irons Cast iron with a graphitic structure that can withstand heat for long periods without crystal growth.

heat-resistant steels Steels with good structural stability and resistance to oxidation at elevated temperatures.

heat sealing Joining films by the application of heat and pressure.

heat-sealing adhesive Thermoplastic film that acts as an adhesive when melted between two surfaces.

heat shock A test used to determine the stability of a material by sudden exposure to a high temperature for a short period of time.

heat shrinkable tubing Polymer tubing that can be joined and sealed by the application of heat.

heat sink A mounting base, usually metallic, that dissipates, carries away, or radiates into the surrounding atmosphere the heat generated within a semiconductor device.

heat toughened glass *See* tempered glass.

heat transfer The transfer or exchange of heat by radiation, conduction, or convection with a substance and between the substance and its surroundings.

heat transfer, coefficient of Heat flow per unit time across a unit area of a specified surface under the driving force of a unit temperature difference between two specified points along the direction of heat flow. Also known as overall coefficient of heat transfer.

heat transfer coefficient, surface or film The conductance of the thin layer of fluid immediately adjacent to the surface.

heat transmission Heat transmitted from one substance to another.

heat treatable alloy An alloy that can be hardened by heat treatment.

heat treated glass *See* tempered glass.

heat treating Term used to cover annealing, hardening, tempering, solutioning, aging, etc.

heat treatment Heating and cooling a solid metal or alloy in such as way as to produce desired conditions or properties. Heating for the sole purpose of hot working is excluded from the meaning of this definition.

heavy oil A viscous fraction of petroleum or coal-tar oil having a high boiling point.

heavy water *1.* Water in which the hydrogen of the water molecule consists entirely of the heavy hydrogen isotope of mass 2 (deuterium). *2.* A liquid compound D_2O whose chemical properties are similar to H_2O (light water), and which occurs in a ratio of 1 part in 6000 in fresh water.

heliarc welding *See* gas tungsten-arc welding.

helical winding A winding in which a filament band is wound in a helical path.

helium bombing A method of testing for leaks in which hermetically sealed units containing an internal volume are subjected to a helium pressure prior to being bell jar tested. If leads are present in the sealed unit, the helium pressure will drive some helium into the internal volume and this may be subsequently detected during bell jar testing.

helix A spiral winding.

Hellige turbidimeter A variable-depth instrument for visually determining the cloudiness of a liquid caused by the presence of finely divided suspended matter.

hematite A common iron mineral; ferric oxide.

henry The inductance of a closed circuit in which an electromotive force of 1 volt is produced when the electric current in the circuit varies uniformly at the rate of one ampere per second.

Henry's law A principle of physical chemistry that relates equilibrium partial pressure of a substance in the atmosphere above a liquid solution to the concentration of the same substance in the liquid; the ratio of concentration to equilibrium partial pressure equals the Henry's-law constant, which is a temperature-sensitive characteristic; Henry's law generally applies only at low liquid concentrations of a volatile component.

hermeticity The effectiveness of the seal of microelectronic and semiconductor devices with designed internal cavities.

hertz (Hz) Unit of frequency equal to one cycle per second.

Hessian matrices Given a real value function of N variables, an N by N symmetric of all second order partial derivatives.

heterochain polymers Polymers in which the carbon atoms in the chain backbone are replaced by other organic elements.

heterogeneous When applied to materials, describes a material with uniquely identifiable elements with regions of different properties.

heterogeneous radiation A beam of radiation containing rays of several different wavelengths or particles of different energies or different types.

heterojunctions Boundaries between two different semiconductor materials, usually with a negligible discontinuity in the crystal structure.

heuristic Pertaining to a method of problem solving in which solutions are discovered by evaluation of the progress made toward the final solution, for example, a controlled trial and error method. Thus, heuristic methods lead to solutions of problems or inventions through continuous analysis of results obtained thus far, permitting a determination of the next step. Contrast with stochastic.

heuristic program A program that monitors its performance with the objective of improved performance.

hex *1.* A number representation system of base 16. The hex number system is very useful in cases where computer words are composed of multiples of four bits (that is, 4-bit words, 8-bit words, 16-bit words, and so on). *2.* A six-sided geometric shape (for example, a hex-shaped bar).

hexa An abbreviation for hexamethylene-tetramine.

hexadecimal A number system using base 16 and the digit symbols from 0 to 9 and A to F. *See* hex.

hexadecimal notation A numbering system using 0, 9, A, B, C, D, E, and F with 16 as a base.

hexadecimal number *See* sexadecimal number.

hexagonal boron nitride (HBN) Boron nitride with hexagonal crystal structure similar to graphite.

hexa-isobutylene-polyvinylidene copolymer A hard fluoropolymer with excellent resistance to abrasion and chemicals.

Hf Chemical symbol for hafnium.

HF *See* high frequency.

Hg Chemical symbol for mercury.

HID *See* high interstitial defect.

hierarchy Specified rank or order of items; thus, a series of items classified by rank or order.

high-alloy steel An iron-carbon alloy containing at least 5% by weight of additional elements.

high aluminum defect (HAD) An aluminum-rich alpha stabilized region containing an abnormally large amount of aluminum which may extend across a large number of beta grains.

high brass A commercial wrought brass containing 65% copper and 35% zinc.

high-carbon steel A plain carbon steel with a carbon content of at least 0.6%.

high chromium steels Alloys with 15% or more chromium and a minimum of carbon.

high cycle fatigue Failure caused by cyclic loading to levels less than the material elastic limit, which results in complete failure in more than 50,000 loading cycles.

high definition radiography Using an x-ray with a very small focal spot, making it possible to increase magnification by moving the recording film further from the specimen.

high density inclusion (HDI) A region with a concentration of elements, usually tungsten or columbium, having a higher density than the matrix. Regions are readily detectable by x-ray and will appear brighter than the matrix.

high density polyethylene (HDPE) A long chain polymer material with good strength, hardness, and resistance to chemicals.

high energy fragments Any fragment that has sufficient energy to pass through a shield of soft aluminum with a thickness of 0.040 in maximum during the test demonstration.

high frequency (HF) The radio-frequency band between 3 and 30 MHz.

high-frequency heating Absorption of heat energy into a material by exposure to an electric field. *See also* electronic heating.

high heating *See* electronic heating.

high-heat value *See* calorific value.

high-impact polystyrene (HIPS) A thermosetting resin with good dimensional stability and impact strength.

high intensity laser *See* high-power laser.

high interstitial defect (HID) Interstitially stabilized alpha phase region of substantially higher hardness than surrounding material.

high-level data link control (HDLC) A type of data link protocol.

high-level human interference (HLHI) A device that allows a human to interact with the total distributed control system over the shared communications facility.

high-level language(s) A programming language whose statements are translated into more than one machine-language instruction. Examples of high-level languages are BASIC, FORTRAN, COBOL, and TELEVENT.

high-low bias test *See* marginal check.

high order Pertaining to the weight or significance assigned to the digits of a number, for example in the number 123456, the highest-order digit is 1, and the lowest-order digit is 6. As another example, one may refer to the three higher-order bits of a binary word.

high power laser Stimulated emission device having high energy flux density outputs.

high pressure laminates Laminates molded and cured at pressures above 7.0 MPa.

high resolution graphics A finely defined graphical display on a computer monitor screen.

high Reynolds number A Reynolds number above the critical Reynolds number of a sphere.

high speed steels Alloy steels that retain their hardness and strength at relatively elevated temperatures. These are commonly used as cutting tools.

high-strength alloy A metallic material having a strength considerably above that of most other alloys of the same type or classification. To be more specific, steels exhibiting a room temperature ultimate tensile strength of at least 180,000 psi.

high-strength alloy conductor A conductor that shows a maximum of 20% increase in resistance and a minimum of a 70% increase in breaking strength over the equivalent construction in pure copper while exhibiting a minimum elongation of 5% in 10 inches.

high-strength low alloy steels (HSLA) Low alloy steels with a fine grain structure with yield strengths between 350 and 550 MPa, approximately 50–80 ksi.

high-temperature alloy A metallic material suitable for use at 930°F (500°C) or above. This classification includes iron-base, nickel-base, and cobalt-base superalloys, and the refractory metals and their alloys, which retain enough strength at elevated temperature to be structurally useful and generally resist undergoing metallurgical changes that weaken or embrittle the material. *See also* heat-resistant alloys.

high temperature superconductor New superconducting material consisting of mixed metal oxide ceramics that maintain their superconductivity at higher temperature ranges (above 24 K) than the more traditional superconductors.

HIP (process) *See* hot isostatic pressing.

HIPS *See* high-impact polystyrene.

hit In data processing, the isolation of a matching record.

hit rate The number of successful matches in a computer search.

HK *See* Knoop hardness number.

HLHI *See* high-level human interference.

hoist A mechanical device consisting of a supporting frame and integral mechanism specifically designed to raise or lower a load.

hold time In any process cycle, an interval during which no changes are imposed on the system. Hold time is usually used to allow a chemical or metallurgical reaction to reach completion, or to allow a physical or chemical condition to stabilize before proceeding to the next step.

hole geometry (mechanics) The sizes, locations, and shapes of perforations created in materials.

holidays Unwanted abnormalities on a coated surface that allows corrosive elements to invade the coated base metal.

Hollerith code A widely used system of encoding alphanumeric information onto cards, hence Hollerith cards are the same as punch cards.

hollow glass fiber Hollow fiberglass fibers used for reinforcements to reduce weight.

homogeneous Of the same or similar nature. Uniform in structure or composition.

homogeneous carburizing A process that converts a low-carbon ferrous alloy to one of substantially uniform and higher carbon content throughout the section, so that a specific response to hardening may be obtained.

homogeneous glass Glass of essentially uniform composition throughout its structure.

homogenizing Soaking at a temperature sufficient to eliminate or decrease chemical segregation by diffusion, usually applied prior to mechanical working.

homopolymer A polymer derived from the polymerization of a single monomer.

hone To remove a small amount of material using fine-grit abrasive stones and thereby obtain an exceptionally smooth surface finish or very close dimensional tolerances.

honeycomb Manufactured product of resin-impregnated sheet material or metal foil, formed into hexagonal-shaped cells.

honeycomb core Lightweight strengthening material of structures resembling honeycomb meshes.

honeycomb sandwich assembly A structural composition consisting of relatively dense, high-strength facings bonded to a lightweight, cellular honeycomb core.

Hookean behavior A condition in liquid expansion in which the fractional change in volume is proportional to the hydrostatic stress, if under such stress it evidences ideal elastic behavior.

Hooke's Law Law stating that stress is always proportional to strain.

hoop stress The radial stress on the surface of a cylindrical form imposed by internal or external loads.

horn antenna *1.* Antenna shaped like a horn. *2.* The flared end of a radar waveguide, the dimensions of which are chosen to give efficient radiation of electromagnetic energy into the surrounding environment.

hot atoms Atoms with high internal or kinetic energy as a result of a nuclear process such as beta decay or neutron capture.

hot-cold working Mechanical deformation of austenitic and precipitation hardening alloys at a temperature just below the recrystallization range to increase the yield strength and hardness by either plastic deformation or precipitation hardening effects induced by plastic deformation or both.

hot corrosion The corrosion at high temperatures as a result of the reduction of protective oxide coatings and scales and the subsequent accelerated oxidation.

hot crack A crack that develops in a casting by internal stress during cooling.

hot densification *See* hot pressing.

hot-dip coating A continuous process of applying coatings to steel in which the steel is immersed in a molten bath of the coating material.

hot dip galvanizing A process for rustproofing iron and steel products by the application of a coating of metallic zinc.

hot dipping A process for coating parts by briefly immersing them in a molten metal bath, then withdrawing them and allowing the metal to solidify and cool.

hot finishing Final finishing of metal above the recrystallization temperature, resulting in no strain hardening.

hot forging Applying pressure on a metal in two or more directions at temperatures higher than the recrystallization temperature.

hot gas welding Joining thermoplastic material using highly focused hot air to soften the areas to be joined.

hot isostatic pressing (HIP) A thermomechanical process for forming metal-powder compacts or ceramic shapes by use of isostatically applied gas pressure in order to achieve high density in the treated material.

hot melt adhesive A molten adhesive that forms a bond after cooling to a solid.

hot pressed silicon nitride (HPSN) A dense engineering ceramic. *See* silicon nitride.

hot pressing Using heat and pressure to form a dense compact.

hot quenching An imprecise term used to cover a variety of quenching procedures in which a quenching medium is maintained at a prescribed temperature above 160°F (71°C).

hot-rolled Describes bars, plates, or sheets that are reduced to required dimensions at temperatures at which scale is formed and, therefore, carry hot-mill oxide.

hot-rolled, pickled Describes a hot-rolled product that has been subsequently pickled to remove the hot-mill oxide.

hot setting adhesive An adhesive that must be heated above 100°C to set.

hot tear A fracture in a metal caused by limiting contraction during solidification.

hot tops *1.* A reservoir, thermally insulated or heated, to hold molten metal on top of a mold to feed the ingot or casting as it contracts on solidifying, to avoid having pipe or voids. *2.* A refractory lined steel or iron casting which is inserted into the top of the mold and is supported at various heights to feed the ingot as it solidifies.

hot working Mechanical working of a metal at temperatures above the recrystallization temperature but below the melting temperature.

HUD *See* head up display.

hue The attribute of color perception by means of which an object is judged to be red, yellow, green, blue, or intermediate between some adjacent pair of these.

humidification Artificially increasing the moisture content of a gas.

humidistat An instrument for measuring and controlling relative humidity.

humidity A measure of the water vapor content of the air.

humidity, absolute The moisture content of air on a mass or volumetric basis.

humidity ratio The amount of water vapor per unit amount of dry air.

humidity, relative The moisture content of air relative to the maximum that the air can contain at the same pressure and temperature.

humidity, specific The weight of water vapor in air expressed in pounds or grains of water vapor per pound of dry air.

humidity test *1.* An exposure to humidity with an evaluation of effects. *2.* A corrosion test for comparing relative resistance of specimens to a high-humidity environment at constant temperature.

Huygens' principle A very general principle applying to all forms of wave motion which states that every point on the instantaneous position of an advancing phase front (wave front) may be regarded as a source of secondary spherical wavelets.

hybrid *1.* A composite laminate consisting of laminae of two or more composite material systems. *2.* A combination of two or more integrated circuits (ICs) in one package. A combination of an analog and digital computer.

hybrid computer *1.* A computer for data processing using both analog representation and discrete representation of data. *2.* A computing system using an analog computer and a digital computer working together.

hybrid laminate A laminate with two or more fiber reinforcement materials.

hydraulic Refers to any device, operation, or effect that uses pressure or flow of oil, water, or any other liquid of low viscosity.

hydraulic fluid A light oil or other low-viscosity liquid used in a hydraulic circuit.

hydride phase The phase formed in titanium when the hydrogen content exceeds the solubility limit. Hydrogen and, therefore, hydrides tend to accumulate at areas of high stresses.

hydride powder Metal powder produced by removing hydrogen from a metal hydride.

hydrocarbon(s) *1.* All-organic materials including unburned fuel and combustion by-products present in the exhaust which are detected by the flame ionization detector. *2.* A chemical compound of hydrogen and carbon.

hydrocarbon plastics Plastics made by the polymerization of carbon and hydrogen monomers.

hydrodynamic coefficients The factors producing motions in floating objects in liquids.

hydrogen blistering Subsurface voids produced by hydrogen absorption in (usually) low-strength alloys with resulting surface bulges.

hydrogen chloride laser *See* HCL laser.

hydrogen damage Any of several forms of metal failure caused by dissolved hydrogen, including blistering, internal void formation, and hydrogen-induced delayed cracking.

hydrogen deuterium oxide *See* heavy water.

hydrogen embrittlement (HE) A loss of ductility in metals resulting from absorption of hydrogen. *See also* hydrogen damage.

hydrogen loss The measure of oxygen content in a test specimen containing oxides that are reducible with hydrogen.

hydrogen stress cracking A failure process that results from the initial presence or absorption of hydrogen in metals in combination with residual or applied tensile stresses. It occurs most frequently with high-strength alloys.

hydrolysis Chemical decomposition of a substance involving the addition of water.

hydrolytic stability The tendency of fluids to be chemically affected by the presence of water. The instability effects are generally accelerated by elevated temperature.

hydrometer An instrument for directly indicating density or specific gravity of a liquid.

hydrophilic Having an attraction for water; capable of adsorbing or absorbing water.

hydrophilic emulsifier A water-soluble detergent concentrate used with the post-emulsifiable penetrants.

hydrophobic Poorly wetted to repelling water.

hydroscopic Refers to any material that easily absorbs and retains moisture.

hydrostatic burst-pressure test A destructive test that determines the minimum static pressure that a filter can sustain without leaking or rupture.

hydrostatic test Determining the burst resistance or leak tightness of a fluid component or system by imposing internal pressure.

hydrothermal stress analysis The evaluation of the combined effects of temperature-humidity cycling.

hydroxyl group A chemical group with one hydrogen and one oxygen atom.

hygral properties The affinity of something for moisture.

hygrometer An instrument for directly indicating humidity.

hygrometry Any process for determining the amount of moisture present in air or another gas.

hygroscopic Able to absorb and retain atmospheric moisture.

hygrothermal effect The change in properties of a material due to absorbed moisture and temperature change.

hygrothermograph An instrument that records both temperature and humidity on the same chart.

hypereutectic alloy An alloy that has an excess of an alloying element contrasted with the eutectic composition.

hypereutectoid alloy An alloy that has an excess of an alloying element contrasted with the eutectoid composition.

hyperon In the classification of subatomic particles according to mass, the heaviest of such particles. Some large and highly unstable components of cosmic rays are hyperons.

hypoeutectic alloy An alloy that has an excess of base metal contrasted to the eutectic composition.

hypoeutectoid alloy An alloy that has an excess of base metal contrasted to the eutectoid composition.

hysteresimeter A device for measuring a lagging effect related to physical change, such as the relationship between magnetizing force and magnetic induction.

hysteresis *1.* The difference between the response of a unit or system to an increasing and a decreasing signal. Hysteretical behavior is characterized in inability to retrace exactly on the reverse swing a particular locus of input/output conditions. *2.* In a transducer, the maximum difference in output, at any measurand value within the specified range, when the value is approached first with increasing and then with decreasing measurand. *3.* The difference in the input pressure between the increasing and decreasing output curve at a given output pressure.

Hz SI symbol for hertz. *See also* hertz.

I

I Chemical symbol for iodine.

IACS Abbreviation for "International Annealed Copper Standard," a method of rating the conductivity of copper.

IAE *See* integral absolute error.

ibrigizing A surface hardening process that produces a hard, corrosion-resistant silicon-rich surface.

IC *See* integrated circuit.

ICE *See* in-circuit emulation.

ICP *See* inductively coupled plasma.

ID *See* inside diameter.

ideal elastic behavior A material characteristic, under given conditions, whereby the strain is a unique straight-line function of stress and is independent of previous stress history.

ideal gas A hypothetical gas characterized by its obeying precisely the equation for a perfect gas, $PV = nRT$.

identifier A symbol used in data processing whose purpose is to identify, indicate, or name a body of data.

idiomorphic crystal A crystal with clearly developed planes.

idle characters Control characters interchanged by a synchronized transmitter and receiver to maintain synchronization during non-data periods.

idle time *1.* The part of available time during which computer hardware is not being used. Contrast with operating time. *2.* The part of uptime in which no job can run because all jobs are halted or waiting for some external action such as I/O data transfer.

IEC *See* ion-exchange chromatography.

ignition loss The loss of weight in a material after burning off binders or sizing.

illuminance (illumination) *1.* The density of luminous flux on a surface. It is equal to the flux divided by the area when the surface is uniformly illuminated. *2.* The total luminous flux received on a unit area of a given real or imaginary surface, expressed in such units as the footcandle, lux, or phot. Illuminance is analogous to irradiance, but is to be distinguished from the latter in that illuminance refers only to light.

illuminants Light oil or coal compounds that readily burn with a luminous flame such as ethylene, propylene, and benzene.

image analysis *1.* Technique for understanding or quantification of digital data as presented in a two-dimensional format. *2.* Electronic scanning of a test specimen to determine the size, shape, and distribution of phases within the microstructure.

image contrast The degree of noticeable differences in intensity.

image processing Conversion of optical images into digital data form for storage and reconstruction by computer techniques.

image reconstruction The reproduction of the original scene from data stored or transmitted after scanning by an electron beam. In reprography, the re-creation of graphic images from digital data stored in a computer.

image resolution In optics, a measure of the ability of an optical instrument to produce separable images of different points on an object.

imaging rate The frequency of renewal of information for a given point expressed in renewals per second or in images per second when all the points of the image are renewed simultaneously.

immersion coating Dipping a material in a chemical to produce a coating.

immersion plating Dipping a material in a metallic liquid to produce a deposit without the use of external electric current.

immersion test Test used to determine the relative abilities of greases to prevent corrosion on metal surfaces when the grease-coated metal object is immersed in water for an extended period of time. *See also* ultrasonic testing.

immiscible The inability of two fluids to attain homogeneity.

immunity An inherent or induced electrochemical condition that enables a metal to resist attack by a corrosive solution.

impact energy Measurement of the amount of energy needed to fracture a test specimen.

impact extrusion *See* cold extrusion.

impact fusion The conversion of the kinetic energy of a fast-moving, initially stationary, macroparticle projectile into the internal energy of fusile material using a particle accelerator. Impact fusion is generally an inertial confinement fusion concept.

impact melts Molten material resulting from hypervelocity impact.

impact sintering High-energy compacting of a powder metal compact which results in heating and fusion of the particles.

impact strength A material property that indicates its ability to resist breaking under extremely rapid loading, usually expressed as energy absorbed during fracture. *See* Charpy test and Izod test.

impact temperature The temperature of a gas after impact with a solid body, which converts some of the kinetic energy of the gas to heat and thus raises the gas temperature above ambient.

impact test A test to determine the behavior of materials when subjected to high rates of loading, usually in bending, tension, or torsion. The quantity measured is the energy absorbed in breaking the specimen by a single blow. *See* Charpy test and Izod test.

impeller *See* diffuser.

imperfection Any quality or state that differs from the quality or state intended for a particular material, part, or product. However, such is generally thought or known to not adversely influence service performance. *See also* defect.

impervious The ability of a material to resist absorption of water or gases.

impingement *1.* The striking of moving matter, such as the flow of steam, water, gas or solids, against similar or other matter. *2.* A method of removing entrained liquid droplets from a gas stream by allowing the stream to collide with a baffle plate.

impingement attack A form of accelerated corrosion in which a moving corrosive liquid erodes a protective surface layer, thus exposing the underlying metal to renewed attack.

implication Contained in the nature of something, although not readily apparent. *See* inclusion.

implosion The rapid inward collapsing of the walls of vacuum systems or devices as the result of failure of the walls to sustain the ambient pressure.

impregnate Saturating a reinforcement fiber with a resin.

impregnated fabric Fabric saturated with a polymer resin.

impulse sealing A pulse of intense thermal energy applied to a sealing surface followed by rapid cooling.

impurity Undesirable elements or compounds in a material.

incandescence *1.* The generation of light caused by heating a body to a high temperature. *2.* Spontaneous radiation of light energy from a hot object.

inch-pound units Units based on the yard and the pound, commonly used in the United States of America and defined by the National Institute of Standard and Technology (NIST). Note that units having the same names in other countries may differ in meaning.

incidence Partial coincidence, as a circle and a tangent line. The impingement of a ray on a surface.

in-circuit emulation (ICE) A development aid for testing the software in computer hardware. It involves an umbilical link between a development system and the target hardware being plugged into the microprocessor socket.

inclusion *1.* Nonmetallic materials in a solid metallic matrix. More specifically,

an undesirable phase in a metal, such as an oxide, sulfide, or silicate particle. *2.* Foreign matter trapped within the glass substrate or the surface by a coating. *3.* A logic operator having the property that if P is a statement and Q is a statement, then P inclusion Q is false if P is true and Q is false; true if P is false; and true if both statements are true. P inclusion Q is often represented by P>Q.

incoherent fiber optics A bundle of fibers in which the fibers are randomly arranged at each end. The pattern may be truly random to achieve uniform illumination, or the manufacturer may simply not bother to align individual fibers. In either case, the fiber bundle cannot transmit an image along its length.

incomplete trim A trimmed surface that does not have all designated material removed.

incompressible Describes a substance whose change in volume due to pressure is negligible; liquids, for example.

incompressible flow Fluid flow under conditions of constant density.

increment The specific amount by which a variable is changed.

incremental *See* incremental representation.

incremental encoder An electronic or electromechanical device that produces a coded digital output based on the amount of movement from an arbitrary starting position; the output for any given position with respect to a fixed point of reference is not unique.

incremental representation A method of representing a variable in which changes in the value of the variables are represented, rather than the values themselves.

indene resins Resins derived from polyindene, used in many coatings.

independent laboratory A laboratory whose systems are not dependent on those of specific material suppliers.

independent variable *1.* A process or control-system parameter that can change only due to external stimulus. *2.* A parameter whose variations, intentional or unintentional, induce changes in other parameters according to predetermined relationships.

index *1.* An ordered reference list of the contents of a computer file or document, together with keys or reference notations for identification or location of those contents. *2.* To prepare a list as in *1.* 3. A symbol or number used to identify a particular quantity in an array of similar quantities, for example, the terms of an array represented by X(1), X(2), … , 100 respectively. *4.* Pertaining to an index register. *5.* To move a machine part to a predetermined position, or by a predetermined amount, on a quantized scale.

index graduations The heaviest or longest division marks on a graduated scale, opposite the scale numerals.

index matching fluid A liquid with refractive index that matches that of the core or cladding of an optical fiber. It is used in coupling light into or out of optical fibers, and can help in suppressing reflections at glass surfaces.

indication In nondestructive testing, any visible sign or instrument reading that must be interpreted to determine whether or not a flaw exists.

indicator *1.* A device that makes information available about a measured characteristic, but does not store the information or initiate a responsive or corrective action. *2.* An instrument that graphically shows a value of a variable. *3.* The pointer on a dial or scale which provides a visual readout of a measurement. *4.* An instrument for diagramming pressure-volume changes during the working cycle of a positive-displacement compressor, engine, or pump.

indium A soft metallic element with good corrosion resistance.

indium-tin-oxide semiconductors *See* ITO (semiconductors).

inductance (L) *1.* The property of an electric circuit whereby a varying current in it produces a varying magnetic field which induces voltage in the same circuit or in a nearby circuit—measured in henrys. *2.* In an electrical circuit, the property that tends to oppose changes in current magnitude or direction. *3.* In electromagnetic devices, generating electromotive force in a conductor by means of relative motion between the conductor and a magnetic field such that the conductor cuts magnetic lines of force.

induction bonding Electromagnetically raising the temperature of electrically conductive materials sufficient to bond under pressure.

induction brazing Brazing an electrically inductive material by electromagnetically raising the temperature of the workpiece.

induction hardening *1.* A surface hardening process in which only the surface layer of a suitable ferrous workpiece is heated by electrical induction to above the upper transformation temperature and appropriately quenched. *2.* More generally, quench hardening in which the heat is generated by electrical induction.

induction heating Heating by electrical induction.

induction welding *See* induction brazing.

inductively coupled plasma (ICP) A plasma source for excitation in mass or atomic emission spectroscopy.

inductor A wire coil that will store energy in the form of a magnetic field.

industrial computer A computer used on-line in various areas of manufacturing, including process industries (chemical, petroleum, etc.), numerical control, production lines, etc. *See* process computer and numerical control.

industrial computer language A computer language for industrial computers. A language used for programming computer control applications and system development, for example, assembly language, FORTRAN, RTL, PROSPRO, BICEPS, and AUTRAN.

inelastic collision A collision between two or more bodies in which there is a net change in internal energy of at least one of the participating bodies and a net change in the sum of their kinetic energies.

inelastic scattering A collision of particles that changes their energy.

inelastic stress A force acting on a solid and producing a deformation such that the original shape and the size of the solid are not restored after the force is removed.

inert atmosphere A gaseous medium that, because of its lack of chemical reaction, is used to enclose tests or equipment.

inert filler A non-chemical-reacting material added to plastic to enhance properties.

inert gas *See* rare gases.

inert gas fusion A method of finding the concentration of oxygen, hydrogen, and nitrogen in a sample.

inertia Inherent resistance of a body to changes in its state of motion.

inertia bonding The joining of materials with friction and pressure.

infiltration Casing molten metal to be drawn into void spaces in a powder-metal compact, formed-metal shape, or fiber-metal layup.

infinity A point, line, or region extending beyond measurable limits.

influence A change in an instrument's indicated value caused solely by a difference in value of a specified variable or condition from its reference value or condition when all other variables are held constant.

information Any facts or data that can be used, transferred, or communicated.

information processing The organization and manipulation of data, usually by a computer. *See* data processing.

infrared The portion of the light spectrum with wavelength greater than the visible.

infrared absorption The taking up of energy from infrared radiation by a medium through which the radiation is passing.

infrared imaging device Any device that receives infrared rays from an object and displays a visible image of the object.

infrared radiation Electromagnetic radiation lying in the wavelength interval from 75 microns to an indefinite upper boundary, sometimes arbitrarily set at 1000 microns (0.01 centimeter).

infrared spectroscopy A technique for determining the molecular species present in a material, and measuring their concentrations, by detecting the characteristic wavelengths at which the material absorbs infrared energy and measuring the relative drop in intensity associated with each absorption band.

ingate *See* gate.

ingot That which is solidified in an ingot mold and is suitable for working or re-melting.

inherent Achievable under ideal conditions, generally derived by analysis.

inherent error The error in quantities that serve as initial conditions at the beginning of a step in a step-by-step set of operations; thus, the error carried over from the previous operation from whatever source or cause.

inherent reliability A measure of reliability that includes only the effect of an item design and its application and assumes an ideal operation and support environment.

inhibitor A chemical substance or combination of substances, which when present in the proper concentration and forms in the environment, prevents or reduces corrosion.

inhomogeneous Consisting of more than one phase, for example, discrete regions of different materials.

initialize In data processing, to send a rest command to clear all previous or extraneous information, as when starting a new operating sequence.

initial modulus The slope of the first part of a stress-strain curve.

initial set The start of a hardening reaction following water addition to a powdery material such as plaster or portland cement.

initial strain The strain in a test specimen under specific load conditions before creep occurs.

initial stress The stress in a test specimen under specific load before stress relaxation occurs.

initial value problem *See* boundary value problem.

initial yield The point at which 0.2% (or sometimes 0.02%) permanent deformation has occurred.

initiator Peroxides used in free-radical polymerization, curing thermosetting resins, and crosslinking agents for elastomers.

injection laser diode A semiconductor device in which lasing takes place within the P-N junction. Light is emitted from the diode edge.

injection molding A forming process in which a heat softened or plasticized material is forced from a cylinder into a relatively cool cavity which gives the product a desired shape. A similar process is used for forming solid propellants from quick-cure ingredients.

injector Any nozzle or nozzle-like device through which a fluid is forced into a chamber or passage.

in line *1.* Centered on an axis. *2.* Having several features, components, or units aligned with each other. *3.* In a motor-driven device, having the motor shaft parallel to the driven shaft of the device and approximately centered on each other.

inoculated iron Molten iron inoculated with an element just prior to casting to enhance the structure of the cast product.

inorganic pigments Metallic oxides that provide color, plus heat and weather resistance, to plastics.

inorganic polymers Materials with excellent stability and heat resistance.

219

input *1.* Signals taken in by an input interface as indicators of the condition of the process being controlled. *2.* Data keyed into a computer or computer peripherals. *See* excitation and measurand.

input channel A channel for impressing a state on a device or logic element.

input interface Any device that connects computer hardware or other equipment for the input of data.

input-output (I/O) *1.* The interface to a unit that provides data or signals used or generated by that unit. *2.* A general term for the equipment used to communicate with a computer and the data involved in the communications.

input-output control system (IOCS) A set of flexible routines that supervises the input and output operations of a computer at the detailed machine-language level.

input-output limited Pertaining to a computer system or condition in which the time for input and output operation exceeds other operations.

input-output (I/O) software The portion of the operational software that organizes efficient flow of data and messages to and from external equipment.

input-output (I/O) statement A statement that controls the transmission of information between the computer and the input/output units.

input port *1.* Any port into which flow is directed for the purpose of a test. *2.* In computer hardware, terminals for connection in external devices which input data to the computer.

input signal A signal applied to a device, element, or system.

input state The state occurring on a specified computer input channel.

input strobe (INSTRB) A signal that enters set-up data into registers.

input work queue A list of summary information of job-control statements maintained by the job scheduler, from which it selects the jobs and job steps to be processed.

inquiry A technique in which the interrogation of the contents of a computer's storage may be initiated at a keyboard.

insert *1.* Any design feature of a cast or molded component that is made separately and placed in the mold cavity prior to the casting or molding step. *2.* A removable part of a die, mold, or cutting tool.

inside diameter (ID) The inner dimension of a circular member such as a pipe or tube.

inside-out corrosion Corrosion that starts from the inside surface of a material specimen and works outward.

in situ composites Composites in which a reinforcing phase grows during solidification.

inspection A deliberate critical examination to determine whether an item meets established standards. Inspection may involve measuring dimensions, observing visible characteristics, or determining inherent properties of an object, but usually does not involve determining operating characteristics. The last is more properly termed testing.

inspection door A small door in the outer enclosure so that certain parts of the interior of the apparatus may be observed.

inspection, nondestructive (NDI) A family of methods for investigating the quality, integrity, properties, and dimensions of materials and components, without damaging or impairing their serviceability through the use of dye penetrant, magnetic eddy current, ultrasonic, radiographic, infrared, etc. devices.

instability Lack of stability. Contrast with stability.

instantaneous recovery The rapid decrease in strain that occurs when reducing the load on a test specimen.

instantaneous strain The strain that occurs immediately when applying a load to a test specimen.

instruction In data processing, a statement that specifies an operation and the values or locations of its operands. In that context, the term instruction is preferable to the terms command or order, which are sometimes used synonymously. Command should be reserved for electronic signals, and order should be reserved for sequence, interpolation, and related usage.

instruction buffer An eight-bit byte buffer in the computer processor that is used to contain bytes of the instruction currently being decoded and to prefetch instructions in the instruction system.

instruction code *See* operation code.

instruction counter *See* location counter.

instruction time The portion of an instruction cycle during which the computer control unit is analyzing the instruction and setting up to perform the indicated operation.

instrument *1.* A device for measuring the value of an observable attribute; the device may merely indicate the observed value, or it may also record or control the value. *2.* Measuring, recording, controlling, and similar apparatus requiring the use of small to moderate amounts of electrical energy in normal operation.

instrumental analysis Any analytical procedure that uses an instrument to measure a value, detect the presence or absence of an attribute, or signal a change or end point in a process.

instrument correction A quantity added to, subtracted from, or multiplied into an instrument reading to compensate for inherent inaccuracy or degradation of instrument function.

instrument loop diagram A diagram that contains the information needed to understand the operation of the loop and also shows all connections to facilitate instrument startup and maintenance of the instruments. The loop diagram must show the components and accessories of the instrument loop, highlighting special safety and other requirements.

Instrument Society of America (ISA) A U.S. society of instrument and controls professionals.

insulation *1.* A material of low thermal conductivity used to reduce heat loss. *2.* A material of specific electrical properties used to cover wire and electrical cable.

insulation resistance (S) *1.* The resistance measured between specified insulated portions of a transducer when a specified d-c voltage is applied at room conditions unless otherwise stated. *2.* The ratio of applied voltage to the total current between two electrodes in contact with a specific insulation.

insulative The characteristic of a material that describes its ability to resist electric current flows when a voltage is applied across the material.

insulator *1.* A dielectric material that does not conduct current. *2.* A high-resistance device that supports or separates conductors to prevent a flow of current between them or to other objects.

integer A whole number signified by a binary "word."

integer programming *1.* In operations research, a class of procedures for locating the maximum or minimum of a function subject to constraints, where some or all variables must have integer values. Contrast with convex programming, dynamic programming, linear programming, and mathematical programming. *2.* Loosely discrete programming.

integral *1.* Describes a control action that will cause the output signal to change according to the summation of the input signal values sampled at regular intervals up to the present time. *2.* Mathematically, the reciprocal of reset.

integral absolute error (IAE) A measure of controller error defined by the integral of the absolute value of a time-dependent error function; used in tuning automatic controllers to respond properly to process transients. *See also* integral time absolute error.

integral-blower burner A burner of which the blower is an integral part.

integral composite structure A composite in which multiple structural elements are "laid up" and cured as one rather than preassembled.

integral control Form of control action that returns the value of the controlled variable to the set point when sustained offset occurs without this action. Also called reset control.

integral orifice A differential-pressure-measuring technique for small flow rates in which the fluid flows through a miniature orifice plate integral with a special flow fitting.

integrals Of or pertaining to an integer.

integral skin foam Foam with a porous core and a harder, nonporous skin.

integral time absolute error (ITAE) A measure of controller error defined by the integral of the product of time and the absolute value of a time-dependent error function; whereas the absolute value prevents opposite excursions in the process variable from canceling each other, the multiplication by time places a more severe penalty on sustained transients.

integrated circuit (IC) A complete electronic circuit containing active and passive elements fabricated and assembled as a single unit, usually as a single piece of semiconducting material, resulting in an assembly that cannot be disassembled without destroying it.

integrated optics Thin-film devices containing tiny lenses, prisms, and switches to transmit very thin laser beams, which serve the same purposes as the manipulation of electrons in thin-film devices of integrated electronics.

integrating accelerometer A device that measures acceleration of an object, and converts the measurement to an output signal proportional to speed or distance traveled.

integrating ADC A type of analog-to-digital converter in which the analog input is integrated over a specific time with the advantages of high resolution, noise rejection, and linearity.

integrator(s) *1.* A device that continually totalizes or adds up the value of a quantity for a given time. *2.* A device whose output is proportional to the integral of the input variable with respect to time.

intensiostatic *See* galvanostatic.

intensity The amount of light incident per unit area. For human viewing of visible light, the usual term is illuminance; for electromagnetic radiation in general, the term is radiant flux.

intensity level The amplitude of a sound wave, commonly measured in decibels.

interactive In data processing, a technique of user/system communication in which the operating system immediately acknowledges and sets upon a request entered by the user at a terminal. Compare with batch.

interactive graphics *See* computer graphics.

intercalation Production of layer-type semiconducting as well as other conducting materials.

intercept method A method for estimating the quantity of particles or number of grains within a unit area of a microscopic image by counting the number intercepted by a series of straight lines through the image. This is one of the standard methods of determining grain size of a polycrystalline metal.

interconnected porosity A type of porosity in a powder metal compact whereby a liquid can pass through the entire mass.

intercrystalline That space between crystals of a metal.

intercrystalline corrosion *See* intergranular corrosion.

interdendritic corrosion *1.* Corrosive attack of cast materials which progresses preferentially along interdendritic paths. *2.* Corrosion due to the different compositions within a material.

interelectrode capacitance *1.* The capacitance between electrodes of a vacuum tube. *2.* A capacitance determined by measuring the short-circuit transfer admittance between two electrodes.

interface *1.* A shared boundary. An interface might be a hardware component to link two devices, or it might be a portion of storage or registers accessed by two or more computer programs. *2.* The surface that forms a boundary between phases in a powder metal compact. *3.* The surface between laminate reinforcements and a resin.

interface activity The determination of the chemical potential between particles in a compact.

interfacial bonding The effective adhesion between a base material and a reinforcing material.

interfacial seal *See* seal.

interference *1.* The waveform resulting from superimposing one wave train on another. *2.* In signal transmission, spurious or extraneous signals that prevent accurate reception of desired signals. *3.* A difference in dimension of two mating materials at ambient temperature.

interference fit *1.* Any combination of pin or shaft diameter and mating hole diameter where the tolerance envelope of the hole overlaps or is smaller than the tolerance envelope of the pin. *2.* The condition in which the diameter of the fastener is larger than the hole that it is to fit in.

interference microscopy Testing technique for the analysis of material surfaces.

interference pattern The pattern of some characteristic of a stationary wave produced by superimposing one wave train on another.

interference suppression Action that reduces or eliminates electrical interference.

interferometer An instrument so designed that the variance of wavelengths and light path lengths within the mechanism allows very accurate measurement of distances.

intergranular beta Beta phase situated between alpha grains. It may be at grain corners, as in the case of equiaxed alpha type of microstructures in alloys having low beta stabilizer content.

intergranular corrosion Preferential corrosion at grain boundaries of a metal or alloy. Also called intercrystalline corrosion.

intergranular cracking Rupture that occurs at grain boundaries.

intergranular stress corrosion cracking Stress corrosion cracking in which the cracking occurs along grain boundaries.

interlaminar Referring to an object or occurrence existing between laminates.

interlaminar shear Shearing force tending to produce a relative displacement between two laminae in a laminate along the plane of their interface.

interlaminar shear strength The maximum shear stress between layers that a laminated material can resist.

interleaving *1.* The act of simultaneously accessing two or more bytes or streams of data from distinct computer memory banks. *2.* The alternating of two or more operations or functions through the overlapped use of a computer facility.

intermediate annealing Annealing wrought metals at one or more stages during manufacture and before final thermal treatment.

intermediate band A mode of recording and playback in which the frequency response at a given tape speed is "intermediate."

intermediate oxides Oxides with both acidic and basic properties.

intermediate phase A distinct compound or solid solution in an alloy system whose composition limits do not extend to any of the pure constituents.

intermediate temperature-setting adhesive An adhesive that sets at temperature ranges from 30 to 100°C.

intermetallic compound A phase in an alloy system which usually occurs at a definite atomic ratio and exhibits a narrow solubility range.

internal absorption factor The ratio of the directed luminous flux absorbed by a transparent body during a single passage from the first surface to the second surface (difference between the flux leaving the first surface and that reaching the second surface) to the flux leaving the first surface.

internal energy Ability of a working fluid to do its work based on the arrangement and motion of its molecules.

internal friction The heat developed in a material when under fluctuating loads.

internal oxidation A form of degradation of a material involving absorption of oxygen at the surface and diffusion of oxygen to the interior, where it forms subsurface scale or oxide inclusions.

internal pressure The pressure inside a portion of matter due to the attraction between molecules.

internal standard In chemical analysis, especially instrumental analysis, a material present in or added to a sample in known amounts to serve as a reference in determining composition.

internal wave In fluid mechanics, wave motions of stably stratified fluids in which the maximal vertical motions occur below the surface of the fluids.

international rubber hardness degrees (IRHD) A standard unit used to indicate the relative hardness of elastomeric materials, where zero represents a material having a Young's modulus of zero, and 100 represents a material of infinite Young's modulus.

International System of Units The metric system of units based on the meter, kilogram, second, ampere, Kelvin degree, and candela. Other SI units are hertz, radian, newton, joule, watt, coulomb, volt, ohm, farad, weber, and tesla.

interpass temperature The lowest temperature reached by weld metal before the next pass is deposited in a multiple-pass weld.

interphase A boundary between a polymer and a base material where the polymer has a strong molecular attraction to the base material.

interply hybrid A composite where adjacent reinforcing laminates have different composition.

interrupted aging Aging at two or more temperatures, by steps, and cooling to room temperature after each step. *See* aging. Compare with progressive aging.

interrupted quenching A quenching procedure in which the workpiece is removed from the first quench at a temperature substantially higher than that of the quenchant and is then subjected to a second quenching system having a different cooling rate than the first.

interrupt service routine In data processing, a unique address that points to two consecutive memory locations containing the start address of the interrupt service routine and priority at which the interrupt is to be serviced.

intersection(s) In Boolean algebra, the operation in which concepts are described by stating that they have all the characteristics of the classes involved. Intersection is expressed as AND.

interstices In materials, the spaces between individual particles.

interstitial element An element with relatively small atomic diameter which can assume position in the interstices of a crystal lattice. Common examples are oxygen, nitrogen, hydrogen, and carbon.

interval timer *See* timer.

intralaminar Descriptive term pertaining to an object, event or potential field existing entirely within a single lamina without reference to any adjacent laminae.

intraply hybrid A composite in which different materials are used within a specific layer or band.

intrinsic joint loss A loss intrinsic to the fiber, caused by fiber parameter mismatches when joining two nonidentical fibers.

introfaction The change in wetting properties of a fluid by the addition of an introfier.

introfier A chemical that modifies a chemical solution to a molecular solution.

invalidity *See* error.

invalid test value A test value considered to be untrue because it does not fit the population of other values from the same sample or because the sample or test instrument did not conform to prescribed standards.

inverse scattering Method of analyzing some classic wave scattering.

inverse-time *See* time-inverse.

inversion temperature In a thermocouple, the temperature of the "hot" junction when the thermoelectric emf of the circuit is equal to zero.

inverter A NOT element. The output signal is the reverse of the input signal.

I/O *See* input-output.

I/O-bound A state of program execution in which all operations are dependent on the activity of an I/O device; for example, when a program is waiting for input from a terminal. *See also* CPU-bound.

IOCS *See* input-output control system.

I/O hardware Computer hardware used to carry signals into and out of the processing hardware.

I/O isolation Usually refers to the electrical separation of field circuits from computer internal circuits. Accomplished by optoelectronic devices. Occasionally refers to the ability to have input or output field wiring on isolated circuits, i.e., with one return for each.

I/O limited *See* input-output limited.

I/O module Basic set of I/O interfaces sharing a common computer unit housing. Can be a set of discrete I/O or a smart control I/O.

ion An electrically charged atom or group of atoms.

ion chamber *See* ionization chamber.

ion chromatography A testing technique that uses an ion exchange material to separate various types of ions in a solution and transfer them to a detector for analysis.

ion density (concentration) In atmospheric electricity, the number of ions per unit volume of a given sample of air; more particularly, the number of ions of a given type (positive small ion, negative small ion, positive large ion, negative large ion) per unit volume of air.

ion exchange A chemical process for removing unwanted dissolved ions from water by inducing an ion-exchange reaction (either cation or anion) as the water passes through a bed of special resin containing the substitute ion.

ion-exchange chromatography (IEC) A testing technique used to separate ionic charged compounds for analysis.

ion-exchange resin(s) *1.* A synthetic organic compound (resin) that can remove unwanted ions from a dilute solution by combining with them or by exchanging them for ions that produce desirable or neutral effects. *2.* Polymers that form salts with ions.

ion gage *See* ionization gage.

ionic bond A bond between atoms due to the attractive forces between positively and negatively charged ions.

ionic crystal A crystal in which the bonds are ionic. *See* ionic bond.

ionic mobility In gaseous electric conduction, the average velocity with which a given ion drifts through a specified gas under the influence of an electric field of unit strength.

ionic strength Effective strength of all ions in a solution; for dilute solutions, equal to the sum of one-half of the product of the individual ion concentration and their ion valence or charge squared.

ion implantation A method of semiconductor doping in which impurities that have been ionized and accelerated to a high velocity penetrate the semiconductor surface and become deposited in the interior.

ionitriding A plasma method for nitriding steel.

ionization The process of splitting a neutral molecule into positive and negative ions, or of detaching one or more electrons from a neutral atom.

ionization chamber An enclosure filled with gas that is ionized when radiation enters the chamber; it contains two or more electrodes that sustain an electric field and collect the charge resulting from ionization.

ionization constant A measure of the degree of dissociation of a polar compound in dilute solution at equilibrium; it equals the product of the concentrations of the dissociated compound (ions) divided by the concentration of the undissociated compound.

ionization counter *See* radiation counter.

ionization gage A pressure transducer based on conduction of electric current through ionized gas of the system whose pressure is to be measured.

ionization potential The energy required to ionize an atom or molecule. The energy is usually given in terms of electron volts.

ionization time In a gas tube, the interval between the time when conduction conditions are established and the time when conduction actually begins at some stated value of tube potential.

ionized plasma *See* plasmas (physics).

ionizing event Any interaction between an atom or molecule and an energy beam, particle, atom, or molecule that causes one or more ions to be generated.

ionizing radiation Any electromagnetic or particulate radiation that can produce ions, either directly or indirectly, when it interacts with matter.

ion laser A laser in which the active medium is an ionized gas, typically one of the rare gases, argon or krypton, or a mixture of the two.

ion nitriding A vacuum process for producing a nitrided case on ferrous materials using a glow discharge.

ionomer resin An ethylene-based polymer with excellent strength and resistance to chemicals.

ion pair The combination of a positive ion and a negative ion having the same magnitude of charge, and formed from a neutral molecule due to absorption of the energy in radiation.

ion-pair chromatography (IPC) A testing technique using mobile ions that combine with sample ions to form neutral ion pairs for subsequent analysis.

ions Charged atoms or molecularly bound groups of atoms; or, sometimes, free electrons or other charged subatomic particles.

ion-scattering spectrometry A testing technique used to define the properties and structure on the surface of a material.

ion spectrometer *See* mass spectrometer.

ion storage Ions within an electromagnetic trap, cooled to sub-Kelvin temperatures with lasers. Potential uses are for frequency standards.

ion stripping A procedure following the focusing of ion beams in the target chamber of a reactor to be used for particle beam pellet fusion.

IP (impact prediction) *See* computerized simulation.

IPC *See* ion-pair chromatography

I/P converter A device that linearly converts electric current into gas pressure.

IRHD *See* international rubber hardness degrees.

iridium A metal in the platinum group.

iron A widely used metallic element. It is the base element in steels.

iron 58 A radioactive isotope of iron.

irradiance *1.* The rate at which energy is incident on a surface, per unit area (W/m^2). *2.* The detection rate per unit area of radiation.

irradiation *1.* Exposing an object or person to penetrating ionizing radiation such as x-rays or gamma rays. *2.* Exposing an object or person to ultraviolet, visible, or infrared energy.

irreversible Chemical reactions of thermosetting resins that cannot be changed, for example, remelting.

ISA *See* Instrument Society of America.

isentrope A line of equal or constant pressure, with respect to either space or time.

isentropic Proceeding at constant entropy.

isentropic exponent A ratio defined by the specific heat at constant pressure divided by the specific heat at constant volume.

isobaric Proceeding at constant pressure.

isobaric process Of equal or constant pressure, with respect to either space or time.

isocratic elution The use of phases with non-changeable compositions throughout the sequence of a liquid chromatography separation process.

isocyanate plastics Plastics derived from organic isocyanates and other compounds.

isokinetic sampling Withdrawing a sample stream at the same velocity as the parent stream. This method is used to obtain a representative sample when particulate matter is of interest.

isolated A metallographic feature that occurs in 3% or less of the microstructure.

isolation A technique used in fault-tolerant systems that removes the effects of a failure or prevents a failure from propagating or affecting the continued operation of the system.

isolation level The functional level to which a failure can be isolated using accessory test equipment at designated test points.

isomer Nuclide having the same mass number A and atomic number Z, but existing for measurable times in different quantum states with different energies and radioactive properties. Molecules having the same atomic composition and molecular weight, but differing in geometrical configuration.

isomerization Process of converting hydrocarbon or other organic compound to an isomer.

isoparametric finite elements The basis for the calculation of physical properties of structural shapes including stress analyses.

isopotential point Point on the millivolt versus pH plot at which a change in temperature has no effect. It is at 7 pH and zero millivolts, unless shifted by the standardization and meter zero adjustments or an electrode asymmetry potential.

isoprene An unsaturated hydrocarbon used in the manufacture of synthetic rubbers.

isostatic pressing Forming metal powder compacts in a press that applies pressure equally in all directions.

isostatic stereoisomerism A polymer molecular structure with a sequence of evenly spaced and similar symmetrical groups in a polymer chain.

isotensoid structure Filamentary structure in which the filaments are uniformly stressed throughout from the design loading conditions.

isothermal Proceeding at constant temperature.

isothermal annealing Austenitizing a ferrous alloy and then cooling to and holding at a temperature at which austenite transforms to a relatively soft ferrite-carbide aggregate.

isothermal process Thermodynamic change of state of a system that takes place at constant temperature.

isothermal transformation A change in phase at constant temperature.

isotherms Lines connecting points of equal temperature.

isotope(s) Any of two or more nuclides that have the same number of protons in their nuclei, but different numbers of neutrons; such atoms are of the same element, and thus cannot be separated from each other by chemical means, but because they have different masses, can be separated by physical means.

isotope effect The effect of nuclear properties other than the number of protons on the nonnuclear physical and chemical behavior of the nuclides.

isotopic enrichment Process by which the relative abundance of the isotopes of a given element are altered in a batch to produce a form of the element enriched in a particular isotope.

isotropic Invariant with respect to direction.

isotropic laminate A laminate in which the strength properties are the same in all directions.

isotropic ply A ply with similar properties in all directions.

isotropy The state of a material having identical properties in all directions.

ITAE *See* integral time absolute error.

iterate To repeatedly execute a loop or series of steps. For example, a loop in a routine.

iterative Describes a procedure or process that repeatedly executes a series of operations until some condition is satisfied. An iterative procedure can be implemented by a loop in a routine.

ITO (semiconductor) Semiconductor device consisting of a layer of tin sandwiched between an indium layer and an oxide layer.

Izod test A pendulum type of single-blow impact test in which the specimen, usually notched, is fixed at one end and broken by a falling pendulum. The energy absorbed, as measured by the subsequent rise of the pendulum, is a measure of impact strength or notch toughness. *See also* Charpy test.

J

J *See* joule.

jack A connecting device to which a wire or wires of a circuit may be attached and which is arranged for the insertion of a plug.

jacket *1.* A plastic layer applied over the coating of an optical fiber, or sometimes over the bare fiber. Used for color coding in optical cables, to make handling easier, or for protection of the fiber against mechanical stress and strain. *2.* Stiff plastic protective material that encases a floppy disk with slots for a disk drive to access the data. *3.* The layer of plastic, fiber, or metal surrounding insulated electrical wires to form a cable. This outer cover may be for mechanical or environmental protection of the wires contained therein.

jacketing Covering or coating wire or cable with a resin for insulation or protection.

Jahn-Teller effect The effect whereby, except for linear molecules, degenerate orbital states in molecules are unstable.

jet Rapid flow of fluid from a nozzle or orifice.

jeweler's rouge A mild abrasive used to produce a fine polish.

J integral A contour energy integral formulated by Rice and used for evaluating fracture toughness of elastoplastic materials.

jitter A computer signal instability. *See also* vibration.

joggle A displacement machined or formed in a structural member to accommodate the base of an adjacent member.

Johansson curved crystal spectrometer A type of spectrometer having a reflecting crystal whose face is concave so that x-rays that diverge slightly after passing through the primary slit are refocused at the detector slit.

joint A separable or inseparable juncture between two or more materials.

joint efficiency The strength of a joint expressed as a percent of the strength of the base material.

Jominy test A test to determine the hardenability of steel. *See* end-quench hardenability test.

joule The unit of energy in the SI system.

Joule-Thomson effect A change of temperature in a gas undergoing Joule-Thomson expansion.

Joule-Thomson expansion The non-reversible expansion of gas flowing through a porous medium or a small opening.

Jovignot test A test to determine the ductility of sheet metal.

joy stick A device by which an individual can communicate with an electronic information system through a cathode ray tube.

jump *1.* An instruction that causes a new address to be entered into a computer program counter; the program continues execution from the new program counter address. *2. See* transfer. *3. See* unconditional transfer.

jumper(s) A temporary wire used to bypass a portion of an electrical circuit or to attach an instrument or other device during testing or troubleshooting.

junctions In semiconductor devices, regions of transition between semiconducting regions of different electrical properties.

justification *1.* The act of adjusting, arranging, or shifting digits to the left or right, to fit a prescribed pattern. *2.* The use of space between words in printed material to cause each line to be of equal length.

justify To align computer data about a specified reference.

jute An organic plant material used as a reinforcement material in some plastics.

K

k *1. See* solubility coefficient. *2. See* Boltzmann's constant.

K *1.* Symbol used to indicate a kilobyte, which is a measure of a computer memory. One kilobyte of memory can store 1024 characters. *2.* Chemical symbol for potassium. *3.* Abbreviation for Kelvin. *4. See* luminous efficacy.

K′ *See* cyclic strength coefficient.

Kalman filter A technique for calculating the optimum estimates of process variables in the presence of noise; the technique, which generates recursion formulas suitable for computer solutions, also can be used to design an optimal controller.

kaolinite A hydrous silicate of aluminum. It constitutes the principle mineral in kaolin.

Kapton See polyimide.

Karl Fischer technique A titration method for accurately determining moisture content of solid, liquid, or gas samples using Karl Fischer reagent—a solution of iodine, sulfur dioxide, and pyridine in methanol.

Karnaugh map A tubular arrangement that facilitates the combination and elimination of logical functions by listing similar logical expressions, thereby taking advantage of the human brain's ability to recognize visual patterns to perform the minimization.

K_b *See* Bernoulli coefficient.

Kbyte 1024 (2^{10}) bytes.

keel block A test casting for high-shrinkable cast alloys.

Kelvin Metric unit for thermodynamic temperature. An absolute temperature scale in which the zero point is defined as absolute zero (the point where all spontaneous molecular motion ceases) and the scale divisions are equal to the scale divisions in the Celsius system. In the Kelvin system, the scale divisions are not referred to as degrees as they are in other temperature-measurement systems. $0°C$ equals approximately 273.16 K.

Kern counter *See* dust counter.

kerogen The substantially organic material in oil shale, consisting primarily of carbon and hydrogen with quantities of sulfur, nitrogen, and oxygen, which upon pyrolysis will yield gas and raw shale oil.

Kerr cell A device in which the Kerr effect is used to modulate light passing through the material. The modulation depends on rotation of beam polarization caused by the application of an electric field to the material. The degree of rotation determines how much of the beam can pass through a polarizing filter.

Kevlar A DuPont synthetic textile material, lightweight and nonflammable, and with high impact resistance.

key characteristics The features of a material or part whose variation has a significant influence on product fit, performance, service life, or manufacturability.

keyhole specimen A test specimen with a keyhole-like notch, used in impact testing.

keying *1.* Applying uniform load on a powder metal compact during compacting. *2.* In composites, the mechanical bond between the reinforcement fibers and the base material.

*k***-factor** *1.* In quantitative analysis, the ratio between the unknown and standard x-ray strength. *2. See* thermal conductivity.

K-factor A measure of insulating properties of various polymeric foams.

kg *See* kilogram.

Kikuchi lines An electron-diffraction pattern that provides microstructural information.

killed steel A type of steel from which there may be only a slight evolution of gases during solidification of the metal. Killed steels have more uniform chemical composition and properties than the other types. However, there may be variations in composition, depending on the steelmaking practices used. Alloy steels are of the killed type, while carbon steels may be killed or may be of the following types: killed steels, rimmed steels, semikilled steels, and capped steels.

kiln An oven or similar heated chamber for drying, curing, or firing materials or parts.

kilo A decimal prefix denoting 1,000.

kilogram Metric unit of mass.

kilohm (kohm) Unit of power equivalent to 1000 ohms.

kilometric waves Electromagnetic waves with wavelengths between 1,000 and 10,000 meters.

kilovolt ampere (kVA) 1000 volt × amperes.

kilovolts (kV) 1000 volts.

kilowatt (kW) A unit of power equal to one thousand watts.

kinematic viscosity Absolute viscosity of a fluid divided by its density.

kinetic energy *1.* The energy of a working fluid caused by its motion. *2.* Energy related to the fluid of dynamic pressure, $1/2\ pV^2$.

kinetic theory The derivation of the bulk properties of fluids from the properties of their constituent molecules, their motions and interactions.

Kirchhoff-Huygens principle *See* diffraction.

Kirchhoff's law of radiation The radiation law that states that at a given temperature the ratio of the emissivity to the absorptivity for a given wavelength is the same for all bodies and is equal to the emissivity of an ideal black body at that temperature and wavelength.

Kirchhoff's Law Law stating that the sum of the voltage across a device in a circuit series is equal to the total voltage applied to the circuit.

Kirkendall effect The movement of the junction between two metals due to different rates of diffusion.

K_{iscc} Abbreviation for the critical value of the plane strain stress intensity factor that will just produce crack propagation by stress corrosion cracking of a given material in a given environment.

kish Free graphite that forms in cast iron as it cools.

knit-line The area of a molded plastic part, formed by the union of two or more streams of plastic flowing together.

Knoop hardness A microhardness test using a pyramid shaped indenter.

Knoop hardness number Numbers related to measurements from the Knoop hardness test.

known defect test standard In nondestructive testing, a test standard containing known defects used to perform system performance checks and classify penetrants.

knuckle area The area between sections of a reinforced plastic with different geometries.

kohm *See* kilohm.

kondo effect Change in superconductivity characteristics resulting from magnetic impurities in the compounds involved.

konimeter A device for determining dust concentration by drawing in a measured volume of air, directing the air jet against a coated glass surface, thereby depositing dust particles for subsequent counting under a microscope.

koniscope An indicating instrument for detecting dust in the air.

Kr Chemical symbol for krypton.

K-radiation The standard x-rays produced by an atom or ion when a vacancy in a K-shell is replaced by an electron from another shell.

kreep A yellow-brown glassy lunar mineral enriched in potassium, rare earth elements, and phosphate.

Krenchel model A technique for estimating the modulus of some laminates.

kriging A method of providing unbiased estimates of variables in regions where the available data exhibit spatial autocorrelation; these estimates are obtained in such a way that they have minimum variance.

krypton fluoride lasers Rare gas halide ultraviolet stimulated emission devices in which krypton fluoride is the active lasing medium.

K-series The standard wavelengths for K-radiation in various materials.

K-shell The innermost shell of electrons surrounding an atomic nucleus.

K_t *See* stress concentration factor.

kurtosis In statistics, the extent to which a frequency distribution is peaked or concentrated about the mean; it is sometimes defined as the ratio of the fourth moment of the distribution to the square of the second moment.

Kurtosis number Figure of merit, K, used for monitoring impulsive-type vibrations of ball bearings.

KV *See* kilovolts.

KVA *See* kilovolt ampere.

KW *See* kilowatt.

kyanite An alumina-silicate mineral used in refractories.

Kynar Tradename for polyvinylidene fluoride, manufactured by Pennwalt Corp.

L

L *1. See* litre. *2. See* length.

La Chemical symbol for lanthanum.

labeled molecule A molecule of a specific chemical substance in which one or more of its component atoms is an abnormal nuclide, that is, a nuclide that is radioactive when the molecules normally are composed of stable isotopes, or vice versa.

laboratory ambient conditions 65–75°F (18–24°C) and 45–55 percent relative humidity.

laboratory sample A portion of material taken to represent the lot sample, or the original material, and used in the laboratory as a source of test specimens.

laboratory standard An instrument that is calibrated periodically against a primary standard.

lacing members Structural truss members at angles to and supporting the chords in a lattice boom. They are open or closed members used to transmit shear loads and to maintain the geometry of the lattice boom.

lack of fill out Characteristic of an area, occurring usually at the edge of a laminated plastic, where the reinforcement has not been wetted with resin.

lacquer Thermoplastic coating material dissolved in a solvent.

ladder diagram Symbolic representation of a control scheme. The power lines form the two sides of a ladder-like structure, with the program elements arranged to form the rungs. The basic program elements are contacts and coils as in electromechanical logic systems.

ladder polymer A polymer with two cross-linked polymer chains.

ladle A receptacle used for transferring and pouring molten metal.

ladle analysis Chemical analysis of the heat of steel as obtained from a sample taken (while the metal is still liquid) just prior to casting into molds. It is the analysis for all the specified elements and is determined by analyzing one or more test samples obtained during the pouring of the steel. When such samples are unobtainable or if it is evident that the samples do not represent the analysis of the melt, additional samples are taken from the solid steel. An analysis based on these representative samples may be used. Note: In the case of remelted products, for example, vacuum arc remelting (VAR) and electroslag remelting (ESR), it is common to have the analysis be obtained from the remelted product.

ladle sampling A common practice in most melting operations to obtain more than one test sample; often three or more are taken representing the first, middle, and last portions of the heat. These samples are used to survey the uniformity of the heat and for control purposes.

lag *1.* A relative measure of the time delay between two events, states, or mechanisms. *2.* In control theory, a transfer function term in the form, 1/(Ts+1). *3.* The dynamic characteristic of a process giving exponential approach to equilibrium. *See also* time lag.

lagging *1.* In an a-c circuit, a condition where peak current occurs at a later time in each cycle than does peak voltage. *2.* A thermal insulation, usually made of rock, wood and magnesia plaster, that is used to prevent heat transfer through the walls of process equipment, pressure vessels, or piping systems.

Lagrangian fit A computer technique used to interpolate polynomial curves generated from a set of data points (calibration points). N data points are required to generate a curve to N − 1 deg. A feature of this technique is that the interpolated curve goes through each data point exactly.

lambda *See* failure rate.

lambert A unit of luminance equal to 1 candela/cm^2, which is equal to 1 stilb. A one-centimeter square surface, all points of which are one centimeter from a one candela source, has a luminance of one lambert.

Lambert's cosine law The radiance of certain surfaces, known as Lambertian reflectors, Lambertian radiators, or Lambertian sources, is independent of the angle from which the surface is viewed.

Lambert's law Law stating that the light emitted in a given direction from a perfectly diffusing surface has an intensity that varies as the cosine of the angle between the emitted ray and the normal to the surface. Such a surface is said to obey Lambert's Law. Since projected area also varies as the cosine of the angle, the luminous intensity per projected area of luminance is constant for all angles. *See also* Bouguer law.

lamella Crystalline materials whose grains are in the form of thin sheets.

lamellar corrosion Localized subsurface corrosion in zones parallel to the surface which results in thin layers of uncorroded metal resembling the pages of a book. Synonymous with exfoliation corrosion.

lamellar pearlite Laminations of layers of ferrite and carbide (cementite) generally found in plain carbon and low alloy steels in the as-rolled (not heat treated) condition. In this condition, tool steel is generally fairly hard, and difficult to machine or to cold form.

lamellar structure A material structure made up of thin plates.

lamellar thickness A morphological dimension showing the average thickness of lamella in a specimen.

lamina *1.* A single layer in a polymer laminate. *2.* A surface containing woven fibers in a metal matrix composite.

laminar flow Fluid flow characterized by the gliding of fluid layers (laminae) past one another in orderly fashion. *See also* Poiseuille flow, streamline flow, and viscous flow.

laminar flow element (LFE) A flow rate measuring device in which there is a linear relationship between flow rate and pressure drop.

laminate *1.* A metal composite with two or more bonded layers. *2.* A polymer composite with one or more reinforcement fibers bonded in a resin.

laminate coordinates A system used to describe the properties of a laminate; a reference coordinate method using the direction of the principle axis.

laminated board A general term describing a board composed of two or more single plies of board, paper, or other sheet materials in any combination, firmly adhered to each other by means of an adhesive between the plies. The adhesion and cohesion of the entire finished structure are such that it will function as a single unit.

laminated glass Two or more pieces of sheet, plate, or float glass bonded together by an intervening layer or layers of plastic material. It will crack and break under sufficient impact, but the pieces of glass tend to adhere to the plastic. If a hole is produced, the edges are likely to be less jagged than would be the case with ordinary annealed glass.

laminated materials *See* laminate.

laminated tape A tape consisting of two or more layers of different materials bonded together.

laminate orientation The angular configuration of cross-plied laminate.

laminate ply One reinforcement layer bonded by resin in a laminate.

laminations Defects aligned parallel to the surface of metal, usually in sheet form, resulting from pipe, blisters, seams, inclusions, or segregation elongated and made directional by working.

lampblack Carbon soot used to produce tungsten carbide and titanium carbide metal powders, and as a filler in plastic materials.

LAN (computer networks) *See* local area networks.

Landau damping The damping of a space charge wave by electrons that move at the phase velocity of the wave and gain energy transferred from the wave.

lanthanum A rare earth element.

Lanz Perlit iron A low-carbon and silicon iron used for higher-quality castings.

lap A variation of a seam caused by folding over hot metal, as in the case of a fin or sharp corner; it may result from improper rolling practices or, in the case of forging, the metal being folded over, but failing to weld into a single piece.

Laplace transform In control theory, a mathematical method for solution of differential equations.

lapping Smoothing or polishing a surface by rubbing it with a tool made of cloth, leather, plastic, wood, or metal in the presence of a fine abrasive.

lapping in A process of mating contact surfaces by grinding and/or polishing.

lap weld A lap joint made by welding.

large core fiber An optical fiber with a comparatively large core, usually a step index type. There is no standard definition of "large" but for the purposes here, diameters of 400 micrometers or more are designated as "large."

large-scale integration (LSI) *1.* The simultaneous achievement of large-area circuit chips and high density of component packaging for the express purpose of cost reduction by maximization of the number of system interconnections made at the chip level. *2.* Monolithic digital ICs with a typical complexity of 100 or more gates or gate-equivalent circuits. The number of gates per chip used to define LSI depends on the manufacturer. The term sometimes describes hybrid ICs built with a number of MSI or LSI chips.

largest particle passed test (absolute particle retention rating) A form of particle retention rating test in which the largest hard particle that will pass through a filter under defined test conditions is determined. The term "largest particle passed" is intended to indicate the largest opening in a filter.

Larmor radius For a charged particle moving transversely in a uniform magnetic field, the radius of curvature of the projection of its path on a plane perpendicular to the field.

laser Device for producing light by emission of energy stored in a molecular or atomic system when stimulated by an input signal (from light amplification by stimulated emission of radiation).

laser annealing Rapid heating of metals and/or alloys with the use of lasers.

laser-beam welding A joining process using the heat from a laser beam.

laser cutting The cutting of material by means of lasers.

laser diode array A device in which the output of several diode lasers is brought together in one beam. The lasers may be integrated on the same substrate, or discrete devices coupled optically and electronically.

laser glass An optical glass doped with a small concentration of a laser material. When the impurity atoms are excited by light, they are stimulated to emit laser light.

laser microscopy The application of a laser microscope having a ceramic tube in which a metal vapor is formed at 1600°C. Copper (or other metal atoms) are excited and amplify light so that when used with a projection microscope, the object to be magnified is illuminated. The power of the emitted beam on the screen remains constant.

laser plasma interactions The results of the actions of laser beams on electrically ducting fluids, such as plasmas or ionized gases.

laser simulator A light source that simulates the output of a laser. In practice, the light source is a 1.06-micrometer LED which simulates the output of a neodymium laser at much lower power levels.

laser spectroscopy The use of lasers for spectroscopic analysis; particularly in Raman spectroscopy.

laser stability Characteristic of a laser beam free from oscillations.

laser welding Microspot welding with a laser beam.

LAS glass ceramics Lithia-alumina-silica glasses with very low thermal expansion and good resistance to thermal shock.

Lason-Miller parameter A system used to predict creep behavior.

latching digital output A contact closure output that holds its condition (set or reset) until changed by later execution of a computer program. *See* momentary digital output.

latching relay Real device or program element that retains a changed state without power. In a computer program element, the power removal is only in terms of the logic power expressed in the diagram. Real power removal will affect the PC outputs according to some scheme provided by the manufacturer.

latent curing agent A non-acid curing agent that provides stability at ambient temperature, and fast curing action at elevated temperatures.

latent heat Heat that does not cause a temperature change.

latent heat of fusion *See* heat of fusion.

latex A dispersion of rubber or plastic in solution.

lattice defects Dislocations in the geometric regularity of a crystal which can affect the material strength.

lattice diffusion Diffusion of atoms or vacancies in a crystal lattice, rather than along grain boundaries.

lattice network An electronic network composed of four branches connected end-to-end to form a mesh, and in which two nonadjacent junctions are the input terminals and the two remaining nonadjacent junctions are the output terminals.

lattice parameter In crystallography, the length of any side of the unit cell in a given space lattice; if the sides are unequal, all unequal lengths must be specified.

Laue diffraction X-ray diffraction used to determine crystal structures.

Lauritsen electroscope An electroscope in which the sensitive element is a metallized quartz fiber.

lawrencium A transuranic radioactive element.

lay The axial length of a turn of the helix made by a helical element of fibers, wires, or roving.

layered laminate Laminate in which two or more plies, either of the same or different materials, are bonded and stacked one on top of the other to act as a single structural layered element.

layer lattice A crystal structure with strong-bonded atoms in parallel planes, and weak-bonded layers in successive planes.

lay-up Production of reinforced plastics by positioning the reinforced material (such as glass) in the mold prior to impregnation with resin.

lb *See* pound.

LCD *See* liquid crystal display.

LCD display A passive display whose operating principle is the change in light transmission or polarization in a liquid crystal under the influence of an electric field.

LCF *See* low-cycle fatigue.

LCP *See* liquid crystal polymer.

LDPE Low-density polyethylene.

lead A soft metallic element with high density and a low melting point.

lead equivalent The radiation-absorption rating of a specific material expressed in terms of the thickness of lead that reduces radiation dose an equal amount under given conditions.

leading A condition in which peak current occurs earlier in each cycle than does peak voltage.

leading edge The first transition of a pulse going in either a positive (high) or a negative (low) direction.

lead-tin alloys Alloys of lead, tin, and other metals used in the manufacture of low-pressure bearings.

lead zirconate titanates Dense ceramics with high piezoelectric coefficients and a high relative permittivity.

leakage *1.* A type of streamline flow most often observed in viscous fluids near solid boundaries, characterized by the tendency for fluid to remain in thin, parallel layers to maintain uniform velocity. *2.* A nonturbulent flow regime in which the stream filaments glide along the pipe axially with essentially no transverse mixing. Also known as viscous or streamline flow. *3.* Flow under conditions in which forces due to viscosity are more significant than forces due to inertia. *4.* Undesirable loss or entry across the boundary of a system.

leakage bubble test A method of test by which the fabrication integrity of an element may be determined. Air pressure is applied gradually to the inside of a filter element, which is submerged in a specified liquid, until the first positive stream of bubbles appears. The pressure is recorded and the point or origin of the bubbles is observed.

leakage rate The amount of leakage across a defined boundary per unit time.

leak detector An instrument such as a helium mass spectrometer used for detecting small cracks or fissures in a vessel wall.

least count In testing, the smallest change that can be determined.

least significant bit (LSB) The smallest bit in a string of bits, usually at the extreme right.

least significant digit (LSD) The rightmost digit of a number.

least squares method Any statistical procedure that involves minimizing the sum of squared differences.

LED An indivisible, discrete light source unit containing a semiconductor junction in which visible light is nonthermally produced when a forward current flows as a result of applied voltage.

ledeburite The eutectic in the iron-carbon system.

Ledoux bell meter A type of manometer whose reading is directly proportional to flow rate sensed by a head-producing measuring device such as a pitot tube.

leg *1.* One of the members of a branched object or system. *2.* The distance between the root of a fillet weld and the toe. *3.* Any structural member that supports an object above the horizontal.

length A fundamental measurement of the distance between two points, measured along a straight or curved path.

leptons In the classification of subatomic particles according to mass, the lightest of all particles; examples of leptons are the electron and positron.

let-go In laminated glass, an area where adhesion has been lost.

level Values of a factor being examined in an experiment. If the process is known to be linear, then a two-level experiment may be adequate. For a nonlinear process a three-level design may be needed. Some test designs allow for mixed level testing.

level (logic) A signal that remains at the "0" or "1" level for long amounts of time.

levigation *1.* Separation of finer powders in a mix by forming a suspension of the fine powders in a liquid. *2.* Determining particle size by the settling rate in a suspension.

levitation melting A metallurgical process in which a piece of metal placed above a coil carrying a high-frequency current can be supported against gravity by the Lorentz force caused by the induced surface currents in the metal. At the same time, the heat produced by Joule dissipation melts the metal.

lexical analysis In data processing, a stage in the compilation of a program in which statements, such as IF, AND, END, etc. are replaced by codes.

LF Low frequency.

LFE *See* laminar flow element.

Li Chemical symbol for lithium.

Lichtenberg figure camera A device for indicating the polarity and approximate crest value of a voltage surge.

life test A laboratory procedure used to determine the period of operation during which a component or assembly will operate until it no longer performs its intended function.

ligand The ion, molecule, or group attached to a central atom in a chelate.

light Visible radiation (about 400 to 700 nm in wavelength), considered in terms of its luminous efficiency, i.e., evaluated in proportion to its ability to stimulate the sense of sight.

light anneal copper An imprecise term used to indicate the formability of cold-rolled and annealed products. The use of this term is discouraged. The desired product is properly described as "fully recrystallized; grain size 0.015–0.035 mm."

light-beam galvanometer A type of sensitive galvanometer whose null-balance point is indicated by the position of a beam of light reflected from a mirror carried in the moving coil of the instrument. Also known as d'Arsonval galvanometer.

light-beam instrument A measurement device that indicates measured values by means of the position of a beam of light on a scale.

light crude Crude oil rich in low-viscosity hydrocarbons of low molecular weight.

light intensity *See* luminous efficiency.

light ions Ions of helium, boron, and other elements used in implantation experiments.

light metals Low-density metals such as aluminum, magnesium, titanium, beryllium, or their alloys.

light oil Any oil whose boiling point is in the temperature range 110 to 210°C, especially a coal tar fraction obtained by distillation.

light pen A device by which an individual can communicate with an information system through a cathode ray tube.

light water Water in which both hydrogen atoms in each molecule are of the isotope protium.

lignite Coal of relatively recent origin; an intermediate between peat and bituminous coal.

likelihood ratio The probability of a random drawing of a specified sample from a population, assuring a given hypothesis about the parameters of the population, divided by the probability of a random drawing of the same sample, assuring that the parameters of the population are such that this probability is maximized.

limit The extreme of the designated range through which the measured value of characteristics may vary and still be considered acceptable.

limit-check The comparison of data from a specific source with pre-established allowable limits for that source.

limit checking Internal program checks for high, low, rate-of-change, and deviation from a reference. These checks are to detect signals indicating undesirable or unsafe plant operation.

limit control A sensing device that shuts down an operation or terminates a process step when a prescribed limiting condition is reached.

limit cycle A sustained oscillation of finite amplitude.

limit frequency The lowest frequency that can be used in eddy-current testing; this value is dependent on the conductivity and size of the component being tested.

limiting oxygen index (LOI) A test system used to determine the flammability of a polymer.

limit of detection In any instrument or measurement system, the smallest value of the measured quantity that produces discernible movement of the indicator.

limit of error In an instrument or control device, the maximum error over the entire scale or range of use under specific conditions.

limit of measurement In any instrument or measurement system, the smallest value of the measured quantity that can be accurately indicated or recorded.

limit of proportionality That stress beyond which the relationship of stress and strain are no longer linear.

limit priority A priority specification associated with every task in a multitask operation, representing the highest dispatching priority that the task may assign to itself or to any of its subtasks.

limits The set of values that describe the established boundaries of acceptance. The same characteristic may have several sets of limits, depending on the time or basis of establishing acceptance (for example, new part limits, overhaul limits, and safety limits).

limits, confidence The values, upper and lower, between which a true value can be expected to fall, with a pre-established level of confidence.

lin *See* linear.

Lindemann electrometer An electrometer in which a metallized quartz fiber mounted on a quartz torsion fiber perpendicular to its axis is positioned within a system of electrodes to produce a visual indication of electric potential.

line In process plants, a collection of one or more associated units and equipment modules, arranged in serial and/or parallel paths, used to make a complete batch of material or finished product.

lineal scale length The distance from one end of an instrument scale to the other, measured along the arc if the scale is curved or circular.

linear The type of relationship that exists between two variables when the ratio of the value of one variable to the corresponding value of the other is constant over the entire range of possible values.

linear absorption coefficient A characteristic of a material related to its ability to absorb radiation.

linear (or nonlinear) analysis Method for analyzing nonlinear systems; a linear approximation of the nonlinear equations is made about a specific point.

linear behavior Motions when the vehicle response has the following (linear) properties: superposition holds, the proportionality between input and output is independent of the input amplitude, the output cannot contain components at frequencies not present in the input, and the question of stability is unaffected by the nature of the input function and the initial conditions.

linear elastic fracture mechanics A method of determining the load required to create fracture in a material with crack flaws of specific size and shape.

linear expansion The increase of a given dimension, measured by the expansion or contraction of a specimen or component subject to a thermal gradient or changing temperature.

linear expansion coefficient The change in dimension of a test specimen subject to varying temperatures.

linearity *1.* The relationship between two quantities when a change in a second quantity is directly proportionate to a change in the first quantity. Also, deviation from a straight-line response to an input signal. *2.* The closeness with which a curve of a function approximates a specified time. *3.* The degree to which the normal output curve conforms to a straight line under specified load conditions.

linearity, differential Expresses the difference of the measured output value of adjacent digital codes from the ideal. Any two adjacent digital codes should result in measured output values that are exactly 1 LSB apart. Any deviation of the measured "step" from the ideal difference is called differential nonlinearity, expressed in multiples of 1 LSB.

linearity error The ratio, in percent, of the maximum difference between the calibration value and the corresponding value read on the straight line defined in sensitivity coefficient at the upper limit of the channel amplitude class (data channel full scale).

linearization The process of converting a nonlinear (nonstraight-line) response into a linear response.

linear meter An instrument whose indicated output is proportional to the quantity measured.

linear optimization *See* linear programming.

linear polarization Light in which the electric field vector points in only a single direction.

linear polymers Thermoplastics with chain patterns that are not crosslinked.

linear position sensing detector An optical detector that can measure the position of a light spot along its length.

linear potentiometer A variable resistance device whose effective resistance is a linear function of the position of a control arm or other adjustment.

linear programming (LP) A method of solution for problems in which a linear function of a number of variables is subject to a number of constraints in the form of linear inequalities.

linear quadratic regulator A type of optimal state feedback controller that does not consider noise; primarily used to control aircraft and spacecraft. *See also* linear regulator.

linear regulator Refers to a special case used for minimization of the performance index to obtain optimal performance. *See also* linear quadratic regulator.

linear strain The percent change in dimension due to applied load on a test specimen.

linear variable differential transformer (LVDT) A type of position sensor consisting of a central primary coil and two secondary coils wound on the same core; a moving-iron element linked to a mechanical member induces changes in self induction that are directly proportional to movement of the member.

linear variable reluctance transducer (LVRT) A type of position sensor consisting of a center-tapped coil and an opposing moving coil attached to a linear probe; the winding is continuous over the length of the core, instead of being segmented as in an LVDT.

linear velocity A vector quantity whose magnitude is expressed in units of length per unit time and whose direction is invariant; if the direction varies in circular fashion with time, the quantity is known as angular or rotational velocity; and if it varies along a fluctuating or noncircular path the quantity is known as curvilinear velocity.

line pair In spectroscopy, the standard line and the analytical line that is compared to the standard line.

line spectra The spontaneous emission of electromagnetic radiation from the bound electrons as they jump from high to low energy levels in an atom.

liquation The partial melting of a material.

liquation temperature The lowest temperature at which the melting of a material will commence.

liquefied compressed gas A gas that, under the charging pressure, is partially liquid at a temperature of 70°F (21°C).

liquid chromatography A separation technique based on sample distribution between a stationary phase and a liquid phase.

liquid crystal display A type of digital display device.

liquid crystal light valve A device used in optical processing to convert an incoherent light image into a coherent light image.

liquid crystal polymer (LCP) A thermoplastic polymer primarily made up of benzene ring chains with excellent tensile strength and heat resistance.

liquid erosion failure See cavitation erosion.

liquid H See polyimide.

liquid knockout See impingement.

liquid-liquid chromatography (LLC) Liquid chromatography in which the stationary phase is a liquid dispersed into some inert material.

liquid metal embrittlement The decrease in ductility of a metal caused by contact with a liquid metal.

liquid metal infiltration The immersion of fibers in molten metal to form a metal matrix composite.

liquid oxygen A light blue, magnetic, transparent or water-like fluid that is produced by the fractional distillation of purified liquid air. When cooled to -182.9°C (at 14.7 psia), oxygen passes from the gaseous to the liquid state.

liquid-partition chromatography (LPC) See liquid-liquid chromatography.

liquid penetrant testing Nondestructive testing in which a penetrating dye is introduced to potential surface discontinuities, then a developing agent is added that will show the flaws.

liquid phase epitaxy A liquid phase transformation during crystal growth.

liquid phase sintering Sintering of a compact in which a liquid phase is present at some point in the sintering cycle.

liquid plus solid zones See mushy zones.

liquid resin An organic, polymeric liquid that becomes a solid when converted to its final state for use.

liquids Substances in a state in which the individual particles move freely with relation to each other and take the shape of the container, but do not expand to fill the container.

liquid-solid chromatography (LSC) Liquid chromatography in which silica or alumina is the stationary phase.

liquid trap *See* air bind.

liquidus *1.* The temperature at which a metal, or phase, begins to freeze on cooling or completes melting on heating. *2.* The maximum temperature at which equilibrium exists between a molten glass and its primary crystalline phase.

literal An element of a programming language that permits the explicit representation of character strings in expressions and command and function elements.

lithium A soft, low-melting and lightweight metal used as an alloying element.

lithium iodates Salts of iodic acid containing the 10 to the third power radical.

lithium oxide Known as lithia, it is used in the manufacture of refractories and lithium-alumina-silica (LAS) glass-ceramics.

litmus A blue, water-soluble powder derived from lichens and used as an acid-base indicator; it is blue at pH 8.3 and above, and is red at pH 4.5 and below. A pH of 7 is considered neutral.

litre Also spelled liter. Abbreviated L. The SI unit of volume; it equals 0.001 m^3 or 1.057 quarts.

lixiscopes Portable lightweight battery-operated low-intensity x-ray imaging systems with medical, industrial, and scientific applications.

LLC *1. See* logical link control. *2. See* liquid-liquid chromatography.

lm *See* lumen.

load *1.* The force in weight units applied to a body. *2.* The weight of the contents of a container or transportation device. *3.* A qualitative term denoting the contents of a container. *4.* The element or circuit driven by the output of a device or circuit. *5.* To store a computer program or data into memory. *6.* To mount a magnetic tape on a device so that the read point is at the beginning of the tape. *7.* To place a removable disk in a disk drive and start the drive. *8.* The amount of force applied to a structural member in service. *9.* The quantity of parts placed in a furnace, oven, or other piece of process equipment. *10.* The power demand on an electrical distribution system. *11.* The amount of power needed to start or maintain motion in a power-driven machine. *12.* (process load) A term to denote the nominal values of all variables in a process that affect the controlled variable. *13.* In a physical structure, the externally applied force, or the sum of external forces and the weight of the structure borne by a single member or by the entire structure.

load cell A transducer for the measurement of force or weight. Action is based on strain gages mounted within the cell on a force beam.

load deflection curve A curve in which loads are plotted on the ordinate axis, and the deflections plotted on the abscissa axis.

load, dynamic A load imposed by dynamic action, as distinguished from a static load.

load factor The ratio of the average load in a given period to the maximum load carried during that period.

load impedance A measure of total opposition to current flow in an alternating-current circuit.

load pressure The pressure reacting to a static or dynamic load.

load, ultimate The load that should cause destructive failure according to stress analysis; the load that causes failure during a test of strength.

local area network (LAN) A communications mechanism by which computers and peripherals in a limited geographical area can be connected. They provide a physical channel of moderate to high data rate (1 to 20 Mbit) which has a consistently low error rate (typically 10^{-9}).

local cell A galvanic cell resulting from inhomogeneities between areas on a metal surface in an electrolyte. The inhomogeneities may be of physical or chemical nature in either the metal or its environment.

local effect The impact of the failure mode on the item that is being analyzed.

localization level The functional level to which a failure can be traced or located without using accessory test equipment.

localized corrosion Corrosion resulting in differential attack across a metallic surface. Localized corrosion is typically of a cosmetic nature, but can lead to catastrophic failure if a primary structural component is the site of severe attack.

localized precipitation Similar to continuous precipitation, except the precipitating particles form at specific locations, such as grain boundaries.

local processing unit Field station with input/output circuitry and the main processor. These devices measure analog and discrete inputs, convert these inputs to engineering units, perform analog and logical calculations (including control calculations) on these inputs, and provide both analog and discrete (digital) outputs.

location counter 1. In data processing, the control-section register that contains the address of the instruction currently being executed. 2. A register in which the address of the current instruction is recorded. Synonymous with instruction counter and program address counter.

locking Pertaining to code extension characters that change the interpretation of an unspecified number of following characters. Contrast with nonlocking.

lockout Any condition that prevents any or all senders or receivers from communicating.

locus of failure Site of failure.

log 1. A record of everything pertinent to a machine run including: identification of the machine run, record of alteration switch settings, identification of input and output tapes, copy of manual key-ins, identification of all stops, and a record of action taken on all stops. 2. To record occurrences in a chronological sequence.

logarithm The power to which a fixed number, called the base, usually 10 or e (2.7182818), must be raised to produce the value to which the logarithm corresponds.

logarithmic amplifier An amplifier whose output is a logarithmic function of its input.

logarithmic decrement In an exponentially damped oscillation, the natural logarithm of the ratio of one peak value to the next successive peak value in the same direction.

logger A device that automatically records physical processes and events, usually chronologically.

logic 1. A mathematical approach to the solution of complex situations by the use of symbols to define basic concepts.

In computers and information-processing networks, the systematic method that governs the operations performed on the information, usually with each step influencing the one that follows. *2.* A means of solving complex problems through the repeated use of simple functions that define basic concepts. Basic logic functions are "AND," "OR," "NOT," etc. *3.* The science dealing with the criteria or formal principles of reasoning and thought. *4.* The systematic scheme that defines the interactions of signals in the design of an automatic data processing system. *5.* The basic principles and application of truth tables and interconnection between logical elements required for arithmetic computation in an automatic data processing system. Related to symbolic logic.

logical block An arbitrarily-defined, fixed number of contiguous bytes used as the standard I/O transfer unit throughout a computer operating system. For example, the commonly used logical block in PDP-11 systems is 512 bytes long. (PDP-11 is a digital minicomputer manufactured by Digital Equipment Corporation [DEC].) An I/O device is treated as if its block length is 512 bytes, although the actual (physical) block length of a device may be different. Logical blocks on a device are numbered from block 0 consecutively up to the last block on the volume.

logical link control (LLC) The upper sublayer of the data link layer (Layer 2) used by all types of IEEE 802 local area networks. LLC provides a common set of services and interfaces to higher layer protocols. Note: IEEE is a U.S. trade association of electronics manufacturers. It has set up a number of special groups to define standards for telecommunications. Group 802, for example, originates standards for local area networks.

logical operation *1.* An operation in which logical (yes or no) quantities form the elements being operated on, for example AND, or OR. *2.* The operations of logical shifting, masking, and other nonarithmetic operations of a computer. Contrast with arithmetic operation.

logical record A logical unit of data within a file whose length is defined by the user and whose contents have significance to the user; a group of related fields treated as a unit.

logical sum A result, similar to an arithmetic sum, obtained in the process of ordinary addition, except that the rules are such that a result of one is obtained when either one or both input variables is a one, and an output of zero is obtained when the input variables are both zero.

logical unit number A number associated with a physical device unit during the I/O operations of a task; each task in the system can establish its own correspondence between logical unit numbers and physical device units.

logical variable A variable that may have only the value true or false. Also called Boolean variable.

logic analyzer A device used to analyze the logical operation of a microcomputer. It is a test device used for debugging systems.

logic device The general category of digital fluidic components which perform logic functions; for example, AND, NOT, OR, NOR, and NAND. They can gate or inhibit signal transmission with

247

the application, removal, or other combinations of control signals.

logic diagram *1.* In data processing, a diagram that represents a logic design and sometimes the hardware implementation. *2.* Graphic method of representing a logic operation or set of operations.

logic levels Electrical convention for representing logic states. For TTL systems, the logic levels are nominally 5V for logic 1, and 0V for logic 0.

LOI *See* limiting oxygen index.

London dispersion forces Very weak intermolecular attraction from transient dipole-dipole action.

long-arc xenon A xenon arc in which the length of the arc between the electrodes is greater than the diameter of the envelope enclosing the arc.

long-chain branching A type of branching in addition polymers resulting from internal transfer action.

longevity The length of useful life of a product to its ultimate wearout requiring complete rehabilitation. This is a term generally applied in the definition of a safe, useful life for an equipment or system under the conditions of storage and use to which it will be exposed during its lifetime.

longitude Angular distance, along a primary great circle, from the adopted reference point; the angle between a reference plane through the polar axis and a second plane through that axis.

longitudinal direction The principal direction of flow in a worked metal.

longitudinal wave(s) A wave in which the medium is displaced in a direction perpendicular to the wave front at all points along the wave.

longos Small-angle helical or longitudinal windings.

long terne Cold-rolled sheet steel coated on both sides with a lead-tin alloy by a continuous hot-dip process.

long transverse In mechanical testing of rectangular bars, the longer of the two transverse directions. Contrast with short transverse.

loop *1.* A sequence of instructions that is executed repeatedly until a terminal condition prevails. *2.* Synonymous with control loop. *3.* The doubled part of a cord, wire, rope, or cable; a bight or noose. *4.* A complete hydraulic, electric, magnetic, or pneumatic circuit. *5.* A length of magnetic tape or motion picture film that has been spliced together, end-to-end, so it can be played repeatedly without interruption. *6.* In data processing, a closed sequence of instructions that are repeated. *7.* All the parts of a control system: process, or sensor, any transmitters, controller, and final control element. *8.* In a computing program, a sequence of instructions that is written only once, but executes many times (iterates) until some predefined condition is met.

loose powder Metal powder that is not compacted.

loose powder sintering Sintering a metal powder with no external pressure.

Lorentz force The force affecting a charged particle due to the motion of the particle in a magnetic field.

loss *1.* Dissipation of power which reduces the efficiency of a machine or system. *2.* Dissipation of material or energy due to leakage.

loss factor In a dielectric material, the product of its dissipation factor and dielectric constant.

lossless materials Dielectric materials that do not dissipate energy or that do not dampen oscillations.

loss modulus The dissipation of energy into heat when a material is under load.

lossy media A material that dissipates energy of electromagnetic or acoustic energy passing through it.

lost cluster A group of one or more disk sectors that are not available for storage use.

lot *See* batch.

low-alloy steel An iron-carbon alloy that contains up to about 1% C, and less than 5% by weight of additional elements.

low brass A binary copper-zinc alloy containing about 20% zinc.

low-carbon steel *1.* An iron-carbon alloy containing about 0.05 to 0.25% C, and up to about 0.7% Mn. *2.* Iron alloys containing carbon in low percentages which display temper and malleability characteristics not found in ordinary carbon steels.

low-cycle fatigue Fatigue that occurs after a small number of cycles—less than 10^4 cycles.

low-energy electron diffraction A method for analyzing the atomic structure of single-crystal surfaces.

low frequency (LF) The frequency band between 30 and 300 kHz.

low-hydrogen electrode A covered welding electrode that provides an atmosphere around a welding arc that is low in hydrogen.

low-pressure laminates Laminate molded and cured at pressures below 2800 kPa.

low-profile resins Polyester resin for reinforced plastics that are made up of thermoset and thermoplastic resins to reduce shrinkage and produce a uniformly flat surface.

low-residual phosphorous copper Copper with very small amounts of residual phosphorous which are insufficient to lower conductivity.

low resolution graphics In data processing, the ability of a dot-matrix printer to reproduce simple forms or pictures.

low Reynolds number A Reynolds number below the critical Reynolds number of a sphere.

low stress grinding Grinding/polishing under controlled conditions to minimize compressive surface stresses and thereby reduce residual stresses and the likelihood of surface damage.

low-temperature hygrometry The measurement of water vapor at low temperatures.

LP *See* linear programming.

LPC *See* liquid-liquid chromatography.

LQR *See* linear quadratic regulator.

L-radiation X-rays produced when the vacancy in the L shell is replaced by an electron from another shell.

Lr Chemical symbol for lawrencium.

LSB *See* least significant bit.

LSC *See* liquid-solid chromatography.

LSD *See* least significant digit.

L-series A set of representative x-ray wavelengths for L-radiation for various elements.

L-shell The second layer of electrons surrounding the nucleus of an atom.

LSI *See* large-scale integration.

lubricant *1.* Any substance used to reduce friction between two surfaces in contact. *2.* A material added to most sizings to improve the handling and processing properties of textile strands, especially during weaving.

lubricant, mold *See* mold lubricant.

Lüder's lines (stretcher strain) *See* stretcher strain.

luggin probe A small tube or capillary filled with electrolyte and terminating close to the metal surface under study, used to provide an ionically conducting path without diffusion between an electrode under study and a reference electrode.

lumen (lm) The unit of luminous flux. It is equal to the flux through a unit solid angle from a point source of one candela. The flux on a unit surface, all points of which are at unit distance from a one candela source, is one lumen.

luminance (brightness) The luminous intensity per projected area normal to the line of observation.

luminance intensity The luminous flux emitted into a given solid angle by a source. It is measured in units of candelas.

luminescence The process whereby light (or visual radiation) is emitted at certain wavebands in excess of that which would be expected from a full radiator at the temperature of the emitter. Particular aspects of luminescence are referred to as fluorescence and phosphorescence.

luminescent intensity *See* luminous intensity.

luminosity Emissive power with respect to visible radiation.

luminosity coefficients The constant multipliers for the respective tristimulus values of any color such that the sum of the three products is the luminance of the color.

luminous Emitting radiation in the form of visible light.

luminous efficacy (K) The ratio of the luminous flux emitted by a source to the radiant power emitted by the source; has units of lumens per watt. The ratio is calculated using a standard spectral response curve for the average human observer.

luminous efficiency The ratio, expressed as a percent, of the actual luminous efficacy of a source to the maximum possible luminous efficacy of 673 lumens per watt for a monochromatic source operating at 555 nm. Luminous efficiency is also used, loosely, for the ratio of luminous flux emitted by the source to its input power in watts. When used in this manner, the units are lumens per watt instead of a percent. Therefore, it is used as a relative figure of merit for the power consumption of various displays. Typical values for current active displays are in the range of 0.05 to 0.5 lumen per watt.

luminous energy (quantity of light) The product of the luminous flux and the time maintained. It is expressed in lumen-seconds.

luminous flux The amount of light energy emitted per unit time. The unit is the lumen.

luminous flux density *See* luminous intensity.

luminous intensity The luminous flux emitted by a source per unit solid angle. The unit is the candela. Practical units for electronic displays are the milli and microcandela.

luster finish A finish produced on ground rolls suitable for decorative painting and plating with additional surface preparation after forming.

lux (lx) The SI unit of illumination equal to one lumen per square meter.

LVDT *See* linear variable differential transformer.

LVRT *See* linear variable reluctance transducer.

lx *See* lux.

Lyman alpha radiation The radiation emitted by hydrogen at 1216 angstroms.

lyotropic liquid crystal A liquid crystal polymer that can only be processed from suspension.

lysimeters Instruments for measuring the water percolating through soils and determining the materials dissolved by the water.

M

m *See* meter.

MAC See media access control.

macerate To shred fabric or fibers for use as a filler in a resin.

machinability A measure or rating of how readily a material can be machined.

machine recognition *See* artificial intelligence.

machining Removing material, in the form of chips, from a workpiece, usually through the use of a machine.

machining allowance A stated minimum amount of material (stock) to be removed before the resulting surface is expected to satisfy a defined quality. *See* allowance. *See also* cleanup.

macroetching Etching a material surface to highlight structural details, such as grain flow or cracks. Examination is typically conducted with the naked eye or low-power magnification.

macrograph A magnified reproduction of a macroetched surface.

macro modeling The representation of a component or device in terms of a netlist description of an equivalent circuit. Standard components, such as resistors or capacitors, are typically employed.

macropore Pores in metal powder compacts that can be seen without magnification.

macroprogramming Programming with macro instructions.

macroscopic stress *1.* Load per unit area distributed over an entire structure or over a visible region of the structure. *2.* Residual stress at a distance that can be detected by x-ray or dissection techniques.

macroshrinkage Voids in castings that can be detected macroscopically.

macrostrain The mean or average strain measured from a finite control volume of material with a gage length or characteristic lineal dimension that is several orders of magnitude greater than the interatomic dimensions (or characteristic length of the material microstructure). Note: Macrostrain can be measured by several methods, including electrical resistance strain gages and mechanical or optical extensometers. Elastic macrostrain can be measured by x-ray diffraction or other nondestructive techniques.

macrostructure The features of a polycrystalline metal revealed by etching and visible at magnifications of 10 diameters or less.

magma Naturally occurring mobile rock materials, generated within the earth and capable of intrusion and extrusion, from which igneous rocks are thought to have been derived by solidification and related processes.

magnesia The oxide of magnesium used in the manufacture of refractories and ceramics.

magnesite The carbonate of magnesium used in the manufacture of refractories.

magnesium A highly combustible, low-density metal.

magnet Anything that attracts; any piece of iron or steel that has the property of attracting iron. This property may be naturally present or artificially induced.

magnetically hard alloy A ferrous alloy that can be permanently magnetized.

magnetically soft alloy A ferrous alloy that loses magnetic strength when a magnetizing field is removed.

magnetic analysis testing A nondestructive test to determine discontinuities and hardness variation in ferrous alloys by measuring changes in magnetic flux.

magnetic bubble memory A high-density information storage device composed of a magnetic film only a few micrometers thick deposited on a garnet substrate; information is stored in small magnetized regions (bubbles) whose magnetic polarity is opposite to that of the surrounding region.

magnetic compass Any of several devices for indicating the direction of the horizontal component of a magnetic field, but especially for indicating magnetic north in the earth's magnetic field.

magnetic compression The force exerted by a magnetic field on an electrically conducting fluid or on a plasma.

magnetic contrast Contrasts that develop from the interaction of electrons in a beam with the magnetic fields of magnetic materials.

magnetic cooling Keeping a substance cooled to about 0.2 K by using a working substance (paramagnetic salt) in a cycle of processes between a high-temperature reservoir (liquid helium) at 1.2 K and a low-temperature reservoir containing the substance to be cooled.

magnetic core *1.* A configuration of magnetic material that is, or is intended to be, placed in a spatial relationship to current-carrying conductors and whose magnetic properties are essential to its use. *2.* A storage device in which binary data are represented by the direction of magnetization in each unit of an array of magnetic material, usually in the shape of toroidal rings, but also forms such as wraps on bobbins.

magnetic damping Progressive reduction of oscillation amplitude by means of current induced in electrical conductors due to changes in magnetic flux.

magnetic fields Regions of space in which magnetic dipoles would experience a magnetic force of torque; often represented as the geometric array of the imaginary magnetic lines of force that exist in relation to magnetic poles.

magnetic hardness comparator A device for determining hardness of a steel part by comparing its response to electromagnetic induction with the response of a similar part of known hardness.

magnetic ink An ink that contains magnetic particles. Characters printed in magnetic ink can be read both by humans and by machines designed to read the magnetic pattern.

magnetic particle display A passive display whose operating principle is the orientation of permanently magnetic particles under the influence of an applied magnetic field.

magnetic particle testing A nondestructive test to determine surface cracks and other imperfections in ferrous alloys by applying magnetic particles to a magnetized material; the particles outline a pattern of discontinuities.

magnetic recorder A device for producing a stored record of a variable electrical signal as a variable magnetic field in a ferromagnetic recording medium.

magnetic resistance *See* reluctance.

magnetic separator A machine that uses strong magnetic fields to remove pieces of magnetic material from a mixture of magnetic material and nonmagnetic or less strongly magnetic material.

magnetic shield A metal shield that insulates the contents from external magnetic fields. Such shields are often used with photomultiplier tubes.

magnetic storms Worldwide disturbances of the earth's magnetic field.

magnetic test coil A coil used in conjunction with a suitable indicating or recording instrument to measure variations or changes in magnetic flux when the coil is linked with a magnetic field.

magnetic variation The angular difference between true north and magnetic north.

magnetic variometer An instrument for measuring variations in magnetic field strength with respect to space or time.

magnet meter An instrument for measuring the magnetic flux of a permanent magnet under specified conditions; it usually incorporates a torque coil or a moving-magnet magnetometer with a unique arrangement of pole pieces.

magnetomechanics (physics) Study of the effects that the magnetization of a material and its strain have on each other.

magnetometer An instrument for measuring the size and direction of a magnetic field.

magnetostriction Changes in dimensions of a material due to exposure to a magnetic field.

magnetostrictive The property of a ferromagnetic material that causes it to change dimensions and vibrate when subjected to a high-frequency magnetic field. In ice detectors, the shift in resonant frequency is used to indicate ice buildup.

magnetostrictive effect An inherent property of some ferromagnetic materials to deform elastically, thereby generating mechanical force, when subjected to a magnetic field.

magnetostrictive resonator A ferromagnetic rod so constructed that an alternating magnetic field can excite it into resonance at one or more frequencies.

main frame The basic processing portion of a computing system, generally containing some basic storage, the arithmetic logic unit and a group of registers.

major defect A defect, other than critical, that is likely to result in failure, or to reduce materially the usability of the material or component for its intended purpose.

major graduations Intermediate graduation marks on a scale which are heavier or longer than other graduation marks but which are not index graduations.

majority A logic operator having the property that if P is a statement, Q is a statement, and R is a statement, then the majority of P, Q, R is true if more than half the statements are true, false if half or less are true.

malfunction *1.* Improper functioning of components, causing improper operation of a system. *2.* The effect of a fault.

malfunction routine Same as diagnostic routine.

malleability The characteristic that permits plastic deformation in compression without rupture.

malleable cast iron Highly annealed white cast iron with reduced cementite.

malleable iron A somewhat ductile form of cast iron made by heat-treating white cast iron to convert the carbon-containing phase from iron carbide to nodular graphite.

malleableizing A process in which the as-cast malleable type (white) iron is thermally treated for the purpose of converting most of all of the carbon in Fe_3C to graphite (temper carbon) to produce a family of products with improved ductility.

mandrel *1.* A rod used to retain the cavity in hollow metal products during working. *2.* A metal bar around which other metal may be cast, bent, formed, or shaped. *3.* A shaft or bar for holding work to be machined.

manganese An important alloying element in steel and some nonferrous metals.

Mannesmann process A process used for piercing tube billets in making seamless tubing. The billet is rotated between two heavy rolls mounted at an angle and is forced over a fixed mandrel.

manometer A gage for measuring pressure or a pressure difference between two fluid chambers. A U-tube manometer consists of two legs, each containing a liquid of known specific gravity.

manometric equivalent The length of a vertical column of a given liquid at standard room temperature which indicates a pressure differential equal to that indicated by a 1-mm-long column of mercury at 0°C.

mantissa *See* floating point. *See also* fixed point.

manual data entry module A device that monitors a number of manual input devices from one or more operator consoles and/or remote data-entry devices and transmits information from them to the computer.

map To establish a correspondence or relationship between the members of one set and the members of another set and perform a transformation from one set to another, for example, to form a set of truth tables from a set of Boolean expressions. Information should not be lost or added when transforming the map from one to another.

MAP/EPA (Manufacturing Automation Protocol/Enhanced Performance Architecture) Part of the EPA architecture, a node that contains both the MAP protocols and the protocols required for communication to MINI-MAP. This node can communicate with both MINI-MAP nodes on the same segment and full MAP nodes anywhere in the network.

mapped system A system that uses the computer hardware memory management unit to relocate virtual memory addresses.

MAP/TOP Manufacturing Automation Protocol/Technical Office Protocol.

maraging A precipitation hardening treatment applied to a special group of iron-base alloys to precipitate one or more intermetallic compounds in a matrix of essentially carbon-free martensite.

Note: The first developed series of maraging steels contained, in addition to iron, more than 10% nickel and one or more supplemental hardening elements. In this series, the aging is done at about 900°F.

maraging steels A family of high-strength steels that are easily worked and machined in a solution-treated condition.

marginal check A preventive-maintenance procedure in which certain operating conditions are varied about their normal values in order to detect and locate incipient defective units, for example, supply voltage or frequency may be varied. Synonymous with marginal test and high-low bias test. *See also* check.

marginal test Same as marginal check.

margin testing Testing in which item environments such as line voltage or temperature are changed to reversibly worsen the performance. Its purpose is to find how much margin is left in the item for its degradation.

marking pointer An adjustable stationary pointer, usually of a color different from that of the indicating pointer, which can be positioned opposite any location on the scale of interest to the user.

mark-sense To mark a position on a punch card with an electrically conductive pencil, for later conversion to machine punching.

mark-sense device An electronic machine that will read mark-sensed forms.

mark sensing A technique for detecting special pencil marks entered in special places on a punch card and automatically translating the marks into punched holes. *See also* sensing, mark.

marstraining A treatment for steel to increase fatigue strength by applying a small amount of plastic strain followed by low-temperature aging.

martempering *1.* A hardening procedure in which an austenitized ferrous workpiece is quenched into an appropriate medium whose temperature is maintained substantially at the M_s of the workpiece, held in the medium until its temperature is uniform throughout, but not long enough to permit bainite to form, and then cooled in air. The treatment is frequently followed by tempering. *2.* When the process is applied to carburized material, the controlling M_s temperature is that of the case. This variation of the process is frequently called marquenching.

martensite A supersaturated solid solution of carbon in iron characterized by an acicular (needle-like) microstructure. *See also* alpha prime.

martensite range The temperature interval between M_s and M_f (M_f and M_s are defined under transformation temperature).

martensitic transformation A phase transformation occurring on cooling in some metals and resulting in formation of martensite.

Martin's diameter (of a particle) *See* particle diameter (Martin's diameter).

mask *1.* A protective face covering that usually provides for filtration of breathing air or for attachment to an external supply of breathing air. *2.* A frame or similar device to prevent certain areas of a workpiece surface from being coated, as a paint. *3.* A frame that conceals the edges of a cathode-ray tube, for example, a television screen. *4. See* filter.

masking *1.* The process of extracting a nonword group or a field of characters from a word or a string of words. *2.* The process of setting internal program controls to prevent transfers that otherwise would occur upon setting of internal machine latches.

mass *1.* A quantity of matter. In the earth's gravitational field, 1 kg of mass weighs about 9.8 N (in the metric, SI system), and 1 slug of mass weighs about 32 lb (in the traditional U.S. system). *2.* A quantity characteristic of a body, which relates the attraction of this body toward another body. Since the mass of a body is not fixed in magnitude, all masses are referred to the standard kilogram, which is a lump of platinum. *3.* Amount of matter an object contains.

mass center The point in a free rigid body at which, if a force is applied, only translational acceleration results. Conversely, if a force is applied to a free body at any point other than the mass center, a rotational acceleration will occur, also, due to the resulting moment about the mass center.

mass flow The amount of fluid, measured in mass units, that passes a given location or reference plane per unit time.

mass number The sum of the number of protons and the number of neutrons in the nucleus of a specific nuclide.

mass spectrograph A mass spectroscope that records intensity distributions on a photographic plate.

mass spectrometer A mass spectroscope that uses an electronic instrument to indicate intensity distribution in the separated ion beam.

mass spectrometry An analytical method to identify chemical structures, analyze mixtures, and perform quantitative analysis.

mass spectroscope An instrument for determining the masses of atoms or molecules, or the mass distribution of an ion mixture, by deflecting them with a combination of electric and magnetic fields which act on the particles according to their relative masses.

mass spectrum In a mixture of ions, the statistical distribution by mass or by mass-to-charge ratio.

mass storage Pertains to a computer device that can store large amounts of data so that they are readily accessible to the central processing unit; for example, disks, Digital Equipment Corporation (DEC) tape, or magnetic tape.

mass velocity Mass flow per unit cross-sectional area.

master *1.* A device that controls other devices in a system. *2.* A precise pattern for making replicate workpieces, as in certain types of casting processes.

master standard The appropriately identified engineering-approved standard sample against which specified like materials are evaluated.

mat Chopped fibers bonded by an adhesive, used for plastic reinforcement.

material certification The certificate of test or conformance issued by a laboratory.

material dispersion Light pulse broadening due to differential delay of various wavelengths of light in a waveguide material. This group delay is aggravated by broad band width light sources.

material noise Extraneous signals caused by the structure of the material being tested.

material specification The document(s) describing which tests are to be performed and the conformance limits required/expected, and (sometimes) the test specification(s) to be used.

materials recovery The treatment of a material to reclaim one or more of its components.

materials science The study of materials used in research, construction, and manufacturing; includes the fields of metallurgy, ceramics, plastics, rubber, and composites.

mathematical check A check that uses mathematical identities or other properties, occasionally with some degree of discrepancy being acceptable; for example, checking multiplication by verifying that $A \times B = B \times A$. Synonymous with arithmetic check.

mathematical logic Same as symbolic logic.

mathematical model The general characterization of a process, object, or concept, in terms of mathematics, which enables the relatively simple manipulation of variables to be accomplished in order to determine how the process, object, or concept would behave in different situations.

mathematical programming In operations research, a procedure for locating the maximum or minimum of a function subject to constraints. Contrast with convex programming, dynamic programming, integer programming, linear programming, nonlinear programming, and quadratic programming.

matrix *1.* In materials, the principal phase or aggregate in which another constituent is embedded. *2.* The binding agent in a composite or agglomerated mass. *3.* In mathematics, a two-dimensional rectangular array of quantities. Matrices are manipulated in accordance with the rules of matrix algebra. *4.* In computers, a logic network in the form of an array of input leads and output leads with logic elements connected at some of their intersections. *5.* By extension, an array of any number of dimensions.

matrix material The ingredients used as binding agents to produce composite materials.

matrix printer A type of computer device that forms letters and symbols by printing a pattern of dots.

matte *1.* A smooth, but relatively nonreflective, surface finish. *2.* An intermediate product in the refining of sulfide ores by smelting.

matte finish The texture produced on sheets by rolls that have been blasted to various degrees of roughness depending upon the end use.

maturing temperature With regard to composites, the temperature that produces a desirable bond in a specific time.

maximum continuous load The maximum load that can be maintained for a specified period.

maximum stress (S_{max}) The stress having the highest algebraic value in the stress cycle, tensile stress being considered positive, and compressive stress negative. In this definition, the nominal stress is used most commonly.

maximum vehicle test weight Either (a) the gross vehicle weight rating (GVWR) of the vehicle or (b) the maximum legal load at the test location, whichever is lower.

Maxwell The CGS unit of magnetic flux.

Maxwellian distribution The velocity distribution of the moving molecules of a gas in thermal equilibrium, as determined by applying the kinetic theory of gases.

Mayer index A hardness index derived from Brinell hardness tests using a uniform size indenter with variable loads.

MB Megabyte; millions of bytes.

Mbit Million bits per second.

MBM *See* magnetic bubble memory.

MBM junctions (metal-barrier-metal junctions) Diode devices using metal-barrier-metal layers.

Mbyte 1,048,576 (2^{20}) bytes.

McQuaid-Ehn test A test to reveal grain size after heating into the austenitic temperature range. Eight standard grain sizes rate the microstructure; No. 8 is finest and No. 1 is coarsest.

mean *1.* The expected value of a random variable. *2.* The first moment of a probability distribution about its origin. As specifically defined and modified, for example, the arithmetic mean (sums), the geometric mean (products), the harmonic mean (reciprocals), logarithmic mean, etc.

mean free path *1.* Of any particle, the average distance that a particle travels between successive collisions with the other particles of an ensemble. *2.* In a gas, liquid, or colloid, the average distance traveled by an individual atom, molecule, or particle between successive collisions with other particles.

mean square values In statistics, values representing the average of the sum of the squares of the deviations from the mean value.

mean stress The algebraic mean of maximum and minimum stress in one loading cycle.

mean-time-to-failure (MTTF) The average or mean-time between initial operation and the first occurrence of a failure or malfunction, as the number of measurements of such time on many pieces of identical equipment approaches infinity.

measurand A physical quantity, property, or condition that is measured.

measure *1.* The quantity or capacity of anything measured. *2.* A unit specified by a scale. *3.* A Device used for measuring. *4.* The act of measuring. *5.* A reference used for quantitative comparison of properties.

measured variable *1.* The physical quantity, property, or condition that is to be measured. Common measured variables are temperature, pressure, rate of flow, thickness, speed, etc. *2.* The part of the process that is monitored to determine the actual condition of the controlled variable.

measured variable modifier The second letter when first-letter is used in combination with modifying letters D (differential), F (ratio), M (momentary), K (time rate of change), Q (integrate or totalize), (A could be used for absolute), or any combination of these is intended to represent a new and separate measured variable, and the combination is treated as a first-letter entity.

measurement *1.* The act of measuring or being measured. *2.* The quantity or extent ascertained by measuring—capacity, size, area, contents, bulk. *3.* A dimension.

measurement system Any set of interconnected components, including one or more measurement devices, that performs a complete measuring function, from initial detection to final indication, recording, or control-signal output.

mechanical hysteresis Energy absorbed in a material in a complete cycle of loading and unloading within the elastic limit.

mechanical properties The properties of a material that reveal its elastic and plastic behavior when force is applied. Examples include tensile test properties, impact test properties, and fracture mechanics properties.

mechanical scale A weighing device in which objects are balanced through a system of levers against a counterweight or counterpoise.

mechanical tubing Tubing intended primarily for structural application and not normally used for the transmission of fluids.

media *1.* The physical interconnection between devices attached to the local area network (LAN). Typical LAN media are twisted pair, baseband coax, broadband coax, and fiber optics. *2.* The plural of medium. *3.* A name for the various materials used to hold or store electronic data, such as printer paper, disks, magnetic tape, or punched cards.

media access control (MAC) The lower sublayer of the data link layer (Layer 2) unique to each type of IEEE 802 local area networks. MC provides a mechanism by which users access (share) the network.

medium-carbon steel An alloy of iron and carbon containing about 0.25 to 0.6% C, and up to about 0.7% Mn.

medium-scale integration (MSI) *1.* A medium level of chip density, lower than for large-scale integration (LSI) circuits,

but more than small-scale integration. *2.* An integrated circuit with 10 to 100 logic gates.

mega Prefix denoting 1,000,000.

megabit One million bits.

megabyte (Mbyte) A unit of computer memory size. One million bytes. *See* Mbyte.

megahertz One million hertz or cycles per second.

megapixel Millions of graphics display picture elements.

Meissner effect *See* superconductivity.

melamine plastics Thermosetting plastics derived from melamine and formaldehyde resins used to produce "unbreakable" products.

melt extrusion An extrusion process in which insulation material is heated above its melting point and forced through a die.

melt index Extrusion rate of a thermoplastic material through an orifice of specified diameter and length under specified condition of time, temperature, and pressure.

melting point The temperature at which a solid substance becomes liquid; for pure substances and some mixtures, it is a single unique temperature; for impure substances, solutions, and most mixtures, it is a temperature range.

melts (crystal growth) Molten substances from which crystals are formed during the cooling or solidifying process.

melt spinning A material process by which polymers such as nylon and polyesters and glass are melted to permit extrusion into fibers through spinnerets.

melt time *See* element melt time.

membrane *1.* A thin tissue that covers organs, lines cavities, and forms canal walls in the body of an animal. *2.* A thin sheet of metal, rubber, or treated fabric used to line cavities or ducts, or to act as a semirigid separator between two fluid chambers.

membrane structures Shell structures, often pressurized, that do not take wall ending or compression loads.

memory map Graphic representation of the general functional assignments of various areas in memory. Areas are defined by ranges of addresses.

memory protect A technique of protecting the contents of sections of memory from alteration by inhibiting the execution of any memory modification instruction upon detection of the presence of a guard bit associated with the accessed memory location.

meniscus The concave or convex surface, caused by surface tension, at the top of a liquid column, as in a manometer tube.

meniscus lens A lens with one concave surface and one convex surface.

mer The recurring structural pattern of a polymer.

mercury A toxic metallic element that is liquid at room temperature.

mercury-cadmium tellurides Compounds of tellurium exhibiting photovoltaic characteristics and used for photodiodes and photodetectors in the 3 to 12 micrometer wavelengths at cryogenic temperatures.

meridian plane Any plane that contains the optical axis.

mesh *1.* A measure of screen size equal to the number of openings per inch along the principal direction of the weave. *2.* The size classification of particles that pass through a sieve of the stated screen size. *3.* Engagement of a gear

with its mating pinion or rack. *4.* A closed path through ductwork in a ventilation survey.

mesh number *1.* The number of a screen's openings per linear inch. *2.* The number of the finest screen that all other metal powders will pass through.

metal An opaque, lustrous elemental chemical substance that is a good conductor of heat and electricity and, when polished, a good reflector of light.

metal-barrier-metal junctions *See* MBM junctions.

metal corrosion *See* corrosion.

metal dusting The catastrophic deterioration of metals in carbonaceous gases at elevated temperatures.

metal flake mottling A puddling of a metallic and non-metallic pigments in a topcoat, giving it a blotchy appearance.

metal flake re-orientation A uniform change, over a sealer, in the apparent color of a metallic topcoat. This difference is independent of the configuration of the sealer bead. It appears to be related to light reflecting off the metallic flakes which have shifted their relative positions from that of the rest of the panel, probably the result of floating, or vertical pigment separation, due to currents set up in a Benard cell.

metal inert-gas welding *See* gas metal-arc welding.

metal-insulator-metal diodes *See* MIM diodes.

metal ion concentration cell A galvanic cell caused by a difference in metal ion concentration at two locations on the same metal surface.

metalization The deposition of a thin-film pattern of a conductive material onto a substrate to provide interconnection of electronic components or to provide conductive pads for interconnections.

metallic Exhibiting characteristics of a metal.

metallic bond The primary bond between metallic atoms.

metallic brilliance Departure from solid color (straight shade) appearance to a highly metallized or opalescent appearance, often accompanied by a change in the angle of viewing (goniochromatic effect).

metallic coating A thin layer of metal applied to an optical surface to enhance reflectivity.

metallic fiber Reinforcement fibers composed of metal, metal coated with plastic, or plastic coated with metal.

metallic friction material A sintered friction material formulated with metallic or metallic-ceramic materials.

metallic glass(es) Amorphous alloys (glassy metals) produced by extremely rapid quenching of molten transition-metal alloys (for example, iron, nickel, and/or cobalt). These metallic glasses exhibit unique mechanical, magnetic, and electrical properties, superconductive behavior, and anticorrosion resistance, depending on the alloys, their formation, and quenching techniques. *See also* glassy alloy.

metallic superoxides The more common metallic superoxides are those of potassium and sodium with KO_2. In addition to serving as oxygen sources, the superoxides remove carbon dioxide, water, and odors. Referred to as superoxides because they occupy a smaller volume than oxygen or lithium hydroxide (for carbon dioxide removal) systems. They

provide a degree of sterilization and additionally weigh less, cost less, and have an excellent shelf life.

metallizing *1.* The coating of a surface with a thin metal layer by spraying, hot dipping, or vacuum deposition. *2.* Applying a conductive metal layer on a non-conductive surface.

metallograph An optical instrument for observing highly magnified solid test specimens.

metallography The science dealing with the study of the structure of metals. The most common techniques are optical microscopy, electron microscopy, and x-ray diffraction analysis.

metalloids Elements that will form bonds with metallic structures, such as boron and carbon.

metallurgy The science and technology of metals. Chemical metallurgy is connected with the extraction of metals from their ores and with the refining of metals; process metallurgy involves control of melting, hot work, and cold work of metals; physical metallurgy deals with the physical and mechanical properties of metals as affected by composition, mechanical working, and heat treatment.

metal-matrix composites Metal composites produced by diffusion bonding of metal fibers and metal matrix.

metal-nitride-oxide semiconductor *1.* Class of semiconductors utilizing silicon nitride and silicon oxide dielectrics. *2.* One type of computer semiconductor memory used in EAROMs.

metal powder Very small metal or metal alloy particles.

metal-semiconductor-metal semiconductors *See* MSM (semiconductors).

metal spraying Coating material by spraying molten metal on its surface.

metal units Concentration units defined as the number of gram-moles per 1000 g of solvent.

metastable A state that is not in equilibrium, yet shows apparent stability.

metastable beta In titanium metallography, a nonequilibrium phase composition that can be partially or completely transformed to martensite, alpha, or eutectoid decomposition products with thermal or strain energy activation during subsequent processing or service exposure.

meter *1.* A device for measuring and indicating the value of an observed quantity. *2.* An international metric standard for measuring length, equivalent to approximately 39.37 inches in the U.S. customary system of units. Spelled metre in the International System of Units (SI).

methanation The conversion of various organic compounds to produce methane.

methane A hydrocarbon represented by the chemical formula CH_4.

methanol or methyl alcohol An alcohol of formula CH_3OH.

methyl cellulose A thickening agent in adhesives and textiles.

methyl methacrylate A volatile liquid resin used in the manufacture of acrylic resins.

methylol melamines *See* melamine plastics.

metre Metric unit of length (SI). *See* meter.

metrication *1.* The conversion on an industry and/or nationwide basis of English units of measurement into the International System of Units, including engineering and manufacturing standards, tools, and instruments, and all affected areas in the government and private sectors. *2.* Any act tending to increase the use of the metric system (SI), whether it be increased use of metric units or of engineering standards that are based on such units.

metric conversion *See* metrication.

metric photography The recording of events by means of photography (either singly or sequentially), together with appropriate metric coordinates to form the basis for accurate measurements.

metric system *See* International System of Units.

MeV Mega-electron-volts; a unit of energy equivalent to the kinetic energy of a single electron accelerated through an electric potential of 1 million volts.

MF Middle or medium frequency. *See* middle frequency.

MFLOPS Millions of floating point operations per second.

M_f temperature The temperature at which the austenite to martensite reaction is complete.

Mg Chemical symbol for magnesium.

M-glass Glass designed with a high modulus of elasticity.

mho A customary unit of conductance and admittance generally defined as the reciprocal of one ohm, or the conductance of an element whose resistance is one ohm; the equivalent SI unit Siemen is preferred.

mica An inorganic material that separates into layers and has high insulation resistance, dielectric strength, and heat resistance.

micelle A submicroscopic structure developed from atoms or molecules.

micro *1.* (μ) Prefix denoting one-millionth. *2.* For composites, the properties of the various elements. *3.* Colloquial or shop expression meaning a microspecimen intended for study of microstructures. *4. See* microsection.

microbalance A small analytical balance for weighing masses of 0.1 g or less to the nearest μg.

microballoons Also called microspheres. Small, hollow glass spheres used as fillers in epoxy and polyester compounds to reduce density.

microbar A unit of pressure equal to one dyne per square centimeter.

micro code *1.* A system of coding making use of suboperations not ordinarily accessible in programming, for example, coding that makes use of parts of multiplication or division operations. *2.* A list of small program steps; combinations of these steps, performed automatically in a prescribed sequence from a macro-operation such as multiply, divide, and square root.

microcrack Cracks of microscopic size.

microcracking Cracks in composite materials that occur when stresses exceed the strength of the matrix resin.

microcurie A unit of radioactivity equal to one millionth (10^{-6}) curie.

microdensitometer A device for measuring the density of photographic films or plates on a microscopic scale; the small scale version of a densitometer.

microdynameter An instrument for measuring force and microscopic changes in a test microetching.

microetching An expression meaning to etch a highly polished surface of metal to examine microstructures, typically using at least 50X power magnification.

microfarad (μf) One-millionth of a farad, a unit of capacitance.

microfaradmeter A capacitance meter calibrated in microfarads.

microfissure A microscopic crack.

micrograph A graphic reproduction of a material surface.

microhardness The hardness of microscopic areas or of the individual microconstituents in a metal, as measured by such means as Tukon, Knoop, or scratch methods.

microinch One millionth (0.000001) part of the U.S. standard linear inch.

micromechanics The study of the constraints, the grain size, and their interrelationship in materials.

micrometer Unit of measurement one millionth of a meter long.

micron A unit of length equal to 10^{-6} meter or approximately 0.00004 inches.

micropore Pore in a metal powder compact that can only be seen through a microscope.

microporosity Porosity visible only with the aid of a microscope.

microprocessor *1.* A usually monolithic, large-scale-integrated (LSI) central processing unit (CPU) on a single chip of semiconductor material; memory, input/output circuits, power supply, etc. are needed to turn a microprocessor into a microcomputer. *2.* A large-scale integrated circuit (IC) that has all the functions of a computer, except memory and input/output systems. The IC thus includes the instruction set, arithmetic and logical unit (ALU), registers and control functions. *3.* Sometimes abbreviated as MPU or μP.

microprogramming A method of operating the control unit of a computer, in which each instruction initiates or calls for the execution of a sequence of more elementary instructions. The microprogram is generally a permanently stored section of nonvolatile storage. The instruction repertory of the microprogrammed system can thus be changed by replacing the microprogrammed section of storage without otherwise affecting the construction of the computer.

microradiography Projecting x-rays through a very thin section of a material onto film to determine the distribution of elements and voids.

microradiometer A device for detecting radiant power which consists of a thermopile supported on and directly connected to the moving coil of a galvanometer.

microscope Instrument used to study microstructures at high magnification.

microscopic stress *1.* Load per unit area over a very short distance, on the order of the diameter of a metal grain, or smaller. A term usually reserved for characterizing residual stress patterns. *2.* Residual stress in a very small distance which can be determined by x-ray diffraction patterns.

microsecond (μs) One-millionth of a second (10^{-6} second).

microsection A specimen prepared for analytical examination. Also called a microspecimen—or simply a micro.

microspheres *See* microballoons.

microstrain The strain measured from a finite control volume of material with a characteristic lineal dimension that is of the same order of magnitude as the materials interatomic distance. Note: These are the strains that are "averaged" by the macrostrain measurement.

Microstrain is not measurable with the commonly employed techniques used for macrostrain measurement, such as finite length foil resistant strain gages. Microstrain distribution is typically measured with x-ray diffraction techniques. The term "microstrain" is often used to signify the macrostrain multiplied by 10^6.

microstructure The crystalline structure of a material.

microwave Electromagnetic radiation having a wavelength of 1 to 300 mm.

microwave spectrum The portion of the electromagnetic spectrum of frequencies lying between infrared waves and radio waves.

microyield strength Stress at which a microstructure (single crystal, for example) exhibits a specified deviation in its stress-strain relationship.

middle frequency (MF) The frequency band between 300 kHz and 3 MHz.

Mie equation Equation that provides the potential energy of two atoms or ions in close proximity.

Mie scattering Any scattering produced by spherical particles without special regard to comparative size of radiation wavelength and particle diameter.

Mie theory *See* Mie scattering.

migration *1.* Movement of particles from the filter assembly into the effluent. *2.* The movement of ions from an area of the same charge to an area of opposite charge.

MIG welding Metal inert-gas welding. *See* gas metal-arc welding.

mil One one-thousandth of an inch.

mild steel Carbon steel with a maximum of 0.25% carbon.

mile A British and U.S. unit of length commonly used to specify distances between widely separated points on the earth's surface; a statute mile, used for distances over land, is defined as 5280 feet; a nautical mile, used for distances along the surface of the oceans, is defined as one minute of arc measured along the equator, which equals 6080.27 feet or 1.1516 statute miles.

mill edge The normal edge produced in hot rolling. This edge is customarily removed when hot-rolled sheets are further processed into cold-rolled sheets.

Miller-Bravais indices An indexing system for identifying crystallographic planes and direction in a hexagonal system.

Miller indices An indexing system for identifying planes and directions in a crystal system.

millimeter Also spelled millimetre. *1.* A unit of length equal to 0.001 meter. *2.* (millimeter of mercury) Abbreviated mm Hg, a unit of pressure equivalent to the pressure exerted by a column of pure liquid mercury one mm high at 0°C under a standard gravity of 980.665 cm/s^2; it is roughly equivalent to 1/760th of standard atmospheric pressure.

millimetre *See* millimeter.

millisecond One thousandth of a second (10^{-3} second).

mill scale The heavy oxide layer formed during hot fabrication or heat treatment of metals.

MIM diodes Junction diodes each consisting of an insulating layer sandwiched between two metallic surface layers and exhibiting a negative differential resistance in its V-1 characteristics conceivably because of stimulated inelastic tunneling of electrons.

min An abbreviation for minute or minimum.

mineral fillers Minerals used as fillers in polymers to improve properties or reduce costs.

miniaturization The design and production of a scaled-down version of a device or mechanism that is capable of performing all of the same functions as the larger-sized original.

MINI-MAP A subset of the Manufacturing Automated Protocols (MAP) extended to provide higher performance for applications whose communications are limited to a single local area network (LAN). A MINI-MAP node contains only the lower two layers (physical and link) of the MAP protocols. It can only communicate directly with Manufacturing Automation Protocol/Enhanced Performance Architecture (MAP/EPA) or MINI-MAP nodes on the same segment.

minimized spangle Refers to galvanized sheet that has very small spangles which are obtained by treating the galvanized sheet during the solidification of the zinc to restrict to the normal zinc spangle formation.

minimum bend radius The smallest radius around which a piece of sheet metal, wire, bar stock, or tubing can be bent without fracture, or in the case of tubing, without collapse.

minimum pressure The lowest transient pressure than can occur temporarily.

minimum stress (S_{min}) The stress having the lowest algebraic value in the cycle, tensile stress being considered positive and compressive stress negative.

minimum thermometer A thermometer that indicates the lowest temperature reached during a given interval of time.

minor defect A defect that is not likely to reduce materially the usability of a material or component for its intended purpose, or is a departure from established standards having little bearing on the effective use or operation of the unit. Sometimes referred to as an imperfection.

minor graduations The shortest or lightest division marks on a graduated scale, which indicate subdivisions lying between successive major graduations or between an index graduation and an adjacent major graduation.

minute *1.* A measure of time equal to 60 seconds. *2.* A measure of angle equal to 1/60 of one degree.

mirror fusion An open-ended configuration that traps low-beta plasmas. It is realized by associating two identical magnetic mirrors having the same axis.

mischmetal A mixture of rare-earth elements (atomic numbers 57 through 71) in metallic form. It contains about 50% cerium, the remainder being principally lanthanum and neodymium.

miscibility The ability of two or more liquid solutions to dissolve into a single-phase solution.

misconvergence The degree to which the midpoint of the line widths of two or three primary colors are misregistered at the phosphor surface of a cathode ray tube (CRT).

mixed oxides Mixture of oxides, particularly of radioactive metals.

Mn Chemical symbol for manganese.

MNOS *See* metal-nitride-oxide-semiconductor (one type of computer semiconductor memory used in EAROMs).

mobility *1.* The average drift velocity of a charged particle induced by a unit electrical potential gradient. *2.* In gases, liquids, solids, or colloids, the relative ease with which atoms, molecules, or particles can move from one location to another without external stimulus.

mode *1.* A particular event (for example, acceleration, deceleration, cruise, or idle) of a test cycle. *2.* A single component in a computer network. *3.* Real or complex (number system). *4.* A stable condition of oscillation in a laser. A laser can operate in one mode (single-mode) or in many modes (multimode).

moderation Reducing the kinetic energy of neutrons, usually by means of successive collisions with hydrogen, carbon, or other light atoms.

moderators Materials that have a high cross section for slowing down fast neutrons with a minimum of absorption, for example, heavy water, beryllium, used in reactor cores.

MODFETS Heterojunction field effect transistor device structures in which only the larger (Al, GaAs) bandgap is doped with donors, while the GaAs layer is left undoped. This results in high electron mobilities due to spatially separated electrons and donors.

modified acrylic A thermoplastic polymer that cures at room temperature.

modularity The degree to which a system of programs is developed in relatively independent components, some of which may be eliminated if a reduced version of the program is acceptable.

modular ratio In metal or polymer reinforced materials, the ratio of the Young's modulus of the reinforcement to the same modulus of the matrix.

modulated wave A radio-frequency wave in which amplitude, phase, or frequency is varied in accordance with the waveform of a modulating signal.

modulation *1.* The process of impressing information on a carrier for transmission; AM, amplitude modulation; PM, phase modulation; FM, frequency modulation. *2.* Regulation of the fuel-air mixture to a burner in response to fluctuations of load on a boiler. *3. See* contrast.

modulation doped fets *See* MODFETS.

modulation doping The process of doping only the larger bandgap of a heterojunction device with donors, while the other layer is left undoped. Since the electrons and donors are spatially separated, ionized impurity scattering is avoided and extremely high electron mobilities are obtained.

module *1.* A combination of assemblies, subassemblies, and parts, contained in one package. *2.* A computer program unit that is discrete and identifiable with respect to compiling, combining with other units, and loading, for example, the input to, or output from, an assembler, compiler, linkage editor, or executive routine.

modulo A mathematical operation that yields the remainder function of division; thus, 39 modulo 6 equals 3.

modulo N check *1.* A check that makes use of a check number that is equal to the remainder of the desired number when divided by n, for example, in a modulo 4 check, the check number will be 0, 1, 2, or 3 and the remainder of the desired number when divided by 4 must equal the reported check number, otherwise an equipment malfunction has occurred. *2.* A method of verification by congruences, for example, casting out nines.

modulus of elasticity *1.* A measure of the rigidity of an alloy. *2.* In any solid, the slope of the stress-strain curve within the elastic region; for most materials, the value is nearly constant up to some limiting value of stress known as the elastic limit; modulus of elasticity can be measured in tension, compression, torsion or shear; the tension modulus is often referred to as Young's modulus. This value in metals is typically 29,000,000 to 30,000,000 at room temperature; it decreases with a rise in temperature.

modulus of resilience The energy that can be absorbed by a solid material without permanent distortion.

modulus of rigidity The ratio of stress to strain for shear or torsional stress.

modulus of rupture in bending (S_b) The value of maximum shear stress in the extreme fiber in a member of cross section loaded to fail by torsion.

modulus, rubber *See* rubber modulus.

Mohs hardness scale A number indicating the scratch resistance of a material.

moiety A part of a molecule with a unique chemical property.

moisture Water in the liquid or vapor phase.

moisture absorption Generally, the amount of moisture, in percentage, that a material will absorb under specified conditions.

moisture barrier A material or coating that retards the passage of moisture through a wall made of more permeable materials.

moisture content The measure of moisture in a material expressed as a percent of the mass of the specimen.

moisture equilibrium The point at which a material no longer accepts or gives up moisture in a specific environment.

moisture-free *See* bone dry.

moisture in steam Particles of water carried in steam usually expressed as the percentage by weight.

moisture loss The loss representing the difference in the heat content of the moisture in the exit gases and that at the temperature of the ambient air.

moisture resistance The ability of a material to resist absorbing moisture.

moisture vapor transmission (MVT) The rate at which a vapor passes through a material under fixed temperature and humidity conditions.

mol *See* mole.

molar conductivity The conductivity of an electrolytic solution with 1 mol solute per liter of solution.

molar units Concentration units defined as the number of gram-moles of a component per liter of solution.

molding compound Polymer material, plus any other fillers, reinforcements, pigments, etc.

mold lubricant The substance used to coat the surfaces of a mold to prevent the elastomer from adhering to the mold cavity surface during vulcanization.

mold, potting An accessory used as a form for containing the potting compound around the terminations of a connector.

mold wash *See* wash.

mole Metric unit for amount of a substance.

molecular attrition *See* fretting.

molecular dissociation *See* dissociation.

molecular fluorescence spectroscopy An analytical method that measures fluorescent emission characteristics of a molecular species rather than an atomic species.

molecular mass *See* molecular weight.

molecular structure The representation of how electrons and nuclei interact to form a molecule.

molecular weight The weight of a given molecule expressed in atomic weight units.

molecule(s) The smallest division of a unique chemical substance which maintains its unique chemical identity.

moles Number of molecular weights, which is the weight of the component divided by its molecular weight.

molten salts High-temperature inorganic salt or mixtures of salts used for thermal energy storage, heat exchangers, high-power electric batteries, heat treatment of alloys, etc.

molybdenum A metallic element used primarily as an alloying agent.

moment *1.* Of force, the effectiveness of a force in producing rotation about an axis; it equals the product of the radius perpendicular to the axis of rotation that passes through the point of force application and the tangential component of force perpendicular to the plane defined by the radius and axis of rotation. *2.* Of inertia, the resistance of a body at rest or in motion to changes in its angular velocity. *3. See* torque.

momentary digital output A contact closure, operated by a computer, that holds its condition (set or rest) for only a short time. *See also* latching digital output.

moment of inertia A measure of resistance of the mass of a body to rotation. It reflects the amount of mass as well as its distribution within the body. Torque applied to a moment of inertia results in a rotational acceleration, analogous to the way that force applied to a mass results in a translational acceleration.

momentum *1.* Quantity of motion. *2.* The product of a body's mass and its linear velocity.

momentum energy *See* kinetic energy.

monitor software The portion of the operational software that controls on-line and off-line events, develops new on-line applications, and assists in their debugging. This software is also known as a batch monitor.

monochromatic A single wavelength or frequency. In reality, light cannot be purely monochromatic, and actually extends over a range of wavelengths.

monochromatic radiation Any electromagnetic radiation having an essentially single wavelength, or in which the photons all have essentially the same energy.

monoclinic system A crystal class in which the three axes are of unequal length; two axes are at right angles to one another, and the third is at an oblique angle to the others.

monolayer *1.* A single layer of atoms on the surface. *2.* A single laminate form upon which other laminates are constructed.

monolithic substrate A unitary catalyst substrate, usually of honeycomb structure.

monomer A single organic molecule usually containing carbon and capable of additional polymerization.

monostable Pertaining to a device that has one stable state.

monotectic alloys Metallic composite materials having a dispersed phase of solidification products distributed within a matrix. The dispersed components can be selected to provide characteristics such as superconductivity or lubricity.

monotonic Describes a digital-to-analog converter in which the output either increases or remains constant as the digital input increases.

monotrophism The ability of a material to appear in two or more crystalline structure forms.

monovalent elements Elements with one valence electron per atom.

Monte Carlo method A trial-and-error method of repeated calculations to discover the best solution to a problem. Often used when a great number of variables are present, with interrelationships so extremely complex as to forestall straightforward analytical handling.

morphology The form of a material structure.

most significant bit (MSB) The bit in a digital sequence that defines the largest value. It is usually at the extreme left.

most significant digit The leftmost nonzero digit.

motion The act, process, or instance of change of position. Also called movement, especially when used in connection with problems involving the motion of one craft relative to another.

mottled iron Cast iron with a structure that is part grey and part white.

movement *See* motion.

MPS A high-level language suitable for the development of microprocessor application software.

MPU *See* microprocessor.

M_s In titanium, the maximum temperature at which a martensite reaction begins upon cooling from the beta phase. *See also* M_s temperature.

MSB *See* most significant bit.

M-shell The third layer of electrons surrounding the nucleus of an atom.

MSI *See* medium-scale integration.

MSM (semiconductors) Semiconductor devices consisting of a semiconductor layer sandwiched between two layers of metal.

M$_s$ temperature In steel, the temperature at which unstable austenite starts to transform into martensite during quenching. *See also* M$_s$.

MTTF *See* mean-time-to-failure.

mud *See* slime.

mult or multiple *See* forging stock.

multidirectional Having multiple ply orientations in a laminate.

multilayer A type of printed circuit board that has several layers of circuit etch or pattern, one over the other and interconnected by electroplated holes.

multilayer coating Optical coatings in which several layers of different thicknesses of different materials are applied to an optical surface.

multilayer glass *See* cased glass.

multilayer structures *See* laminate.

multimeter *See* volt-ohm-milliammeter.

multimode fiber An optical fiber capable of carrying more than one mode of light in its core.

multioriented ply laminate A laminate made from multioriented plies.

multiphoton absorption Ionization and dissociation of a molecule under the action of powerful laser radiation.

Laser-flux dependent light intensities are emitted by different excited states of the molecule to indicate the various absorption processes.

multiple-layer adhesive An adhesive film with different adhesive compositions on each side.

multiple-ply laminates Laminates with two or more laminar plies.

multiple-purpose meter *See* volt-ohm-milliammeter.

multipliers Devices that have two or more inputs and whose output is a representation of the product of the quantities represented by the input signals.

multiprocessor A machine with multiple arithmetic and logic units for simultaneous use.

multiprogramming A computer processing method in which more than one task is in an executable state at any one time.

Munsell system A system of classifying and designating color attributes of object in terms of perceptually uniform color scales for the three variables; hue, value, and chroma.

muscovite An important mineral of the mica group.

mushy zones Regions of liquid plus solid phases in alloys that solidify over a range of temperatures.

MVT *See* moisture vapor transmission.

N

N *1.* Chemical symbol for nitrogen. *2. See* newton. *3.* A composite sample size when the sample sizes of a number of groups being combined are not equal.

n' *See* cyclic strain hardening exponent.

Na Chemical symbol for sodium.

NAND *1.* A logical operator having the property that if P is a statement, Q is a statement, R is a statement, …, then the NAND of P, Q, R, … is true if at least one statement is false; false if all statements are true. Synonymous with NOT-AND. *2.* Logical negation of AND. Supplies a logic 0 when all inputs are at logic 1.

nano A prefix that means one billionth.

nanometer One billionth of a meter.

nanosecond (ns) One billionth of a second (10^{-9} second).

naphtha A distillate of crude oil used for developing monomers such as styrene and vinyl chloride.

narrowband Describes a frequency measurement whose frequency band of energy is smaller relative to the rest of the band.

natural aging Spontaneous aging of a supersaturated solid solution at room temperature. *See* aging. Compare with artificial aging.

natural gas A mixture of gaseous hydrocarbons trapped in rock formations below the earth's surface. The mixture consists chiefly of methane and ethane.

natural language A language, the rules of which reflect and describe current usage rather than prescribed usage. Contrast with artificial language.

natural lasers *See* lasers.

natural radioactivity Spontaneous radioactive decay of a naturally occurring nuclide.

natural rubber (NR) An elastomer made from natural latex.

nautical mile A unit of distance equal to 6076.11549 feet or 1852 meters.

Navier-Stokes equation The equation of motion for a viscous fluid.

NBR *See* nitrile rubber.

Nd Chemical symbol for neodymium.

NDE Nondestructive evaluation. *See* nondestructive testing.

NDI Nondestructive inspection. *See* inspection, nondestructive and nondestructive testing.

NDIR analyzer *See* nondispersive infrared analyzer.

NDRO *See* non-destructive read out.

NDT Nondestructive testing. *See* nondestructive testing.

NDUV analyzer *See* nondispersive ultraviolet analyzer.

Ne Chemical symbol for neon.

neat resin Resins with no added elements, such as fillers.

nebulizer An apparatus for converting a solution into a gas-liquid aerosol.

necking *1.* Stretching a flat material to reduce a specific cross-section area. *2.* Stretching a shell or tube to reduce the diameter.

necking down Reduction of a cross-sectional area of a test specimen. During a tensile test this occurs within the gage length and becomes an expression of ductility.

negative acceleration *See* deceleration.

negative ions Ions singly or in groups which acquire negative charges by gaining one or more electrons.

negatron(s) A negatively charged beta particle. *See also* electron.

neodymium A rare earth element.

neodymium oxide A sintering additive for silicon nitride ceramics.

neon An inert gas.

neoprene A synthetic rubber made by polymerization of chloroprene (2-chlorobutadiene-1,3). Its color varies from amber to silver to cream. It exhibits excellent resistance to weathering, ozone, flames, various chemicals, and oils.

nephelometer *1.* Instrument that measures, at more than one angle, the scattering function of particles suspended in a medium. *2.* Instrument for chemical analysis that measures the light scattering properties of a suspension. *3.* A general term for instruments that measure the degree of cloudiness or turbidity.

nephelometry The application of photometry to the measurement of the concentration of very dilute suspensions.

nepheloscope An apparatus for making clouds in the laboratory by expanding moist air or by condensing water vapor.

neptunium A radioactive element.

nest *1.* To embed a subroutine or block of data into a larger routine or block of data. *2.* To evaluate an nth degree polynomial by a particular algorithm which uses (n − 1) multiply operations and (n − 1) addition operations in succession. *See also* crimp anvil.

nested Describes subroutines that are used or called with another subroutine.

nested DO loop A FORTRAN statement that directs the computer to perform a given sequence repeatedly.

nested laminate In reinforced plastics, the placing of plies of fabric so that the yarns of one ply lie in the valleys between the yearns of the adjacent ply.

nesting *1.* In certain reinforced polymers, the placement of ply fabric so that the yarns of one ply lie in the valleys of the yarns of the second ply. *2.* In computer software, a program that has loops within loops.

net power Forward power minus reflected power at the same location on the transmission line.

netting analysis Stress analysis of a filament-wound material structure that assumes the filaments support the entire stress load.

network structure A type of alloy microstructure in which one phase occurs predominantly at grain boundaries, enveloping grains of a second phase.

neutral In aqueous solutions, refers to a substance having a pH of 7.

neutral atmosphere An atmosphere that tends neither to oxidize nor reduce immersed materials.

neutral atoms Atoms in which the number of electrons surrounding the nucleus equals the number of protons in the nucleus, resulting in no net electric charge.

neutral filter A filter that focuses the radiant power reaching a detector by the same factor.

neutralization Hydrogen ions of an acid combined with hydroxyl ions of a base to form water ($H^+ + OH^- = HOH$).

neutral point Point on the titration curve at which the hydrogen ion concentration equals the hydroxyl ion concentration.

neutral zone *See* deadband.

neutrinos Sub-atomic particles of zero, or near zero, rest mass, having no electric charge.

neutron A nuclear particle with a mass number of one and exhibiting zero (neutral) charge.

neutron absorber A material that absorbs neutrons without re-emitting them.

neutron activation analysis Analysis in which a specimen is bombarded with neutrons and identification is determined by measuring the emitted radioisotopes.

neutron cross-section A measure of the likelihood that a specific interaction will occur between a nucleus and a neutron.

neutron diffraction The interference phenomenon that occurs when neutrons are scattered by atoms within a material.

neutron embrittlement Embrittlement from neutron bombardment, usually associated with nuclear reactors.

neutron flux The number of neutrons passing through an area in a specific time.

neutron radiography Nondestructive testing and inspection utilizing neutron beams from nuclear reactors, particle accelerators, and/or radioisotopes. Imagery displaying structural defects utilizes neutron image recorders or screens.

neutron spectroscopy A method of determining the energy distribution of neutrons.

neutron spectrum The energy distribution of neutrons bombarding a surface.

newton A unit of force in the SI system; the force that gives to a mass of 1 kilogram an acceleration of 1 meter per second squared.

Newtonian flow Fluid characteristics adhering to the linear relation between shear stress, viscosity, and velocity distribution.

Newtonian fluids Fluids whose viscosities are shear independent and time independent. The shear rate of a Newtonian fluid is directly proportional to the shear stress.

Ni Chemical symbol for nickel.

nibble A word with four bits, or one-half a byte.

nick A void created in a material after molding.

nickel A metal that offers a combination of corrosion resistance, formability, and tough physical properties. For these reasons, nickel is used for alloying purposes and as a coating for copper.

NIST National Institute of Standards and Technology

nickel steels Steels containing nickel as a main alloying element. Not to be confused with nickel-base alloys.

nickel terne Cold-rolled sheet steel electrolytically nickel flash plated and then coated on both sides with a lead-tin alloy by a continuous hot-dip process. Corrosion resistance is superior to standard long terne.

nimonic alloys A family of nickel-chromium alloys with excellent creep resistance at high temperatures.

niobium A metallic element used as an alloying agent to increase high temperature strength of metals.

nitinol alloys Shape memory alloys of titanium and nickel.

nitride-bonded ceramics A porous, coarse-grain composite material used as refractories.

nitriding A case-hardening process in which a ferrous-base material is heated to approximately the iron-nitrogen eutectoid temperature in either a gaseous or a liquid medium containing active nitrogen, thus causing absorption of nitrogen at the surface and, by diffusion, creating a concentration gradient. Within the capabilities of the particular material, slow cooling produces full hardness of the case. In *conventional nitriding*, a hardened and tempered alloy steel or tool steel is treated for sufficient time to produce highly saturated nitrides in the case. In an important variation of the process, sometimes called *ductile nitriding*, applied to any ferrous-base material, the amount of active nitrogen and the time of exposure are so controlled as to produce a case of lower nitrogen content which, within the capabilities of the material, is fully hard on a micro scale, but lower in hardness on a macro scale and relatively ductile.

nitrile A generic term for the copolymers of butadiene and acrylonitrile.

nitrile rubber A copolymer of acrylonitrile and butadiene that has excellent tear and abrasion resistance.

nitrocarburizing Hardening ferrous alloys by exposing a surface layer to a nitrogen and carbon-rich material at a specific temperature.

nitrogen A colorless, odorless gas that does not sustain higher forms of life or combustion. Also available in liquid form at $-195.6°C$ ($-320°F$), which vaporizes into gaseous nitrogen.

nitrogen lasers Stimulated emission devices in which the nitrogen molecule is the lasing medium.

NMR Nuclear magnetic resonance.

noble gases *See* rare gases.

noble metal A metal that is not very reactive, such as silver, gold, and copper, and may be found naturally in metallic form on earth.

node bonds The area of the honeycomb core in which the cell walls are adhesively bonded.

node(s) (standing waves) Points, lines, or surfaces in standing waves where some characteristic of the wave field has essentially zero amplitude.

nodular cast iron *See* ductile iron.

nodular iron *See* ductile iron.

nodular powder Metal powders with irregular shapes.

nol ring A parallel filament- or tape-wound hoop test specimen developed by the Naval Ordnance Laboratory for measuring various mechanical strength properties of the material, such as tension and compression, by testing the entire ring or segments of it.

nomex Aramid fiber or paper; low smoke and flame. The paper form is used to make honeycomb.

nominal composition Expressing the composition of an alloy in general percentages rather than exact composition.

nominal size *1.* The standard dimension closest to the central value of a toleranced dimension. *2.* A size used for general identification.

nominal specimen thickness The nominal ply thickness multiplied by the number of plies.

nominal stress The stress calculated by dividing nominal load by nominal cross-sectional area, ignoring the effect of stress raisers, but taking into account localized variations due to general part design.

nominal surface The imaginary true surface that would result if all surface irregularities (peaks, waves, ridges, and hollows) were leveled off to zero value, or were non-existent. It is this nominal surface or "mean line" from which the surface irregularities deviate.

nominal value A value assigned for the purpose of a convenient designation.

nonadiabatic conditions In thermodynamics, changes in volume, temperature, flow, etc., accompanied by a transfer of heat.

non-blackbody A term used to describe the thermal emittance of real objects, which emit less radiation than blackbodies at the same temperature, which may reflect radiant energy from other sources, and which may have their emitted radiation modified by passing through the medium between the body and a temperature-measuring instrument.

noncondensable gas The portion of a gas mixture (such as vapor from a chemical processing unit or exhaust steam from a turbine) that is not easily condensed by cooling.

nonconforming A product or service that does not match expected criteria, but is not necessarily defective.

non-conforming test result A test result that does not conform to the material specification.

nondestructive evaluation (NDE) *See* nondestructive testing.

nondestructive inspection (NDI) *See* nondestructive testing.

non-destructive read out A method of reading from memory whereby the stored value is left intact by the reading process; plated wire and modern semiconductor random-access memory (RAM) are examples of NDRO memory.

nondestructive testing Any testing method that does not damage or destroy the sample. Usually, it consists of stimulating the sample with electricity, magnetism, electromagnetic radiation, or ultrasound, and measuring the sample's response.

nondispersive infrared (NDIR) analyzer An instrument to determine carbon monoxide, carbon dioxide, nitric oxide, and hydrocarbons in exhaust gas. Now primarily being used for carbon monoxide and carbon dioxide determinations.

nondispersive ultraviolet (NDUV) analyzer An analytical instrument used to measure NO_2 concentration in exhaust gas.

nonhygroscopic Lacking the ability to absorb moisture.

non-interlaced In raster graphics display tubes, describes the method whereby every scanning line is refreshed for each sweep down the display. This is contrasted with the interlace method used in standard television whereby every other scanning line is refreshed during each sweep, thereby requiring two sweeps to completely refresh the display.

nonisothermal process In thermodynamics, compression or expansion of substances at nonuniform temperatures.

nonisotropy *See* anisotropy.

nonlinear optics Study of the interaction of radiation with matter in which certain variables describing the response of the matter are not proportional to variables describing the radiation.

nonlinear optimization *See* nonlinear programming.

nonlinear programming *1.* In operations research, a procedure for locating the maximum or minimum of a function of variables that are subject to constraints, when either the function or the constraints, or both, are nonlinear. Contrast with convex programming and dynamic programming. *2.* Synonymous with nonlinear optimization.

nonliquefied compressed gas A gas not in solution which, under the charging pressure, is entirely gaseous at a temperature of 70°F (21°C), and which is maintained in the liquid state at absolute pressure by maintaining the gas at a temperature less than 70°F (21°C).

nonlocking Pertains to code extension characters that change the interpretation of one or a specified number of characters. Contrast with locking.

nonmetallic inclusions Insoluble impurities, such as oxides, aluminates, sulfides, and silicates, which are trapped mechanically, or are formed during solidification.

non-repeatability *See* precision.

nonresonant forced and vibration technique A method of dynamic mechanical measurement whereby the sample is oscillated at a fixed rate.

non-return-to-zero (NRZ) Coding of digital data for serial transmission or storage whereby a logic ONE is represented by one signal level and a logic ZERO is represented by a different signal level.

non-symmetric laminates Laminates in which individual plies are not oriented in a particular direction.

nonwoven fabric A planar textile structure produced by loosely compressing together fibers, yarns, rovings, etc., with or without a scrim cloth carrier.

normal axis *See* vertical axis.

normality Concentration units defined as the number of gram-ions of replaceable hydrogen or hydroxyl groups per liter of solution. A shorter notation of gram-equivalents per liter is frequently used.

normalize *1.* In heat-treating ferrous alloys, to heat to 50 to 100°F above the upper transformation temperature, then to cool in still air to a temperature substantially below the transformation range. *2.* In programming, to adjust the exponent and fraction of a floating-point quantity such that the fraction lies in a prescribed normal standard range. *3.* In mathematical operations, to reduce a set of symbols or numbers to a normal or standard form. Synonymous with standardize.

normalized stress Stress calculated by multiplying the raw stress value by the ratio of measured fiber volume to the nominal fiber volume. This ratio is often approximated by the ratio of the measured specimen thickness to the nominal specimen thickness.

normalizing Heating a ferrous alloy to a suitable temperature above the transformation range (austenitizing), holding a temperature for a suitable time, and then cooling in still air to a temperature substantially below the transformation range.

normalizing gas (span gas) A single calibrating gas blend routinely used in calibration of each analytical instrument.

normal-phase chromatography (NPC) Liquid-solid or bonded-phase chromatography with a polar stationary phase and a nonpolar mobile phase.

normal segregation Concentration of low-melting alloys in the portions of a casting that quickly solidify.

normal solution A solution with 1 mol of reagent in 1 liter of solution.

normal stress Stress that is perpendicular to the plane on which the load acts.

normal variable A random variable that is normally distributed. In situations where the random variable represents the total effect of many "small" independent causes, each with mutually independent errors, the central limit theorem leads to the prospect that the variable will be normally distributed.

normative Prescribing or directing a norm or standard; used in standards to indicate text that poses requirements.

notation 1. The act, process, or method of representing facts or quantities by a system or set of marks, signs, figures, or characters. 2. A system of such symbols or abbreviations used to express technical facts or quantities, as mathematical notation. 3. An annotation.

notch 1. A V-shaped indentation in an edge or surface. 2. An indentation of any shape that acts as a severe stress raiser.

notch bar test See Charpy and Izod tests.

notch brittleness The propensity of a material to fracture at stress concentration points.

notch depth The vertical distance from the surface of the test specimen to the bottom of the notch.

notched specimen A test specimen with a notch, usually V-shaped, that functions as a stress raiser.

notch factor The ratio of the strength of a plain specimen versus a notched specimen.

notching Cutting out various shapes from the edge of a metal strip, blank, or part.

notch rupture strength See notch tensile strength.

notch sensitivity A measure of the sensitiveness of a material to the presence of stress concentration caused by notches in the form of threads or grooves, scratches, and other stress raisers.

notch tensile strength The maximum load before rupture on a notched tensile specimen divided by the original notch cross-sectional area of the test specimen. See also stress rupture test.

no-test See replaced test.

novolac A thermoplastic phenolic resin that reacts with some cross-linking resin groups to form thermoset phenolic.

NO$_x$ See oxides of nitrogen.

nozzle A device used to control the flow of air, gas, or liquids. May be designed with a constricting throat section and/or a divergent section.

Np Chemical symbol for neptunium.

NPC See normal-phase chromatography.

NR See natural rubber.

NRZ See non-return-to-zero.

N-shell The fourth layer of electrons surrounding the nucleus of an atom.

N-type material A crystal of pure semiconductor material to which has been added an impurity so that electrons serve as the majority charge carriers.

nuclear Pertaining or relating to a nucleus; having the character of a nucleus; using nuclear energy.

nuclear emulsion(s) A photographic emulsion specially designed to record the tracks of ionizing particles.

nuclear fluorescence thickness gage A device for determining the weight of an applied coating by exciting the coated material with gamma rays and measuring low-energy fluorescent radiation that results.

nuclear magnetic resonance (NMR) An instrument with a magnet, radio-frequency accelerator, sample holder, and detector that is capable of producing an oscilloscope image of a NMR spectra.

nuclear radiation Corpuscular emissions, such as alpha and beta particles, or electromagnetic radiation, such as gamma rays, originating in the nucleus of the atom.

nucleating agent A substance added to a polymer to increase its solidification rate.

nuclei The positively charged cores of atoms which are associated with practically the whole mass of each atom, but only a minute part of its volume.

nucleons In the classification of subatomic particles according to mass, the second heaviest type of particles. The mass of a nucleon is intermediate between that of the meson and the hyperon.

nucleus *1.* The central, positively charged core of an atom. *2.* The first stable particle in a material capable of starting recrystallization of a phase. *3.* A number of atoms or molecules bound together with interatomic forces sufficiently strong to make a small particle of a new phase stable in a mass otherwise consisting of another phase; creating a stable nucleus is the first step in phase transformation by a nucleation-and-growth process. *4.* The portion of the control program that must always be

present in main storage. Also, the main storage area used by the nucleus and other transient control program routines.

nucleus counter An instrument that measures the number of condensation or ice nuclei in a sample volume of air.

nuclide(s) *1.* An atom that is a radioactive isotope. *2.* A species of atom characterized by a unique combination of charge, mass number, and quantum state of its nucleus.

null A condition (typically a condition of balance) that results in a minimum absolute value of output. Often specified as the calibration point at which the least error can be tolerated by the associated control system.

null point method A means of locating the center of gravity based on the principle that the center of gravity of a balanced body is in the vertical plane through the line of support. The machine is successively balanced on three or more lines and the respective planes containing the center of gravity are established. Intersection of these planes locates the center of gravity.

number *1.* A mathematical entity that may indicate quantity or amount of units. *2.* Loosely, a numeral. *3. See* binary number and random numbers.

number system A systematic method for representing numerical quantities whereby any quantity is represented as a sequence of coefficients of the successive powers of a particular base with an appropriate point. Each succeeding coefficient from right to left is associated with and usually multiplies the next higher power of the base. The following are names of the number systems with bases 2 through 20: 2, binary; 3, ternary; 4, quaternary; 5, quinary; 6, senary; 7, septenary; 8, octal or octonary; 9, novenary; 10, decimal; 11, undecimal; 12, duodecimal; 13, terdenary; 14, quaterdenary; 15, quindenary; 16, sexadecimal—or hexadecimal; 17, septendecimal; 18, octodenary; 19, novemdenary; 20, vicenary. Also 32, duosexadecimal—or duotricinary; and 60, sexagenary. The binary, octal, decimal, and sexadecimal systems are widely used in computers. *See* decimal number and binary number. Clarified by octal digit and binary digit.

numerical analysis The study of methods of obtaining useful quantitative solutions to mathematical problems, regardless of whether an analytic solution exists or not, and the study of the errors and bounds on errors in obtaining such solutions.

numerical control Automatic control of a process performed by a device that makes use of all or part of numerical data generally introduced as the operation is in process.

numerical differentiation Approximate estimation of a derivative of a function by numerical techniques.

NURBS Non-uniform rational B-spline surfaces.

Nusselt number A number expressing the ratio of convective to conductive heat transfer between a solid boundary and a moving fluid.

nylon A plastics material used to make filaments, fibers, fabric, sheet, and extrusions; a generic name for a type of long-chain polymer containing recurring amide groups within the main chain.

Nyquist frequency One-half of the sampling frequency in a sampled data system.

O

O Chemical symbol for oxygen.

object code *1.* Output from a compiler or assembler, which is itself executable machine code or is suitable for processing to produce executable machine code. *2.* A relocatable machine-language code.

object language *1.* The language to which a statement is translated. *2.* A language that is the output of an automatic coding routine. Usually object language and machine language are the same, however, a series of steps in an automatic coding system may involve the object language of one step serving as a source language for the next step, and so forth.

object layer The portion of a communication protocol specifying how to handle and how to buffer entire messages and their respective control bits, and how to interface the actual application.

object program A fully compiled or assembled program that is ready to be loaded into the computer. *See also* target program.

obscuration *See* occultation.

obsolescent Lower in physical or functional value due to changes in technology rather than to deterioration.

obsolete No longer suitable for the intended use because of changes in technology or requirements.

obstacle *See* barrier.

occasional A metallographic feature that occurs in 10% or less of the microstructure.

occlusion Specifically, the trapping of undissolved gas in a solid during solidification. *See also* adsorption.

occultation The disappearance of a body behind another body of larger apparent size.

OC curve *See* operating characteristic curve.

octal Pertaining to eight; usually describing a number system of base or radix eight; for example, in octal notation, octal 214 is 2 times 64, plus 1 times 8, plus 4 times 1, and equals decimal 140.

octal digit The symbol 0, 1, 2, 3, 4, 5, 6, or 7 used as a digit in the system of notation that uses 8 as the base or radix. Clarified by number systems.

octal number A number of one or more figures, representing a sum in which the quantity represented by each figure is based on a radix of eight. The figures used are 0, 1, 2, 3, 4, 5, 6, and 7. Clarified by octal.

octet A group of eight bits treated as a unit. *See* byte.

OD *See* outside diameter.

odd-even check Same as parity check.

odometer An instrument for measuring and indicating distance traveled.

oersted The CGS unit of magnetic field strength; the SI unit, ampere-turn per meter, is preferred.

offal The material trimmed from blanks or formed panels.

offgassing *1.* The emanation of volatile matter of any kind from materials into habitable areas. *2.* The relative high mass loss characteristics of many non-metallic materials upon initial vacuum exposure.

off-line maintenance information Data to indicate long-term engine degradation through trend monitoring and tracking of engine usage history; for example, engine hours, starts, LCF counts, and hot section usage.

offset *1.* The distance between the original segment of a stress-strain curve and a parallel line that intersects the stress-strain curve that is used to determine yield strength. The popular offset is 0.2%, but sometimes 0.02% or even 0.01% is specified. *2.* A sustained deviation of the controlled variable from set point. This characteristic is inherent in proportional controllers that do not incorporate reset action. *3.* Offset caused by load changes. *4.* The steady-state deviation. The count value output from an A/D converter resulting from a zero input analog voltage. Used to convert subsequent nonzero measurements. *5.* A short distance measured perpendicular to a principal line of measurement in order to locate a point with respect to the line. *6.* A constant and steady state of deviation of the measured variable from the set point.

offset (programming) The difference between a base location and the location of an element related to the base location; the number of locations relative to the base of an array, string, or block.

ohm The units of resistance. One ohm is the value of resistance through which a potential of one volt will maintain a current of one ampere.

ohmmeter A device for measuring electrical resistance.

ohms per volt A standard rating of instrument sensitivity determined by dividing the instrument's electrical resistance by its full-scale voltage.

oil Any of various viscous organic liquids that are soluble in certain organic solvents, such as naphtha or ether, but are not soluble in water; may be of animal, vegetable, mineral, or synthetic origin.

oil can (oil canning) Refers to an area in a formed panel that, when depressed slightly, will recover its original contour after the depressing force is removed.

oil canning *See* oil can.

oil gas A heating gas made by reacting petroleum oil vapors and steam.

oil hardening steel A type of tool steel containing a range of alloying elements requiring the quench rate of oil to harden. More alloying elements could provide air hardening characteristics; less would mandate a more rapid quench, for example, water or brine.

oil impregnation Filling a sintered compact with oil by capillary action or pressure.

oil permeability The measure of oil flow through a sintered compact.

oil resistance The measure of an elastomer's ability to withstand the deteriorating effect of oil on its mechanical properties.

oil shale A sedimentary rock containing a relatively high (5 to 30%) content of kerogen, from which shale oil can be produced by heating.

oil swell The change in volume of a rubber material due to absorption of oil.

olefinic based thermoplastic elastomers (OLTPE) Thermoplastic elastomers with a better resistance to temperature than other synthetic elastomers.

olefinics A group of polymers produced by the polymerization of monomers with the C=C group.

olemeter *1.* A device for measuring the specific gravity of oil. *2.* A device for measuring the proportion of oil in a mixture.

oligomer A polymer containing a small number of monomer units.

Olsen ductility test A method for determining relative formability of metal sheet. A sheet metal sample is deformed at the center by a steel ball until fracture occurs; the cup height at fracture indicates relative ease of forming of deep-drawn or stamped parts.

OLTPE *See* olefinic based thermoplastic elastomers.

omega A nonequilibrium, submicroscopic phase that can be formed either athermally or isothermally preceding the formation of alpha from beta. It occurs in metastable beta alloys, alpha-plus-beta alloys rich in beta content, and unalloyed titanium, and leads to severe embrittlement.

one-component adhesive An adhesive with a hardener or catalyst that can be activated by heat.

one-line data-reduction The processing of information as rapidly as information is received by the computing system or as rapidly as it is generated by the source.

one-line diagnostics *1.* Describes the state of a control system, subsystem, or piece of computer equipment that is operable and actively monitoring or controlling the process. *2.* A program to check out systems and subsystems, providing error codes and alarms if errors are detected. This diagnostic program runs in background while the control system is in the operating mode.

on-line *1.* Describes the state of a subsystem or piece of computer equipment that is operable and currently connected to the main system. 2. Pertains to a computer that is actively monitoring or controlling a process or operation, or to a computer operation performed while the computer is monitoring or controlling a process or operation. *3.* Describes coincidence of the axis of a drilled or bored hole and its intended axis, without measurable lateral or angular deviation. *4.* Directly controlled by, or in continuous communication with, the computer (for example, on-line storage). 5. Done in real time.

on-line debugging The act of debugging a program while time sharing its execution with an on-line process program.

on-line equipment A computer system and the peripheral equipment or devices in the system in which the operation of such equipment is under control of the central processing unit, and in which information reflecting current activity is introduced into the data processing system as soon as it occurs, thus, directly in-line with the main flow of transaction processing. Clarified by on-line.

opacity *1.* The fraction of light transmitted from a source that is prevented from reaching the observer or instrument receiver, in percent (opacity = [1 − transmittance] × 100). *2.* The fraction of the light emanating from a source that is not transmitted by a smoke-containing path, in percent (opacity = [1 − transmittance] × 100. *3.* The inability of light to pass through a material.

open-cell foam Cellular materials with interconnected cells.

open hearth furnace A reverberatory melting furnace with a shallow hearth and a low roof. The flame passes over the charge on the hearth, causing the charge to be heated both by direct flame and by radiation from the roof and sidewalls of the furnace. In the ferrous industry, the furnace is regenerative.

open loop *1.* A system operating without feedback, or with only partial feedback. *See also* closed loop. *2.* Pertaining to a control system in which there is no self-correcting action for misses of the desired operational condition, as there is in a closed-loop system. *See also* feed-forward control action.

open pore A pore on the surface of a material.

open porosity The volume of open space within a material divided by the volume of the material, expressed as a percent.

open system A system consisting of nodes interconnected by a common communications medium (signal bus) according to established standards which will support temporary connections to manufacturing networks, diagnostics, and other local area networks.

open system interconnection (OSI) A connection between one communication system and another using a standard protocol.

open time The interval of time between application of the adhesive on the adherent and completion of assembly of parts for bonding. During this period, the adhesive-coated surfaces are exposed to air before being assembled together.

open type check valve A check valve, normally open to fluid flow in both directions, which closes when fluid flow in one direction exceeds a predetermined value.

operable condition That mechanical condition of the observed machine which permits the machine and/or its components to perform the intended function at a required performance level and in a manner not detrimental to the machine or the operator.

operating characteristic curve (OC curve) A curve showing the relationship between the probability of acceptance and either lot quality or process quality, whichever is applicable.

operating lift The specified minimum length of time over which the specified continuous and intermittent rating of a transducer applies without change in transducer performance beyond specified tolerances.

operating stress The load that a structural unit must withstand in service.

operating system Computer programs for expediting, controlling, and/or recording computer use by other programs.

operation *1.* A defined action; namely, the act of obtaining a result from one or more operands in accordance with a rule that completely specifies the result for any permissible combination of operands. *2.* A set of tasks or processes, usually performed at one location.

operational readiness The probability that, at any time, the system is either operating satisfactorily or is ready to be placed in operation on demand when used under stated conditions, including stated allowable warning time. Thus, total calendar time is the basis for computation of operational readiness.

operation code A code that represents specific operations.

operator's console A device that enables the operator to communicate with the computer. It can be used to enter information into the computer, to request and display stored data, to actuate various preprogrammed command routines, etc. *See also* programmer's console.

operator station Station that serves as the interface between the operator and other devices on the data highway. The operator can observe and control several devices.

optical character reader A scanning device that can recognize some typewritten characters.

optical computers Computers that use light rather than electricity for all or part of their operation. They perform multiple tasks in parallel as opposed to electronic computers, which would perform those tasks sequentially.

optical data channel A system composed of an image-taking device (for example, camera and lenses), a recording medium for these images (film, disc, magnetic tape), an optical path (for example, fiber-optic cable), and a system for analyzing the images, including any analysis procedure that may modify the content of the data.

optical depth *See* optical thickness.

optical disk A large electronic storage device that uses laser beam patterns to store data.

optical-emission spectrometry Measurement of the wavelength(s) and intensities of visible light emitted by a substance following stimulation.

optical fiber Fine glass stands that transmit data using light signals.

optical filter The portion of the NDIR instrument that eliminates wavelength regions in which interference signals are obtained.

optical flat A transparent disk, usually made of fused quartz, having precisely parallel faces, one face polished for clear vision and the other face ground optically flat; when placed on a surface and illuminated under proper conditions, interference bands can be observed and used to either assess surface contour (relative flatness) or determine differences between a reference gage or gage block and a highly accurate part or inspection gage.

optical gage A gage that measures the image of an object without touching the object itself.

optical glass Glass free of imperfections, such as bubbles, chemical inhomogeneity, or unmelted particles, which degrade its ability to transmit light.

optical grating *1.* Diffraction grating usually employed with other appropriate optics to fabricate a monochromator. Commonly referred to as a Ronchi grating. *2.* A highly accurate device used in precision dimensional measurement which consists of a polished surface, commonly aluminum coating on a glass substrate, onto which close, equidistant and parallel grooves have been ruled.

optical indicator An instrument that plots pressure variations as a function of time by making use of magnification in an optical system coupled with photographic recording.

optical mark reader Using light sensing, a device that reads marks made on special forms.

optical material Any material that is transparent to visible light or to x-ray, ultraviolet, or infrared radiation.

optical plastic Any plastics material that is transparent to light and can be used in optical devices and instruments to take advantage of the superior physical or mechanical properties of the plastics material compared to glass, or the lower cost of the plastics material.

optical pyrometer An instrument that determines the temperature of an object by comparing its incandescent brightness with that of an electrically heated wire; the current through the wire is adjusted until the visual image of the wire blends into the image of the hot surface, and temperature is read directly from the calibrated dial attached to the current adjustment.

optical scanner A light source and phototube combined as a single unit for scanning moving strips of paper or other materials in photoelectric side-register control systems.

optical thickness Specifically, in calculations of the transfer of radiant energy, the mass of a given absorbing or emitting material lying in a vertical column of unit cross-sectional area and extending between two specific levels. Also called optical depth.

option module Any additional device that expands a computer's capability.

optogalvanic spectroscopy A method of obtaining absorption spectra of atomic and molecular species in flames and electrical discharges by measuring voltage and current changes upon laser irradiation.

OR A logic operator having the property that if P is an expression, Q is an expression, R is an expression ..., then the OR of P, Q, R ... is true if at least one expression is true, false if all expressions are false. P OR Q is often represented by $P + Q$, $P \vee Q$.

orange peel A coarse-textured or pebbly surface which may develop as a result of forming metal having coarse grains.

ordered structure The orderly or periodic arrangement of solute atoms on the lattice sites of the solvent.

ore A natural mineral that may be mined and treated for the extraction of any of its components, metallic or otherwise, at a profit.

organic Designating or composed of matter originating in plant or animal life or composed of chemicals of hydrocarbon origin, either natural or synthetic.

organic charge transfer salts Organic compounds exhibiting temperature-dependent electrical, magnetic, and heat transfer properties.

organic coatings Coatings, primarily paints, applied to metallic or other substrates typically to provide corrosion protection and to improve aesthetic characteristics of the material.

organic fiber A fiber derived or composed of matter originating in plant or animal life, or composed of chemicals of hydrocarbon origin, either natural or synthetic.

organic friction material A friction material, having organic binders, substantially formulated with nonmetallic fibers.

organic peroxides Organic compounds containing radical groups combined with oxides in which two atoms of oxygen are linked together, for example diethyl peroxide.

organic solids Solid materials composed of organic materials.

organic zinc-rich paint Coating containing zinc powder pigment and an organic resin.

organosol A suspension of resin in an organic fluid that will not dissolve at room temperature, but will form into a plastic material at elevated temperatures.

orient To place an instrument, particularly one for making optical measurements, so that its physical axis is aligned with a specific direction or reference line.

orientation Alignment with a specific direction or reference line.

orientation alpha A nonuniform alpha structure that results from colonies or domains of platelets or wormy alpha lying at different angles, and having no significance to crystallographic orientation, such that different areas exhibit different aspect ratios and alpha grain outlines.

orientation anneal A final, high-temperature anneal applied principally to flat-rolled electrical steel to develop secondary grain growth and directionality of magnetic properties.

oriented materials Materials whose molecules are aligned in a specific direction or reference line.

orsat A gas-analysis apparatus in which certain gaseous constituents are measured by absorption in separate chemical solutions.

orthorhombic A crystal class in which the three crystal axes are unequal in length and inclined at right angles to each other

orthotropic Having three mutually perpendicular planes of elastic symmetry.

orthotropic laminate Laminates created from orthotropic materials.

orthotropic material Material with three perpendicular planes of elastic balance.

OS *See* operating system.

oscillation(s) Fluctuations or vibrations on each side of a mean value or position. One oscillation is half an oscillatory cycle, consisting of a fluctuation or vibration in one direction; half a vibration.

oscillator An electronic device that generates alternating current power at a frequency determined by the values of certain constants in its circuits.

oscillator crystal A piezoelectric crystal device used chiefly to determine the frequency of an oscillator.

oscillatory instability A condition that exists if a small temporary disturbance or control input causes an oscillatory vehicle response of ever-increasing amplitude about the initial trim.

oscilloscope Instrument for producing visual representations of oscillations or changes in an electric current.

OSF/motif Open Software Foundation X-window based user interface product.

OSI *See* open system interconnection.

osmosis The diffusion of a solvent through a semi-permeable membrane until there is an equal concentration on both sides of the membrane.

ounce A U.S. unit of weight; one ounce (avoirdupois) equals 1/16 pound, and is used for most commercial products; one ounce (troy) equals 1/12 pound, and is used for precious metals.

outgassing *1.* The release of adsorbed or occluded gases and water vapor, usually during evacuation or subsequent heating of an evacuated chamber. *2. See* gassing.

outliers (statistics) In sets of data, values so far removed from other values in the distribution that their presence cannot be attributed to the random combination of change causes.

out-of-round A dimensional condition in which diameter measurements taken in different directions across a nominally circular object are unequal. The difference between the largest and smallest measurement is the amount of out-of-roundness.

output *1.* The yield or product of an activity furnished by man, machine, or system. *2.* The information transferred from the internal storage of a computer to secondary or external storage or to any device outside the computer. *3.* The transfer from internal storage to an external media.

output area *See* output block.

output block *1.* A block of computer words considered as a unit and intended or destined to be transferred from an internal storage medium to an external destination. *2.* A section of internal storage reserved for storing data that are to be transferred out of the computer. Synonymous with output area. *3.* A block used as an output buffer. *See* buffer.

output port Any port from which flow exits for the purpose of a test.

outside diameter (OD) The outer dimension of a circular member such as a rod, pipe, or tube.

outside-in corrosion Corrosion that starts from the outside surface of a material specimen and works inward.

oven A vessel at atmospheric pressure used to provide a controlled and uniform temperature.

oven dry *1.* The condition of a material after heating when there is no further change in size. *2. See* bone dry.

overaging Aging at any combination of time and temperature in excess of that required to obtain the maximum strength and hardness.

overall coefficient of heat transfer *See* heat transfer, coefficient of.

overcuring The beginning of thermal decomposition resulting from too high a temperature or too long a molding time.

overflow *1.* The condition that arises when the result of an arithmetic operation exceeds the capacity of the storage space allotted in a digital computer. *2.* The digit arising from this condition if a mechanical or programmed indicator is included, otherwise the digit may be lost.

overheat To raise the temperature above a desired or safe limit; in metal heat treating, to reach a temperature that results in degraded mechanical or physical properties.

overheating Heating a metal or alloy to such a high temperature that its properties are impaired. When the original properties cannot be restored by further heat treating, by mechanical working, or by a combination of working and heat treating, the overheating is known as burning.

overlay *1.* A thin surface layer of soft bearing material on a harder lining material which, in turn, becomes an intermediate layer of high load capacity. Normally, overlays are deposited by electroplating, and have a nominal thickness of 0.025 mm (0.001 in.) or less. The result is then a trimetal bearing. *2.* The technique of repeatedly using the same blocks of internal storage during different stages of a problem. When one routine is no longer needed in storage, another routine can replace all or part of it.

overlay sheet A smooth, fibrous mat used as the top layer of a laminate to provide a better surface finish.

overload The maximum magnitude of measurand that can be applied to a transducer without causing a change in performance beyond specified tolerance.

overstress A condition in which the severity levels of operation are more than usual or more than specification.

overvoltage The difference between the given and equilibrium electrode potentials of an electrode due to net current flow.

oxalate conversion A process for forming oxalate coatings on steel.

oxazole Compounds that contain a five-membered heterocyclic ring containing one nitrogen and one oxygen atom.

oxidation A reaction in which there is an increase in valence resulting from a loss of electrons. Oxygen is added and decarburization will occur.

oxidation catalyst A catalyst that promotes the oxidation of HC and CO to form water vapor and carbon dioxide.

oxidation-corrosion stability The extent to which a fluid in the presence of oxygen will tend to corrode various metals, or produce products of degradation that will corrode metals.

oxidation degradation Polymer oxidation that reduces the strength of thermoplastics and elastomers by the formation of chain crosslinks.

oxidation of composites A temperature-limiting factor for fiber-reinforced composites whereby the matrix material can withstand higher temperatures before oxidation than the reinforcement material.

oxidation of iron or ferrous alloys A reaction resulting in rust.

oxidation-reduction reactions An oxidizing chemical change, where an element's positive valence is increased (electron loss), accompanied by a simultaneous reduction of an associated element (electron gain).

oxidative wear The formation of oxide films on intermetallic surfaces which results in sliding (wear) rather than bending.

oxide border *See* oxidized surface.

oxide conversion A process that modifies a surface for improved corrosion resistance or coating adhesion.

oxides of nitrogen (NO_x) The sum total of the nitric oxide and nitrogen dioxide in a sample, expressed as nitrogen dioxide.

oxidized surface Surface having a thin, tightly adhering (discolored from straw to blue) oxidized skin extending in from the edge of a coil or sheet. Sometimes called "annealing border."

oxidizer(s) Specifically, substances (not necessarily containing oxygen) that support the combustion of a fuel or propellant.

oxidizing agent A substance that promotes oxidation.

oxidizing atmosphere An atmosphere that tends to promote the oxidation of materials. In steel, the result will include decarburization of the surface.

oxyacetylene welding A welding (or cutting) process in which acetylene is the fuel gas.

oxygen A colorless, tasteless, odorless gas, constituting one-fifth of the atmosphere; also available in liquid form ($-183°C$ [$-297.3°F$]), which vaporizes into gaseous oxygen.

oxygen 17 An isotope of oxygen.

oxygenate An oxygen-containing organic compound, which may be used as a fuel or fuel supplement; for example, various alcohols and ethers.

oxygenated fuel A class of fuels with chemically bound oxygen, for example, oxygenates such as alcohols, vegetable oils, esters, or ethers.

oxygenation The saturation of a substance with oxygen, either by chemical combination, chelation, or by mixture.

oxygen cutting Using the chemical reaction of oxygen and heated metal to effect a melting, or separation of a material.

oxygen meter A device for measuring either the fraction of oxygen or the partial pressure of oxygen in air or in a mixture of oxygen with other gases.

oxyhydrogen welding A welding (or cutting) process in which hydrogen is the fuel gas.

oxynitrides Base for a broad field of nitrogen ceramics utilizing silicon, aluminum, and other elements to produce high-temperature refractory materials.

oxypropane welding A welding (or cutting) process in which propane is the fuel gas.

ozonator An electrical device that generates ozone from oxygen or air.

ozone O_3 An allotropic form of oxygen. It is a gas that has a characteristic odor and is a powerful oxidizing agent.

ozone resistance The ability of a material to withstand the deteriorating effects of ozone (surface cracking).

P

p *See* pressure.

P *See* poise.

Pa *See* pascal.

PA *See* polyamide.

pachymeter An instrument used to measure the thickness of a material such as paper.

packing *1.* A method of sealing a mechanical joint in a fluid system. A material such as oakum or treated asbestos is compressed into the sealing area (known as a packing box or stuffing box) by a threaded seal ring. *2.* In data processing, the compression of data to save storage space.

packing material Material into which powder metal granules are imbedded before sintering.

PAI *See* polyamide-imide resins.

palette In data processing, the range of display colors that will show on a screen.

PAN *See* polyacrylonitrile.

panel *1.* A sheet of material held in a frame. *2.* A section of an equipment cabinet or enclosure, or a metallic or nonmetallic sheet, on which operating controls, dials, instruments, or subassemblies of an electronic device or other equipment are mounted.

paper Felted or matted sheets of cellulose fibers, bonded together and used for various purposes, especially involving printed language, artwork, or diagrams.

paperboard A generic term for a sheet manufactured primarily from cellulosic fibers, produced by conventional pulping and papermaking process and equipment.

paraffin A hydrocarbon having a chain structure, and the general formula C_nH_{2n+2}.

parallax The apparent relative movement of objects in the field of vision as the point of view is shifted laterally. Objects nearer to the observer than the point fixated seem to move against the direction of the shift; objects beyond the point fixated move with the shift. Also refers to the apparent difference in rate of movement of two objects actually moving at the same velocity, but at different distances from the observer.

parallel computer *1.* A computer having multiple arithmetic or logic units that are used to accomplish parallel operations or parallel processing. Contrast with serial computer. *2.* Historically, a computer, some specified characteristic of which is parallel, for example, a computer that manipulates all bits of a word in parallel.

parallel laminate A series of flat cloth-resin layers stacked one upon the other.

parallel operation *1.* The organization of data manipulating within circuitry wherein all the digits of a word are transmitted simultaneously on separate lines in order to speed up operation. *2.* The performance of several actions, usually of a similar nature, simultaneously through provision of individual similar or identical devices for each such action. Particularly flow or processing of information. Parallel operation is performed to save time.

parallel processing Pertaining to the simultaneous execution of two or more sequences of instructions by a computer having multiple arithmetic or logic units.

parallel search storage A storage device in which one or more parts of all storage locations are queried simultaneously. Contrast with associative storage.

parallel task execution Concurrent execution of two or more programs. Also, simultaneous execution of one program and I/O.

parallel transfer A method of data transfer in which the characters of an element of information are transferred simultaneously over a set of paths.

parallel transmission A method of transmitting digitally coded data in which a separate channel is used to transmit each bit making up a coded word.

paramagnetic material A material whose specific permeability is greater than unity and is practically independent of the magnetizing force.

parameter *1.* A variable that is given a constant value for a specific purpose or process. *See* independent variable. *2.* A quantity in a subroutine whose value specifies or partly specifies the process to be performed; it may be given different values when the subroutine is used in different main routines, or in different parts of one main routine, but

usually remains unchanged throughout any one such use. *3.* A quantity used in a generator to specify machine configuration, designate subroutines to be included, or otherwise to describe the desired routine to be generated. *4.* A definable characteristic of an item, device, or system. *See also* measurand.

parameterize To set up for variable execution depending on run-time parameter.

parametric analysis Analysis of the impact on circuit performance of changes in the individual parameters, such as component values, process parameters, temperature, etc.

parent metal The part of welded materials that is outside the heat affected zone.

parity A code that is used to uncover data errors by making the sum of the "1" bits in a data unit either an odd or even number.

parity bit A check bit appended to an array of binary digits to make the sum of all the binary digits, including the check bit, always odd or always even.

parity check The technique of adding one bit to a digital word to make the total number of binary ones or zeros either always even or always odd. This type of checking will indicate a single error in data, but will not indicate the location of the error.

Parr turbidimeter A device for determining the cloudiness of a liquid by measuring the depth of the turbid suspension necessary to extinguish the image of a lamp filament of fixed intensity.

parse To break a command string into its elemental components for the purpose of interpretation.

parsing algorithms Computer routines for the syntactic and/or semantic analysis and restructuring of natural language instructions or data for internal processing.

part *See also* component.

partial annealing aluminum and aluminum alloys An imprecise term used to denote a treatment given cold worked material to reduce the strength to a controlled level or to effect stress relief. To be meaningful, the type of material, the degree of cold work it has undergone, and the time-temperature cycle used must be stated.

partial decarburization Decarburization with loss of carbon sufficient to cause—when examined metallographically—a lighter shade of tempered martensite than that of the immediately adjacent base metal, but insufficient carbon loss to show clearly defined ferrite grains.

partial failure Failures resulting from deviation in characteristics beyond specified limits, but not such as to cause complete lack of the required function.

particle A minute piece of matter with observable length, width, and thickness; usually measured in micrometers.

particle accelerator *1.* Any of several different types of devices for imparting motion to charged atomic particles. *2.* Specifically, devices for imparting large kinetic energy to charged particles, such as electrons, protons, deuterons, and helium ions.

particle concentration The number of individual particles per unit volume of liquid.

particle counter *See* radiation counter.

particle detector *See* radiation counter.

particle diameter (Feret's diameter) The distance between two tangents, on opposite sides of the particle profile, that are parallel to some fixed direction.

particle diameter (Martin's diameter) The length of the line that bisects the image of the particle; the line may be drawn in any direction that is to be maintained constant for all the image measurements.

particle diameter (Sieve diameter) The width of the minimum square aperture through which the particle will pass.

particle diameter (Stoke's diameter) The diameter of a sphere having the same density as the particle and the same free-falling speed as the particle in a laminar flow region (Re 0.2) in a fluid of the same density and viscosity as the fluid concerned.

particle flux *See* flux.

particle-induced x-ray emission Trace element analysis in which beams of ions are directed at a thin foil on which the sample has been deposited. The energy spectrum of the emitted x-rays is then measured.

particle morphology The form and structure of individual particles.

particle retention rating test A test that determines the ability of a filter to separate and retain particles of a specified size. It is usually conducted as a single pass test with classified contaminant of a narrow size range, and results are expressed as percent efficiency.

particles Elementary subatomic particles such as protons, electrons, or neutrons. Very small pieces of matter.

particle size *1.* A measure of dust size, expressed in microns or percent passing through a standard mesh screen. *2.* The size of a particular powder metal element determined by screening.

particle size analysis The whole of the operation by which a particle size distribution is determined.

particle-size distribution The percent by weight of each fraction that a powder metal has been classified.

particle surface diameter The diameter of a sphere having the same surface area as the particle.

particle volume diameter The diameter of a sphere having the same volume as the particle. Remark: This dimension is the basis of measurement of a Coulter Counter.

particulate composite Metal or polymer particles suspended in a matrix.

particulate composition tests A quantitative and qualitative measurement of effluent solids from an inflator.

particulate matter Any dispersed solid or liquid in which the individual aggregates are larger than 0.001 μm in diameter, excluding uncombined water.

parting agent Material used to prevent sealant from sticking to a surface. Also known as release agent.

partition A contiguous area of computer memory within which tasks are loaded and executed.

parts per million Describing fractional defective, obtained by multiplying percent defective by 10,000; for example, $0.01\% = 100$ ppm.

parts per million carbon (ppmC) The mole fraction times 10^6 of hydrocarbon measured on a methane equivalence basis. Thus 1 ppm of methane is indicated as 1 ppmC. To convert ppm concentration of any hydrocarbon to an equivalent ppmC value, multiply ppm concentration by the number of carbon atoms per molecule of the gas. For example, 1 ppm propane translates as 3 ppmC hydrocarbon; 1 ppm hexane as 6 ppmC hydrocarbon.

part temperature The temperature obtained by locating a temperature-sensing device in or on the specimen or workpiece. In most cases, temperature gradients that develop within flexing rubber specimens make it necessary to define the precise points and techniques to measure temperature.

parylene A polymer used to produce dielectric films.

PAS *See* polyarylsulfone.

pascal Metric unit for pressure or stress.

pass *1.* A single circuit through a process, such as gases through a boiler, metal between forging rolls, or a welding electrode along a joint. *2.* In data processing, the single execution of a loop. *3.* The shaped open space between rolls in a metal-rolling stand. *4.* A confined passageway, containing heating surface, through which a fluid flows in essentially one direction. *5.* A single circuit of an orbiting satellite around the earth. *6.* A transit of a metal-cutting tool across the surface of a workpiece with a single tool setting.

passivating A process for the treatment of stainless steel in which the material is subjected to the action of an oxidizing solution which augments and strengthens the normal protective oxide film, providing added resistance to corrosive attack.

passivation A reduction of the anodic reaction rate of an electrode involved in corrosion.

passivation of metal The chemical treatment of a metal to improve its resistance to corrosion.

passive Refers to a metal corroding under the control of a surface reaction product.

passive-active cell A cell, the emf of which is due to the potential difference between a metal in an active state and the same metal in a passive state.

passive AND gate An electronic or fluidic device that generates an output signal only when both of two control signals appear simultaneously.

passive element An element that is not active, that is, does not control energy; for example, a resistor, a capacitor, or an inductor.

passive metal A metal that has a natural or artificially produced surface film which makes it resistant to electrochemical corrosion.

passive transducer A transducer that has no source of power other than the input signal(s), and whose output signal-power cannot exceed that of the input. Note: The definition of a passive transducer is a restriction of the more general passive network; that is, one containing no impressed driving forces.

paste Mixtures with characteristic soft or plastic consistencies.

paste solder Finely divided solder alloy combined with a semisolid flux.

patenting In wire making, a heat treatment applied to medium-carbon or high-carbon steel before the drawing of wire or between drafts. This process consists of heating to a temperature above the transformation range and then cooling to a temperature below A_{e1} in air or in a bath of molten lead or salt.

patina The green coating that forms on the surface of copper and copper alloys exposed to the atmosphere. Also used to describe the appearance of a weathered surface of any metal.

pattern recognition The recognition of shapes or other patterns by a machine system.

PAW *See* plasma arc welding.

Pb Chemical symbol for lead.

PB *See* polybutylene.

PBB *See* polybrominated biphenyls.

PBI *See* polybenzimidazole.

PBT *See* polybutylene terephthalate.

PBW *See* proportional bandwidth.

PC *See* polycarbonate.

PC/ABS A polymer derived from polycarbonate and ABS with the high impact strength of ABS and superior processing attributes to those of polycarbonate.

PCBN Polycrystalline boron nitride used to produce the cutting edge of long-lasting cutting tools.

PCM *See* phase change materials.

PCM Abbreviation for plug-compatible manufacturer. Abbreviation for pulse code modulation.

PC/PBT A polymer of polycarbonate and PBT with greater resistance to chemical attack than polycarbonate.

Pd Chemical symbol for palladium.

PDD *See* programmable data distributor.

PDU *See* protocol data unit.

PE *See* polyethylene.

peak aging Aging at whatever combination of time and temperature produces maximum strength or hardness.

peak amplitude The maximum value of sound pressure attained, expressed in pounds per square inch (kilopascals). It may also be expressed as sound pressure level in decibels (dB).

peak detector A detector, the output voltage of which is the peak value of the applied signal.

peak-to-peak amplitude *See* double amplitude.

peak-to-peak ripple The greatest variations due to ripple above and below the nominal DC level are called the upper peak and lower peak, respectively. Peak-to-peak ripple is the difference between the upper peak and lower peak voltages.

pearlite An aggregate in steel of ferrite and cementite.

peat Dark brown or black residuum produced from the partial decomposition and disintegration of mosses, hedges, trees, and other plants that grow in marshes and other wet places.

pebbles *See* orange peel.

Peclet number A nondimensional number arising in problems of heat transfer in fluids.

PEEK A class of semicrystalline polymers called polyacrylene ethers, used as molding compounds and as composite matrix materials.

peel strength *See* peel test.

peel test A test in which a sealant is cured on a selected substrate, then peeled from it using one of several types of testing machines.

peening Surface hardening of metal by impingement with loose shot.

peening media Spherical or quasi-spherical material used in the controlled shot peening process.

PEI *See* polyetherimide.

pel *See* pixel.

pellet A compacted unit of gas generant that is used in the inflator.

pelleted substrate A catalyst substrate having such forms as pebbles, beads, small cylinders, or small spheres.

pellicle An extremely thin, tough membrane that is stretched over a frame. Because of its thinness, it transmits some light and reflects other light, and hence can serve as a beam splitter. Its thinness avoids the problem of ghost reflections sometimes produced by other beam splitters. Usually found as beam splitters in interferometers.

Peltier effect The principle in solid-state physics that forms the basis of thermocouples—if two dissimilar metals are brought into electrical contact at one point, the difference in electrical potential at some other point depends on the temperature difference between the two points.

penalty function In mathematics, a function used in treating maxima and minima problems subject to restraints.

penetrameter A stepped piece of metal used to assess density of exposed and developed radiographic film, and to determine relative ability of the radiographic technique to detect flaws in a workpiece.

penetrant A liquid that penetrates. For liquid penetrant inspection, a liquid with low surface tension.

penetrant, fluorescent An inspection penetrant that is characterized by its ability to fluoresce when excited by black light.

penetrant, post emulsifiable A penetrant that requires the application of a separate emulsifier to render the surface penetrant water-washable.

penetrant, water-washable A penetrant with built-in emulsifier which makes it directly water-washable.

penetrating particles *See* corpuscular radiation.

penetration *1.* Distance from the original base metal surface to the point where weld fusion ends. *2.* A surface defect on a casting where molten metal filled surface voids in the sand mold. *3.* The diffusion of an adhesive into a base material.

penetration number A measure of the consistency of materials such as waxes and greases, expressed as the distance that a standard needle penetrates a sample under specified ASTM test conditions.

penetrometer An instrument for determining penetration number.

Penning discharge A direct-current discharge in which electrons are forced to oscillate between two opposed cathodes and are restrained from going to the surrounding anode by the presence of a magnetic field.

Penning effect An increase in the effective ionization rate of a gas due to the presence of a small number of foreign metastable atoms.

pentode An electron tube containing five electrodes—an anode, a cathode, a control electrode, and two others, which usually are grids.

percent conductivity Conductivity of a material expressed as a percentage of that of copper.

percent defective The number of defective pieces in a lot or sample, expressed as a percent.

percent elongation An expression of the ductility of a metal derived from tensile tests.

percent of dilution The amount of diluent in the mixture expressed in whole numbers.

perfect combustion The complete oxidation of all the combustible constituents of a fuel, utilizing all the oxygen supplied.

perfect diffusion Diffusion in which light is scattered uniformly in all directions by the diffusion medium.

perfect gas *See* ideal gas.

perfect vacuum A reference datum analogous to a temperature of absolute zero that is used to establish scales for expressing absolute pressures.

perfluoroalkoxy (PFA) A fluorocarbon resin that offers excellent electrical characteristics, high temperature resistance, chemical inertness, and flame resistance.

perforation corrosion Penetration of a material due to corrosion. Perforation corrosion is usually associated with inside-out corrosion.

performance Degree of effectiveness of operation.

performance characteristic A qualitative or quantitative measurement unique to a piece of equipment or a system, that is evident only during its test or operation.

performance data Information on the way a material or device behaves during actual use.

perfuse *1.* To sprinkle, cover over, or permeate with a liquid. *2.* To pour or spread (as with a liquid) through or over something.

perfusion A perfusing or being perfused. *See* diffusion and suffusion.

period *1.* Of a periodic function, the smallest increment of the independent variable that can be repeated to generate the function. *2.* Of an undamped instrument, the time between two successive transits of the pointer through the rest position in the same direction following a step change in the measured quantity.

periodic function An oscillating quantity whose values repeatedly recur for equal increments of the independent variable.

periodic processes *See* cycles.

periodic vibration Periodic vibration exists in a system when recurring cycles take place in equal time intervals.

peripheral *1.* A supplementary piece of equipment that puts data into, or accepts data from, the computer (printers, floppy disc memory devices, videocopiers). *2.* Any device, distinct from the central processor, that can provide input or accept output from the computer.

peritectic An isothermal reaction in which a solid and liquid phase react to produce a second solid phase during cooling.

peritectoid An isothermal reaction where two solid phases react to produce a third solid phase during cooling.

permanence The characteristic of a material to resist change in its properties over time and environmental exposure.

permanent magnet A shaped piece of ferromagnetic material that retains its magnetic field strength for a prolonged period of time following removal of the initial magnetizing force.

permanent mold A metal mold (other than an ingot mold) of two or more parts that is used repeatedly for the production of many castings of the same shape. Liquid metal is poured in by gravity.

permanent pressure drop The unrecoverable reduction in pressure that occurs when a fluid passes through a nozzle, orifice, or other throttling device.

permanent set The deformation that remains after a specimen has been under load.

permeability *1.* The relationship of flow per unit area to differential pressure across a filter medium. *2.* The measure of how much better a given material is than air as a path for magnetic lines or force. It is equal to the magnetic induction (B) in gausses, divided by the magnetizing force (H) in oersteds.

permeameter *1.* A device for determining the average size or surface area of small particles; it consists of a powder bed of known dimensions and degree of packing through which the particles are forced under pressure. Particle size is determined from flow rate and pressure drop across the bed; surface area, from pressure drop. *2.* A device for determining the coefficient of permeability by measuring the gravitational flow of fluid across a sample whose permeability is to be determined. *3.* An instrument for determining magnetic permeability of a ferromagnetic material by measuring the magnetic flux or flux density in a specimen exposed to a magnetic field of a given intensity.

permittivity Preferred term for dielectric constant.

perovskites Minerals with a close-packed lattice and the general formula ABX_3, where A and B are metals and X is a nonmetal, usually O.

persistence The continuation of luminance of a phosphor after electron excitation has been removed.

PERT *See* program evaluation and review technique.

perturbation generator An instrument that simulates typical data link perturbations such as blanking, noise, bit rate jitter, baseline offset, and wow.

PESV *See* polyether sulfone.

PET *See* polyethylene terephthalate.

petri nets Abstract, formal models of the information flow in systems with discrete sequential or parallel events. The major use has been the modeling of hardware systems and software concepts of computers.

petroleum Naturally occurring mineral oil consisting predominately of hydrocarbons.

petroleum engineering A branch of engineering that deals with drilling for and producing oil, natural gas, and liquifiable hydrocarbons.

petroleum products Materials derived from petroleum, natural gas, and asphalt deposits. Includes gasolines, diesel and heating fuels, lubricants, waxes, greases, petroleum coke, petrochemicals, and sulfur.

PEX—PHIGS+ Extensions to X-window.

pf *See* picofarad.

PF *See* phenol formaldehyde.

PFA *See* perfluoroalkoxy.

ph *See* phot.

pH The symbol for the measurement of acidity or alkalinity. Solutions with a pH reading of less than 7 are acid; solutions with a pH reading of more than 7 are alkaline on the pH scale of 0 to 14, where the midpoint of 7 is neutral.

phase *1.* A portion of a physical system (solid, liquid, or gas) that is homogeneous throughout, has definable boundaries, and can be separated physically from the other phases. The act of separating different phases is commonly termed "phase separation." *2.* The relationship between voltage and current waveforms in a-c electrical circuits. *3.* A microstructural constituent of an alloy that is physically distinct and homogeneous. *4.* For a particular value of the dependent variable in a periodic function, the fractional part of a period that the independent variable differs

from some arbitrary origin. *5.* In batch processing, an independent process-oriented action within the procedural part of a recipe. The phase is defined by boundaries that constitute safe and logical points where processing can be interrupted.

phase angle *1.* A measure of how the output response of a system lags or leads a sinusoidal input to the system. *2.* The difference between the phase of current and the phase of voltage in an alternating-current signal, usually determined as the angle between current and voltage vectors plotted on polar coordinates. *3.* A measure of the propagation of a sinusoidal wave in time or space from some reference instant or position on the wave. *See also* phase shift.

phase angle firing A method of operation for an SCR stepless controller in which power is turned on for the proportion of each half-cycle in the a-c power supply necessary to maintain the desired heating level.

phase change The conversion of one material state to another, such as liquid to solid.

phase change materials Materials undergoing solid/liquid phase transformations, whose latent heat of fusion properties are used to store and deliver thermal energy, usually solar energy.

phase shift *1.* The time difference between the input and output signal, or between any two synchronized signals, of a control unit, system, or circuit, usually expressed in degrees or radians. *2.* A change in phase angle between the sinusoidal input to an element and its resulting output.

phase shift keying *1.* A form of phase modulation in which the modulating function shifts the instantaneous phase of the modulated wave among predetermined discrete values. *2.* A form of PCM achieved by shifting the phase of the carrier, for example, ±90 degrees to represent "ones" and "zeros."

Phelps vacuum gage A modified hot-filament ionization gage useful for measuring pressures in the range 10^{-5} to 1 torr.

phenol formaldehyde (PF) An early thermoset polymer more commonly called bakelite.

phenolic A thermosetting resin produced by the condensation of an aromatic alcohol with an aldehyde, particularly of phenol with formaldehyde.

phenolic resin A thermosetting resin used in many elevated temperature applications.

phenoxy resins A thermoplastic resin used for adhesives and coatings.

phenylamine *See* aniline.

phenylsilane resins A thermosetting copolymer of silicone and phenolic resins.

Philips gage An instrument that measures very low gas pressure (vacuum) indirectly by determining current flow from a glow discharge device.

pH meter An instrument for electronically measuring electrode potential of an aqueous chemical solution and directly converting the reading to pH (a measure of hydrogen ion concentration, or degree of acidity).

phon A unit of loudness level equivalent to a unit pressure level in decibels of a 1000-Hz tone.

Phong shading A computer-intensive shading technique used in the display of objects whereby the light intensity for every point on the object is calculated to obtain a very realistic image.

phosphate coatings Protective coatings formed by reaction of a metallic substrate with an acidic phosphate-containing solution. The primary role of the phosphate coating is to enhance adhesion of the primer (electrocoat or other) to the metal. Phosphate coatings are typically Zn, Fe, Zn-Ni, or Zn-Ni-Mn phosphates.

phosphatizing Forming an adherent phosphate coating on metal by dipping or spraying with a solution to produce an insoluble, crystalline coating of iron phosphate which resists corrosion and serves as a base for paint.

phosphazene A ring or chain polymer that contains alternating phosphorus and nitrogen atoms, with two substituents on each phosphorus atom.

phosphor *1.* Material that gives off visual radiant energy when bombarded by electrons or ultraviolet light. *2.* A phosphorescent material.

phosphor bronze A hard copper-tin alloy, deoxidized with phosphorus, and sometimes containing lead to enhance its machinability.

phosphorescence One form of luminescence. The ability of certain substances to continue to emit light long after the source of excitation energy has been removed.

phosphors Phosphorescent substances such as zinc sulfide, which emit light when excited by radiation, as on the scope of a cathode ray tube.

phot (ph) The CGS unit of illuminance, equal to one candela per square centimeter.

photoacoustic spectroscopy An optical technique for investigating solid and semisolid materials, in which the sample is placed in a closed chamber filled with a gas and illuminated with monochromatic radiation of any desired wavelength, and with intensity modulated at some acoustic frequency. Absorption of radiation results in a periodic heat flow from the sample, which generates sound detectable with a sensitive microphone.

photocathodes Electrodes used for obtaining photoelectric emission.

photocell (photoelectric cell) *1.* A solid-state photosensitive electron device in which use is made of the variation of the current-voltage characteristic as a function of incident radiation. *2.* A device exhibiting photovoltaic or photoconductive effects.

photochemical oxidants Any of the chemicals that enter into oxidation reactions in the presence of light or other radiant energy.

photochemical reactions Chemical reactions that involve either the absorption or emission of radiation.

photochemistry *See* photochemical reactions.

photoconductive cell *1.* Photoelectric cell whose electrical resistance varies with the amount of illumination falling upon the sensitive area of the cell. *2.* A transducer that converts the intensity of EM radiation, usually in the IR or visible bands, into a change of cell resistance.

photoconductor A type of conductor that changes its resistivity when illuminated by light; the changes in resistance can be measured to determine the amount of incident light.

photodegradation Increased oxidation of polymer materials due to exposure to ultraviolet light.

photodissociation The dissociation (splitting) of a molecule by the absorption of a photon. The resulting components may be ionized in the process (photoionization).

photoelastic stress analysis A visual full-field technique for measuring stresses in parts and structures. When a photoelastic material is subjected to forces and viewed under polarized light, the resulting stresses are seen as color fringe patterns. Interpretation of the colorful pattern will reveal the overall stress distribution, and accurate measurements can be made of the stress directions and magnitudes at any point.

photoelectric cell *1.* Transducer that converts electromagnetic radiation in the infrared, visible, and ultraviolet regions into electrical quantities such as voltage, current, or resistance. *2. See* photocell.

photoelectric effect The emission of an electron from a surface as the surface absorbs a photon of electromagnetic radiation. Electrons so emitted are termed photoelectrons.

photoelectric hydrometer A device for measuring specific gravity of a continuously glowing liquid, in which a weighted float, similar to a hand hydrometer, rises or falls with changes in liquid density, changing the amount of light that is permitted to fall on a sensitive phototube whose output is calibrated in specific gravity units.

photoelectric photometer A device that uses a photocell, phototransistor, or phototube to measure the intensity of light. Also known as an electronic photometer.

photoelectric pyrometer An instrument that measures temperature by measuring the photoelectric emission that occurs when a phototube is struck by light radiating from an incandescent object.

photoelectric threshold The amount of energy in a photon of light that is just sufficient to cause photoelectric emission of one bound electron from a given substance.

photoelectrochemical devices Electrochemical devices powered by light or other incident radiation to produce electricity and/or chemical fuels (for example, hydrogen).

photoelectroluminescence The use of light or other electromagnetic energy to create an electric current which, in turn, induces electroluminescence.

photoelectrons Electrons that have been ejected from their parent atoms by interaction between those atoms and high-energy photons. *See* photoelectric effect.

photoemissive Emitting electrons when illuminated, as in electrical currents.

photoemissive tube photometer A device that uses a tube made of photoemissive material to measure the intensity of light; it is very accurate, but requires electronic amplification of the output current from the tube; it is considered chiefly a laboratory instrument.

photogrammetry The art or science of obtaining reliable measurements by means of photography.

photoionization The ionization of an atom or molecule by the collision of a high-energy photon with the particle.

photoluminescence *1.* Nonthermal emission of electromagnetic radiation that occurs when certain materials are excited by absorption of visible light.

2. The re-emission of light power absorbed by an atom.

photometer An instrument used for the measurement of photometric quantities. Photometers may be used for luminance or luminous intensity measurements. Spectral and time response are important characteristics.

photometry *1.* The study of the measurement of the intensity of light. *2.* Any of several techniques for determining the properties of a material or for measuring a variable quantity by analyzing the spectrum or intensity, or both, of visible light.

photon(s) *1.* According to the quantum theory of radiation, the elementary quantities of radiant energy. They are regarded as discrete quantities having a momentum equal to hv/c, where h is the Planck constant, v is the frequency of the radiation, and c is the speed of light in a vacuum. *2.* A quantum of electromagnetic radiation.

photon counting A measurement technique used for measuring low levels of radiation, in which individual photons generate signals that can be counted.

photophoresis Production of unidirectional motion in a collection of very fine particles, suspended in a gas or falling in a vacuum, by a powerful beam of light.

phthalate esters A variety of widely used plasticizers.

physical catalyst Energy that is capable of starting or changing a chemical reaction.

physical input *See* measurand.

physical properties The properties other than mechanical properties, that pertain to the physics of a material; for example, density, electrical conductivity, thermal expansion, etc. Often improperly used to express mechanical properties.

physiochemical instability Change from an initial material bulk property, such as strength, resiliency, volume, composition, etc., as a result of age, pressure, temperature, etc.

PI *See* polyimide.

piano wire Carbon steel wire (0.75 to 0.85% C) cold drawn to high tensile strength and uniform diameter.

PIC *See* pressure-impregnation-carbonization.

pick count The number of yarns per linear inch of a fabric.

pickle A solution or process used to loosen or remove corrosion products such as scale or tarnish.

pickle liquor Spent pickling solution.

pickle patch A tightly adhering oxide or scale not removed during the pickling process.

pickle stain Discoloration present after pickling.

pickling The removal of surface oxides from metal by chemical or electrochemical reaction, commonly at an elevated temperature, i.e., not room temperature.

pickup A mixture of aluminum fines, lubricant, and aluminum oxide generated from roll coating during the deformation process. Metal particles adhering to a work roll or tool which cause a series of dents, scratches, or pits on a sheet or part.

picofarad (pF) A measure of capacitance (10^{-12}) farads.

PID Proportional, integral, and derivative.

piezoelectric The property of certain crystals which: (a) produce a voltage when subjected to a mechanical stress, (b) undergo mechanical stress when subject to a voltage.

piezoelectric ceramics Ceramic materials with piezoelectric properties similar to those of some natural crystals.

piezoelectric effect The generation of an electric potential when pressure is applied to certain materials or, conversely, a change in shape when a voltage is applied to such materials. The changes are small, but piezoelectric devices can be used to precisely control small motions of optical components.

piezoelectricity The property exhibited by some asymmetrical crystalline materials whereby, when subjected to strain in suitable directions, they develop polarization proportional to the strain.

piezoelectric transducer *See* electrostriction transducer.

piezoelectro polymers Polymers that produce an electrical charge when stressed or produce a mechanical response under an electrical field.

piezoid A piezoelectric crystal adapted for use by attaching electrodes to its surface or by other suitable processing.

piezometer *1.* An instrument for measuring fluid pressure. *2.* An instrument for measuring compressibility of materials.

pig *1.* A crude metal casting, usually of primary refined metal, intended for remelting to make alloys. *2.* An in-line scraper for removing scale and deposits from the inside surface of a pipeline; a holder containing brushes, blades, cutters, swabs, or a combination is forced through the pipe by fluid pressure.

pig iron *1.* High-carbon iron made by reduction of iron ore in a blast furnace. *2.* Cast iron in the form of pigs.

pill *See* pellet.

pilot plant A test facility, built to duplicate or simulate a planned process or full-scale manufacturing plant, used to gain operating experience or evaluate design alternatives before the full-scale plant is built.

pimple A small imperfection on the surface of a material.

pinchers Fernlike ripples or creases usually diagonal to the rolling direction.

pinch-off See cutoff.

pin expansion test Forcing a tapered pin into the open end of tubing material to determine expandability or any structural defects.

pin hole A microscopic hole through an insulating (glass) layer. A defect.

pinhole detector A photoelectric device that can detect small holes or other defects in moving sheets of material.

pin holes Very small cavities on the surface of a material.

pinholing Small holes, the size of a pinpoint, in the surface of the topcoat that result when small bubbles burst as the paint cures.

pinning Sites within a superconducting material that are produced by localizing inclusions, dislocations, voids, etc., which provide a means of resisting flux motion (flux jumps) due to Lorentz forces.

PIN photodiode A semiconductor diode light detector in which a region of intrinsic silicon separates the p- and n-type materials. It offers particularly fast response and is often used in fiber-optic systems.

pipe 1. A tubular structural member used primarily to conduct fluids, gases, or finely divided solids; it may be made of metal, clay, ceramic, plastic, concrete, or other materials. 2. A general class of tubular mill products made to standard combinations of diameter and wall thickness. 3. An extrusion defect caused by the oxidized metal surface flowing toward the center of the extrusion at the back end. 4. In sheets, appears as a separation midway between the surfaces containing oxide inclusions. 5. In ingots, a cavity formed by contraction and associated bridging of the metal during solidification. The cavity usually occurs at the top section of the ingot because the liquid metal flows into the cavity in all but the last portion that solidifies.

pipet A tube used to deliver accurate volumes of a liquid during titration.

pit Localized corrosion of a metal surface, confined to a small area which takes the form of a cavity.

pitot-static tube A combination of a pitot tube and a static port arranged coaxially or otherwise parallel to one another and mounted externally on an aircraft (generally on the wing, the nose, or the vertical stabilizer) in a position to sense the air flow and pressure undisturbed by the flow over or around other structures of the aircraft. Used principally to determine airspeed from the difference between impact and static pressures.

pitot tube Open-ended tube or tube arrangement that, when pointed upstream, may be used to measure the stagnation pressure of the fluid for subsonic flow; or the stagnation pressure behind the normal shock wave of the tube for supersonic flow.

pitting A concentrated attack by oxygen or other corrosive chemicals, producing a localized depression in the metal surface.

pitting corrosion A type of perforation corrosion. Pitting corrosion is highly localized corrosion resulting in deep penetration at only a few spots.

pitting factor The ratio of the depth of the deepest pit resulting from corrosion divided by the average penetration as calculated from weight loss.

pixel (pel) *1.* Shortened form of "picture elements." Pixels are image resolution elements in vidicon-type detectors. *2.* The smallest addressable picture element of a graphics display.

PLA *See* programmable logic array.

plain weave A weaving pattern in which the warp and fill fibers alternate; that is, the repeat pattern is warp/fill/warp/fill.

planar Lying essentially in a single plane.

Planck's constant, *h* The ratio of the energy of a photon to its intensity.

plane of polarization Plane of incidence of a beam polarized by reflection at the polarizing angle.

plane of vibration Plane parallel to the electric field of a polarized beam of radiant energy and perpendicular to the plane of polarization.

plane polarized wave An electromagnetic wave in a homogeneous isotropic medium that has been generated, or modified by the use of filters, so that the electric field vector lies in a fixed plane which also contains the direction of propagation.

plane strain A deformation of a body in which the displacements of all points in the body are parallel to a given plane, and the displacement values are not dependent on the distance perpendicular to the plane.

plane strain fracture toughness, K_{IC} *1. See* stress-intensity factor. *2.* An important mechanical property describing a measure of sensitivity to pre-existing surface defects or cracks and thereby tolerances of same.

plane stress A stress condition in which the stress in the thickness direction is zero.

plane stress fracture toughness, K_c *See* stress intensity factor.

planimetric method A method of measuring grain size in which the grains within a definite area are counted.

planishing Producing a smooth surface finish on metal by a rapid succession of blows delivered by highly polished dies or by a hammer designed for the purpose, or by rolling in a planishing mill.

plasma A gas made up of charged particles. Note: Usually plasmas are neutral, but not necessarily so, as for example, the space charge in an electron tube.

plasma arc cutting Use of plasma torches for cutting hard materials at extremely high temperatures.

plasma arc welding (PAW) Welding in which metals are heated with a constricted arc between an electrode and the workpiece (transferred arc), or the electrode and the constricting nozzle (nontransferred arc). Shielding is obtained from the hot, ionized gas issuing from the orifice, which may be supplemented by an auxiliary source of shielding gas.

plasma etching Removal of material by use of a focused plasma beam.

plasma focus A highly compressed plasma.

plasma jet excitation The use of hot ionized gas jet to excite the elements in a sample for analysis.

plasmas (physics) Electrically conductive gases which are composed of neutral particles, ionized particles, and free electrons but which, taken as a whole, are electrically neutral. Plasmas are further characterized by relatively large intermolecular distances, large amounts of energy stored in the internal energy levels of the particles, and the presence of plasma sheaths at all boundaries of the plasma. Plasmas are sometimes referred to as a fourth state of matter.

plasma spraying Propelling molten coating material against the base metal by use of a hot, ionized gas torch.

plasma torches Burners that attain 50,000°C temperatures by the use of plasma gas injected into an electric arc. Plasma torches are used for welding, spraying molten metal, and cutting hard rock or hard metals.

plastic An imprecise term generally referring to any polymeric material, natural or synthetic. Its plural, plastics, is the preferred term for referring to the industry and its products.

plastic-clad silica A step index optical fiber in which a silica core is covered by a transparent plastic cladding of lower refractive index. The plastic cladding is usually a soft material, although hard-clad versions have been introduced.

plastic deformation The changes in dimensions of items caused by stress, that are retained after the stress is removed.

plastic fibers Optical fibers in which both core and cladding are made of plastic material. Typically their transmission is much poorer than that of glass fibers.

plastic flow *1.* Material deformation under load. *2.* The flow of semi-solids under pressure.

plastic foam *See* expanded plastic.

plasticity The degree or rate at which unvulcanized elastomer and elastomeric compounds will flow when subjected to forces of compression, shear, or extrusion. *See also* plastic properties.

plasticizer A material that, when incorporated in an elastomer or a polymer, will change its hardness, flexibility, processability, and/or plasticity.

plastic memory The tendency of a material that is stretched or bent to return to its original shape when heated.

plasticorder A laboratory device for measuring temperature, viscosity, and shear-rate in a plastics material, which can be used to predict its performance.

plastic properties The tendency of a loaded body to assume a deformed state other than its original state when the load is removed.

plastisols A group of vinyl resins and plasticizers that can be cast, molded, or converted to film.

plastometer An instrument for determining flow properties of a thermoplastic resin by forcing molten resin through a fixed orifice at specified temperature and pressure.

plate A flat rolled metal product of some minimum thickness arbitrarily dependent on the type of metal. For example, if the metal is steel, the minimum thickness is 3/16 inch.

platelet alpha A relatively coarse acicular alpha, usually with low aspect ratios. This microstructure arises from cooling alpha or alpha-beta alloys at a slow rate from temperatures at which a significant fraction of beta phase exists.

plating The electrolytic application of one metal over another.

platinum A precious metal used as a catalyst, and in the manufacture of jewelry and thermocouples.

PLC *See* programmable logic controller.

plenum *1.* A condition in which air pressure within an enclosure is greater than barometric pressure outside the enclosure. *2.* An enclosure through which gas or air passes at relatively low velocities.

plotter *1.* A device for automatically graphing a dependent variable on a visual display or flat board, in which a movable pen or pencil is positioned by one or more instrument control signals. *2.* A hardware device that plots on paper the magnitudes of selected data channels, as related to each other or to time.

plotter/printer A plotter that can also print alphanumeric data from the computer.

plus sieve The amount of a metal powder that remains after passing through a standard sieve.

plutonium A radioactive element produced by neutron adsorption in uranium-238.

ply *1.* In general, fabrics or felts consisting of one or more layers. *2.* The layers that make up a stack. 3. Yarn resulting from twisting operations. 4. A single layer of prepreg. 5. A single pass in filament winding. 6. A sheet or layer that is considered to be one discrete piece of manufactured material such as fabric, tape, or adhesive film.

ply bond *See* cohesive strength.

plymetal Metal consisting of one or more bonded layers of different metals.

ply orientation The arrangement of bonded layers comprising laminated materials to obtain optimal strength or other characteristics.

ply rating Index of tire strength; does not necessarily represent the actual number of plies in the tire. It is used to relate a given size tire to its load and inflation limits, as does load range. Ply rating has been replaced by "load range."

ply separation Parting of rubber compound between adjacent plies.

ply steer force The component of lateral force offset that does not change sign (with respect to the tire axis system) with a change in direction of rotation (positive along Y axis). The force remains positive when it is directed away from the serial number on the right side tire and toward the serial number on the left side tire.

PM *See* modulation.

PMMA *See* polymethyl methacrylate.

PMR polyimides A group of elevated temperature-resistant polymers.

PNPN diode A semiconductor device that may be regarded as a two-transistor structure with two separate emitters feeding a common collector.

PNP transistor A transistor consisting of two p-type regions separated by an n-type region.

Po Chemical symbol for polonium.

Pockel's cell A device in which the Pockel's effect is used to modulate light passing through the material. The modulation relies on rotation of beam polarization caused by the application of an electric field to a crystal; the beam then has to pass through a polarizer, which transmits a fraction of the light dependent on its polarization.

Pockel's effect *See* birefringence.

point defects Defects that occur in a crystal lattice causing strain.

pointer *1.* A needle-shaped or arrowhead-shaped element whose position over a scale indicates the value of a measured variable. *2.* (data processing) (a) A data string that tells the computer where to find a specific item. (b) Similar to or the same as a cursor on a computer screen.

point-to-point numerical control A simple form of numerical control in which machine elements are moved between programmed positions without particular regard to path or speed control. Also known as positioning control.

poise The CGS unit of dynamic viscosity, which equals one dyne-second per cm^2; the centipoise (cP) is more commonly used.

Poiseuille flow Laminar flow of gases in long tubes at pressures and velocities such that the flow can be described by Poiseuille's equation. *See also* laminar flow.

Poisson process *See* stochastic processes.

Poisson's ratio The value of the ratio of the transverse strain to the axial strain in a body exposed to uniaxial stress.

POL *See* problem-oriented language.

polar Describes an unsymmetrical molecule such as water or sulfur dioxide, in which the mean center of all the electronic charges does not coincide with the mean electrical center of the nuclei.

polarimeter Instrument for determining the degree of polarization of electromagnetic radiation, specifically the polarization of light.

polarimetry Chemical analysis in which the amount of substance present in a solution is estimated from the amount of optical rotation (polarization) that occurs when a beam of light passes through the sample.

polariscope Instrument for detecting polarized radiation and investigating its properties.

polarity The sign of the electric discharge associated with a given object as an electrode or an ion.

polarization *1.* The deviation from the open circuit potential of an electrode resulting from the passage of current. *2.* The state of electromagnetic radiation when transverse vibrations take place in some regular manner, for example, all in one plane, in a circle, in an ellipse, or in some other definite curve.

polarization curve A plot of current density versus electrode potential for a specific electrode-electrolyte combination.

polarizer A material used to generate polarized light from a non-polarized source. A polarizer may select linear or circularly polarized light. Typical uses include liquid crystal displays and contrast enhancement filters for high ambient applications.

polarizing coating Coatings that influence the polarization of light passing through them, typically by blocking or reflecting light of one polarization and passing light that is orthogonally polarized.

polarizing element A device for producing or analyzing polarized light.

polarographic analysis A method of determining the amount of oxygen present in a gas by measuring the current in an oxygen-depolarized primary cell.

polarography A method of chemical analysis that involves automatically plotting the voltage-current characteristic between a large, non-polarizable electrode and a small polarizable electrode immersed in a dilute test solution;

a curve containing a series of steps is produced, the potential identifying the particular cation involved and the step height indicating cation concentration; actual values are determined by comparing each potential and step height with plots generated from test solutions of known concentrations.

poling A stage in fire-refining of copper during which green-wood poles are thrust into the bath of molten metal; the wood decomposes and forms reducing gases that react with oxygen in the bath.

polished plate glass Glass whose surface irregularities have been removed by grinding and polishing, so that the surfaces are approximately plane and parallel.

polling A method of sequentially observing each channel to determine if it is ready to receive data or is requesting computer action.

polonium A radioactive element.

polyacetylene An aliphatic organic polymer that has high semiconductor properties which can be enhanced by doping.

polyacrylate A type of elastomer that is characterized by an unsaturated chain and is a copolymer of alkyl acrylate and some other monomer such as chloroethyl vinyl ether or vinyl chloroacetate.

polyacrylate elastomers Elastomers with excellent resistance to oxidation.

polyacrylonitrile (PAN) A base material used in the manufacture of carbon fibers.

polyamide (PA) *1.* Nylon. A polymer containing a characteristic amide linkage which is derived from the condensation products of diacids and diamines, or amino acids. *2.* A fiber-forming nylon thermoplastic.

polyamide-imide resins (PAI) A group of resins used for laminates, prepregs and electrical parts.

polyarylsulfone (PAS) An elevated temperature-resistant thermoplastic.

polybenzimidazole (PBI) A condensation polymer that is resistant to very high temperatures and is used to manufacture laminates and adhesives.

polybrominated biphenyls A group of 209 chemicals whose toxicity varies and includes principally one fire retardant called firemaster.

polybutylene (PB) Resin with good creep resistance and toughness.

polybutylene terephthalate (PBT) A resin with high strength, resistance to heat and chemicals, and good electrical characteristics.

polycarbonate (PC) A transparent, very high impact-resistant thermoplastic.

polycased glass *See* cased glass.

polychloroprene Chemical name for neoprene. A synthetic elastomer rubber material.

polychromator A spectrometer that has multiple detectors for measuring light from spectral lines.

polycrystalline A material made up of many crystals.

polydispersed A suspension containing a mix of particle sizes.

polyester Resin formed by condensation of polybasic and monobasic acids with polyhydric alcohols.

polyester yarn Yarns spun from polyethylene terephthalate.

polyetheretherketones *See* PEEK.

polyetherimide (PEI) A porous thermoplastic with excellent thermal properties.

polyether sulfone (PESV) A transparent engineering thermoplastic with excellent flame-resistance.

polyethylene A thermoplastic material composed of polymers of ethylene, and derived from the polymerization of ethylene gas.

polyethylene terephthalate (PET) A thermoplastic resin that is very hard, wear resistant, and resistant to chemicals.

polyimide Materials having good heat-resistant properties. Prepared from diamines and polycarboxylic acids, half esters, anhydrides (alkylene pyromellitimide from mellitic acid derivatives).

polymer *1.* Generic term for an organic compound of high molecular weight and consisting of recurrent structural groups. *2.* A macromolecular material formed by the chemical combination of monomers having either the same or different chemical composition.

polymer blend *See* alloy.

polymerization The ability of certain organic compounds to react together to form a single molecule of higher atomic weight.

polymer matrix composites Materials consisting of reinforcing fibers, filaments, and/or whiskers embedded in polymeric bonding matrices for increased mechanical and physical properties.

polymer quenchant A water solution of polymer used to minimize distortion and residual stresses.

polymethyl methacrylate (PMMA) A transparent thermoplastic with excellent optical properties.

polymorphism The ability of a material to exist in more than one crystallographic structure.

polynomial fit A technique of generating a calibrating curve from a set of points.

polynuclear organic compounds Hydrocarbon molecules with two or more nuclei and with or without oxygen, nitrogen, or other elements.

polynucleotides Linear sequences of esters of nucleotides and phosphoric acid.

polyolefin Any of the polymers and copolymers of the olefin family of hydrocarbons such as ethylene, propylene, butylene, etc.

polypeptides In organic chemistry, chains of amino acids linked by peptide bonds but with lower molecular weights than proteins; obtained by synthesis or by partial hydrolysis of proteins.

polyphenylene sulfide (PPS) A chemical-resistant thermoplastic.

polyphenylene sulfone (PPSU) A thermal- and chemical-resistant thermoplastic.

polypropylene (PP) A tough, lightweight, rigid plastic made by the polymerization of high-purity propylene gas.

polystyrene (PS) A brittle thermoplastic characterized by excellent electrical properties.

polystyrene foam Readily formed foam used in food packaging.

polysulfide A chemical- and gas-resistant elastomer.

polysulfone (PSU) An elevated temperature-resistant thermoplastic with resistance to oxidation and creep.

polytetrafluoroethylene (PTFE) A fluoropolymer with excellent thermal and chemical resistance and a low coefficient of friction. PTFE is usually compounded with fillers such as molybdenum disulfide, graphite, pigments, and glass fibers to improve wear characteristics and other properties.

polytropic expansion An expansion process in which changes of pressure and density are related.

polytropic rate The rate that results when there is limited heat transfer to or from the gas during spring deflection. Polytropic rate results during spring deflections that produce neither adiabatic nor isothermal rate.

polyurethane (PUR) A widely used polymer that can be thermosetting or thermoplastic.

polyvinyl acetals Thermoplastics used in adhesives, coatings, and films.

polyvinyl acetate (PVAC) A thermoplastic resin used in adhesives and coatings.

polyvinyl alcohol (PVAL) A thermoplastic resin soluble in water and used in adhesives and coatings.

polyvinyl butyryl (PVB) A thermoplastic resin used as an interlayer of safety glass.

polyvinyl chloride (PVC) A family of insulating compounds whose basic ingredient is either polyvinylchloride or its copolymer with vinyl acetate.

polyvinyl fluoride (PVF) A crystalline polymer with resistance to staining, chemical attack, and flammability.

polyvinylidene fluoride (PVDF) Thermoplastic resin, characterized by good mechanical, electrical, and chemical properties. Radiation crosslinking improves heat resistance.

PONA analysis Determination of amounts of paraffins (P), olefins (O), naphthalenes (N), and aromatics (A) in gasoline in ASTM standard tests.

pooled Describes the combination of data from different data sources.

pooled standard deviation The square root of averaged sample variances (pooled variance), which gives an estimate of overall population standard deviation.

pooled variance A weighted average of k variances, where the degrees of freedom are used as weights.

pope cell A type of relative humidity (RH) sensor that employs a bifilar conductive grid on an insulating substrate whose resistance varies with relative humidity over a range of about 15 to 99% RH.

pop-off The loss of small fragments of a coating.

popping Usually refers to a fairly uniform distribution of small blisters caused by the volatilization of entrapped solvent beneath a topcoat.

population The totality of a set of items, units, measurements, etc., real or conceptual, that is under consideration.

population mean The average of all potential measurements in a given population weighed by their relative frequencies in the population.

population median That value in the population such that the probability of exceeding it is 0.5 and the probability of being less than it is 0.5.

population variance A measure of dispersion in a population.

porcelain enamel *1. See* enamel. *2. See* vitreous enamel.

pore *1.* Very small voids in a material. *2.* A small perforation in electroplated material.

pore-forming material A volatile substance that produces a desired porosity during sintering.

pore size distribution The rate of the number of pores of given sizes to the total number of pores, per unit of area.

porosimeter A test instrument to measure the porosity of a sintered compact.

porosity *1.* A multitude of minute cavities in a material. *2.* The gas permeability of the bag fabric.

port *1.* The entry or exit point from a computer for connecting communications or peripheral devices. *2.* An aperture for passage of steam or other fluids.

portability The ability of test procedures to be used by more than one test equipment configuration.

portable *1.* Describes a self-contained, battery-operated instrument that can be carried. *2.* Capable of being carried, especially by hand, to any desired location.

portable standard meter A portable instrument used primarily as a reference standard for testing or calibrating other instruments.

ports Data channels dedicated to input or output.

positioning control *See* point-to-point numerical control.

positive displacement Refers to any device that captures or confines definite volumes of fluid for purposes of measurement, compression, or transmission.

positive-displacement flowmeter Any of several flowmeter designs in which volumetric flow through the meter is broken up into discrete elements and the flow rate is determined from the number of discrete elements that pass through the meter per unit time.

positive feedback *1.* A closed loop in which any change is reinforced until a limit is eventually reached. *2.* Returning part of an output signal and using it to increase the value of an input signal.

positive ion Atom or group of atoms that has acquired a positive electric charge by the loss of one or more electrons.

positron(s) *1.* Subatomic particles that are identical to electrons in atomic mass, theoretical rest mass, and energy, but opposite in sign. *2.* A positively charged beta particle.

postcure A secondary high-temperature cure of polymers to improve properties or decrease the amount of volatiles.

postforming The forming, bending, or shaping of fully cured, C-stage thermoset laminates that have been heated to make them flexible.

postheating Heating weldments immediately after welding, for tempering or stress relieving, or for providing a controlled rate of cooling to prevent formation of a hard or brittle structure.

post inspection cleaning The removal of penetrant material residues and developer from a test part after the penetrant inspection process is completed.

pot annealing *See* box annealing.

potassium An alkali metal.

potassium hydroxide The caustic material that is mixed with pure water to formulate the electrolyte solution used in nickel-cadmium cells.

potential *See* electrochemical potential.

potential energy *1.* Energy possessed by a body by virtue of its position in a gravity field, in contrast with kinetic energy, energy possessed by virtue of the body's motion. *2.* Energy related to the position or height above a place to which fluid could possibly flow or a solid could fall or flow.

potential gradients In general, the local space rate of change of any potential, such as the gravitational potential gradient or the velocity potential gradient.

potentiometer *1.* Instrument for measuring differences in electric potential by balancing the unknown voltage against a variable known voltage. If the balancing is accomplished automatically, the instrument is called a self-balancing potentiometer. *2.* A variable electric resistor.

potentiometric titration A technique of automatic titration in which the end point is determined by measuring a change in the electrochemical potential of the sample solution.

potentiostat An instrument for automatically maintaining an electrode at a constant potential or controlled potential with respect to a reference electrode.

pot life The time that a polymer resin retains sufficient flow for processing.

potting form *See* mold, potting.

poultice corrosion Corrosion on outdoor vehicles caused by the collection of road salts and debris on underside areas.

pound The U.S. unit of mass or weight; it is equal to 0.45 kilograms.

poundal A unit of force in the English system of measurement; it is defined as the force necessary to impart an acceleration of one ft/s^2 to a body having a mass of one pound.

pour point *1.* Lowest temperature at which a fluid will flow under specified conditions. *2.* Temperature at which molten metal is cast. *3.* The temperature at which a petroleum-base lubricating oil becomes too viscous to flow, as determined in a standard ASTM test.

pour test Chilling a liquid under specified conditions to determine its ASTM pour point.

powder The particles used to create powder metal compacts.

powder coating A painting process in which finely ground dry plastic is applied to a part using electrostatic and compressed air transfer mechanisms. The applied powder is heated to its melting point and flows out, forming a smooth film, and cures by means of a chemical reaction.

powder metallurgy *1.* The science of creating solid shapes from metal powders. *2.* The art of producing metal powders.

powder pattern An x-ray diffraction pattern consisting of a series of rings on a flat film or a series of lines on a circular strip film which results when a monochromatic beam of x-rays is reflected from a randomly oriented polycrystalline metal or from powdered crystalline material.

power/energy meter An instrument that measures the amount of optical power (watts) or energy (joules). It can operate in the visible, infrared, or ultraviolet region, and detect pulsed or continuous beams.

power spectral density (PSD) A type of frequency analysis on data which can be done by computer using special software, or by an array processor, or by a special-purpose hardware device.

PP *See* polypropylene.

ppm *See* parts per million.

ppmC *See* parts per million carbon.

PPS *See* polyphenylene sulfide.

PPSU *See* polyphenylene sulfone.

Prandtl number A dimensionless number representing the ratio of momentum transport to heat transport in a flow.

precious metal One of the relatively scarce and valuable metals: gold, silver, and the platinum-group metals.

precipitate Particles separated from a fluid as a result of a difference in densities and the action of gravitational force.

precipitation The removal of solid or liquid particles from a fluid.

precipitation hardening *1.* Hardening caused by the precipitation of a constituent from a supersaturated solid solution. *2.* The removal of solid particles from a fluid onto a surface to be hardened. *3. See* age hardening.

precipitation heat treatment Artificial aging in which a constituent precipitates from a supersaturated solid solution.

precision *1.* The degree of mutual agreement between individual measurements, namely repeatability and reproducibility. Contrast with accuracy. *2.* The quality of being exactly or sharply defined or stated. A measure of the precision of a representation is the number of distinguishable alternatives from which it was selected, which is sometimes indicated by the number of significant digits it contains. *See also* stability.

precision index The estimated standard deviation based on n measurements.

precision measurement equipment Test and measurement equipment used to measure, calibrate, gauge, test, inspect, diagnose, or otherwise examine material, supplies, and equipment to determine whether they comply with the established specifications.

precleaning The removal of surface contaminant from a test part so that it cannot interfere with the penetrant inspection process.

pre-coated A material that has been coated prior to the manufacture of the ware or part.

precure A term frequently used to designate the first cure of a material that is given more than one cure in its manufacture.

precursor For carbon or graphite fiber, the rayon, pan, or pitch fibers from which carbon and graphite fibers are derived.

predemonstration phase A period of time immediately prior to commencement of formal maintainability demonstration during which the test team, facilities, and support material are assembled.

predicted That which is expected at some future time, postulated on analysis of past experience and tests.

prediction techniques Method for estimating future behavior of material on the basis of knowledge of its parts, functions, operating environments, and their interrelationships.

preferred orientation A condition in which crystals are aligned in a specific direction within the material.

preform *1.* A cylinder of glass that is made to have a refractive index profile that would be desired for an optical fiber. The cylinder is then heated and drawn out to produce a fiber. *2.* Brazing metal foil cut to the exact outline of the mating parts and inserted between the parts prior to placing in a brazing furnace. *3.* In forging or heading, a shape intermediate to the final shape, which subsequently is deformed to the final shape.

pregel An unintentional, extra layer of cured resin on part of the surface of a reinforced plastic.

preheater air Air at a temperature exceeding that of the ambient air.

preheating *1.* An imprecise term meaning heating to an appropriate temperature in preparation for mechanical work, for welding, or for further thermal treatment. *2.* Heating to an appropriate temperature immediately prior to austenitizing when hardening highly alloyed constructional steels, many of the tool steels, and heavy sections; this treatment is employed to minimize distortion.

preimpregnation The practice of mixing resin and reinforcement and effecting partial cure before use or shipment to the user.

preload An external static load producing a strain in a test specimen. For example, a preload is imposed prior to forced vibration testing. Preload is usually expressed in units of force instead of units of deflection.

premix A molding compound prepared prior to and apart from the molding operations and containing all components required for molding.

premixing The mixing of ingredients prior to a specified action (for example, mixing of fuel and air prior to ignition in combustion).

premodulator filter A lowpass filter at the input to a telemetry transmitter; its purpose is to limit modulation frequencies and thereby limit radiated frequencies outside the desired operating spectrum.

prepolymers High-molecular weight bifunctional molecules which, when catalyzed, produce chain extension.

prepregs The reinforcing materials containing or combined with the full complement or resin before molding operations in the production of composite materials.

preprocessor *1.* A hardware device in front of a computer, capable of making certain decisions or calculations more rapidly than the computer can make them. *2.* The first of the two compiler stages. At this stage, the source program is examined for the preprocessor statements, which are then executed, resulting in the alteration of the source program text. More generally, a program that performs some operation prior to processing by a main program.

presintering The heating of a powder metallurgy compact to a temperature lower than the normal temperature for final sintering, usually to increase the ease of handling or forming the compact or to remove a lubricant or binder before sintering.

pressed density The density of a powder-metal compact after pressing and before sintering.

pressing *1.* Compressive force used to form sheet, plate, or powder metal materials. *2.* A hot forging operation.

pressure Measure of applied force compared with the area over which the force is exerted, psia.

pressure, absolute True pressure, relative to a complete vacuum.

pressure bonding Heat and pressure bonding of materials without the use of another bonding material.

pressure, design *See* design pressure.

pressure drop The difference in fluid pressures as measured between the inlet and the outlet of the heat exchanger or heat exchanging systems.

pressure gage An instrument for measuring pressure by means of a metallic sensing element or piezoelectric crystal.

pressure gas welding Oxyfuel gas welding process using pressure rather than filler material to join materials.

pressure-impregnation-carbonization (PIC) A hot isostatic press operation to increase the density of carbon-carbon composites.

pressure level In acoustic measurement, $P = 1 \log (P_s/P_r)$, where P is the pressure level in bels, P_s is the sound pressure, and P_r is a reference pressure, usually taken as 0.002 dyne/cm^2.

pressure measurement Any method of determining internal force per unit area in a process vessel, tank, or piping system due to fluid or compressed gas; this includes measurement of static or dynamic pressure, absolute (total) or gage (total minus atmospheric), in any system of units.

pressure pulse fatigue test A laboratory test that determines the ability of a filter to withstand lubricating system cyclic pressures encountered by the filter during engine operation.

pressure rating The maximum allowable internal force per unit area of a pressure vessel, tank, or piping system during normal operation.

pressure ratio The relationship of a force to the deformation of a system whose deformation varies in some proportion to the force.

pressure-sensitive adhesive A permanently tacky material that will adhere to nearly any smooth surface.

pressure sintering Heat and pressure sintering of powder metals.

pressure, static *See* static pressure.

pressure vessel A metal container designed to withstand a specified bursting pressure; it is usually cylindrical with hemispherical end closures (but may be of some other shape, such as spherical), and is usually fabricated by welding.

Preston tubes *See* pitot tube.

pre-treatment The treatment of a surface prior to the process of interest, for example, a phosphate coating is a pre-treatment for electrocoat or painting.

prevaporization The phase transformations of liquids to gases prior to some physical or chemical reaction.

primary alpha The allotrope of titanium with hexagonal, close-packed crystal structure which is retained from the last high-temperature alpha-beta heating.

primary billet Compacted powder metal product composed of the specified aluminum matrix material and the properly dispersed silicon carbide.

primary creep The initial stage of deformation.

primary crystal The first type of crystal that separates from a melted material when cooling.

primary element *1.* Detector. *2.* The first system element that responds quantitatively to the measured variable and performs the initial measurement operation. *3.* Device that performs the initial conversion of measurement energy. *4.* Any device placed in a flow line to produce a signal for flow rate measurement. *5.* The component of a measurement or control system that first uses or transforms energy from a given medium to produce an effect that is a function of the value of the measured variable. *6.* The portion of the measuring means that first either utilizes or transforms energy from the controlled medium to produce an effect in response to change in the value of the controlled variable. The effect produced by the primary element may be a change of pressure, force, position, electrical potential, or resistance. *See* sensor.

primary feedback A signal that is a function of the controlled variable and that is used to modify an input signal to produce an actuating signal.

primary instrument An instrument that can be calibrated without reference to another instrument.

primary insulation The layer or layers of nonconducting material that are designed to act as electrical insulation, excluding cosmetic top coatings.

primary measuring element A component of a measuring or sensing device that is in direct contact with the substance whose attributes are being measured.

primary standard dosimetry system The direct measurement of energy deposition for the interpretation of the radiation absorption process.

primer First coat of paint applied to a surface. Formulated to have good bonding and wetting characteristics; may or may not contain inhibitive pigments.

primes Metal sheet or plate free of any visible defects or imperfections.

principle stress Stress on three perpendicular planes where there is no shear stress.

priority *1.* The relative importance attached to different phenomena. *2.* Level of importance of a program or device.

priority interrupt The temporary suspension of a program currently being executed in order to execute a program of higher priority.

prismatic glass Clear glass into whose surface is fabricated a series of prisms, the function of which is to direct the incident light in desired directions.

privilege A characteristic of a user, or program, that determines what kind of operations a user or program can perform; in general, a privileged user or program is allowed to perform operations that are normally considered the domain of the monitor or executive, or that can affect system operation as a whole.

probability *1.* (classical) If an event can occur in N equally likely and different ways, and if n of these ways have an attribute A, then the probability of occurrence of A, denoted Pr(A), is defined as n/N. *2.* (frequency) If an experiment is conducted N times, and outcome A occurs n times, then the limit of n/N as N becomes large is defined as the probability of A, denoted as Pr(A). *3.* (subjective) Denoted Pr(A), a measure of the degree of belief one holds in a specified proposition A.

probability distribution A mathematical function with specific properties which describes the probability that a random variable will take on a value or a set of values.

probability of acceptance Probability that an item under test will be accepted by that test.

probe *1.* A small, movable capsule or holder that allows the sensing element of a remote-reading instrument, usually an electronic instrument, to be inserted into a system or environment and then withdrawn after a series of instrument readings has been taken. *2.* A small tube, movable or fixed, inserted into a process fluid to take physical samples or pressure readings.

probe-type consistency sensor A device in which forces exerted on a cylindrical body in the direction of flow are detected by a strain-gage bridge circuit.

problem-oriented language A programming language designed for the convenient expression of a given class of problems.

procedure *1.* A precise step-by-step method for effecting a solution to a problem. *2.* In data processing, a smaller program that is part of a large program. *3.* In batch processing, the part of a recipe that defines the generic strategy for producing a batch of material.

process *1.* The collective functions performed in and by industrial equipment, exclusive of computer and/or analog control and monitoring equipment. *2.* A series of continuous or regularly recurring steps or actions intended to achieve a predetermined result, as in refining oil, heat treating metal, or manufacturing paper. *3.* A general term covering such terms as assemble, compile, generate, interpret, and compute. *4.* The functions and operations utilized in the treatment of material. *5.* A progressive course or series of actions. *6.* Any operation or sequence of operations involving a change of energy, state, composition, dimension, or other properties that may be defined with respect to a datum. *7.* An assembly of equipment and material that relates to some manufacturing sequence.

process annealing An imprecise term used to denote various treatments used to improve workability. For the term to be meaningful, the condition of the material and the time-temperature cycle used must be stated.

process computer A computer that, by means of inputs from and outputs to a process, directly controls or monitors the operation of elements in that process. *See also* on-line, control computer, and industrial computer.

process control The determination and control of the key manufacturing processes that determine the intrinsic properties of component parts.

process heat Increase in enthalpy accompanying chemical reactions or phase transformations at constant pressure (for example, heat of crystallization and heat of sublimation).

program *1.* In data processing, a series of instructions that tell the computer how to operate. *2.* Any series of actions proposed in order to achieve a certain result. *3.* To design, write, and test a program. *4.* A unit of work for the central processing unit, from the standpoint of the executive program. *See* task.

program address counter Same as location counter.

program control Descriptive of a system in which a computer is used to direct an operation or process and automatically hold or make changes in the operation or process on the basis of a prescribed sequence of events.

program counter A register that contains the address of the next instruction to be executed. At the end of each instruction execution the program counter is incremented by 1 unless a jump is to be carried out, in which case the address of the jump label is entered.

program evaluation and review technique (PERT) A management control tool for managing complex projects. According to PERT, project milestones are defined and interrelated; then, using a flowchart or computer, progress is measured against the milestones. Deviations from the integrated plan are used to trigger decisions or preplanned alternative actions to minimize adverse effects on the overall goal.

programmable data distributor (PDD) An optional module for the telemetry frame synchronizer which causes it to send certain predefined words from each telemetry frame out through a separate port (as to a special buffer area).

programmable logic array (PLA) An integrated circuit that employs ROM matrices to combine sum and product terms of logic networks.

programmable logic controller (PLC) *1.* A microcomputer-based control device used to replace relay logic. *2.* A solid-state control system that has a user-programmable memory for storage of instructions to implement specific functions such as I/O control, logic, timing, counting, three-mode (proportional, integral and derivative [PID]) control, communication, and arithmetic, data, and file manipulation. *3.* Consists of a central processor, input/output interface, and memory. Designed as an industrial control system.

programmer's console A man-machine interface, consisting of various information entry/retrieval devices, arranged as a packaged unit. It is used by the programmer of a computer control system for a manufacturing process, to monitor, modify, and control the internal behavior of the digital controller. *See also* operator's console.

programming The design, writing, and testing of a program. *See* convex programming, dynamic programming, linear programming, mathematical programming, nonlinear programming, quadratic programming, macroprogramming, microprogramming, and multiprogramming.

program parameter A parameter incorporated into a subroutine during computation. A program parameter frequently comprises a word stored relative to either the subroutine or the entry point and dealt with by the subroutine during each reference. It may be altered by the routine and/or may vary from one point of entry to another. Related to parameter.

program storage A portion of the internal storage reserved for the storage of programs, routines, and subroutines. In many systems, protection devices are used to prevent inadvertent alteration of the contents of the program storage. Contrasted with working storage.

progressive aging Aging by increasing the temperature in steps or continuously during the aging cycle. *See* aging. Compare with interrupted aging and step aging.

projection welding A welding process in which the arc is localized by projections, embossments, or intersections.

PROM An abbreviation for programmable read only memory. A memory chip that can be programmed once but cannot be reprogrammed. *See also* EPROM.

promoter A chemical that increases the activity of a catalyst.

proof *1.* To test a component or system at its peak operating load or pressure. *2.* The ethyl alcohol content of a liquid at 15.56°C (60°F) stated as twice the percent of ethyl alcohol by volume.

proof load *1.* Static test load to ensure mechanical integrity at critical points. A proof load shall not cause permanent deformation or affect continued use. *2.* A specific load at which a test specimen is tested before a material is accepted for use.

proof pressure The maximum pressure that can be applied that will not cause any permanent set or deformation. Normally, this is never less than 200% designated pressure.

proof pressure test A test pressure that a filter shall withstand without permanent deformation, external leakage, or other malfunction.

proof stress A specific load at which a structure or member is tested to verify service life.

propagation *1.* In materials, the progression of a crack, for example, a small crack propagates to ultimate breakage. *2.* The spreading abroad or sending forward, as of radiant energy.

propagator Any agent of propagation.

propargyl groups Crosslinking agents for certain aromatic polyamides used as matrix resins in fiber composites.

propellant *See* double base powder.

proportional bandwidth (PBW) In FM telemetry, refers to the condition in which each subcarrier is deviated a fixed percentage of center frequency (and therefore an amount proportional to the center frequency) by data.

proportional, integral, and derivative (PID) *1.* Describes a three-mode controller. *2.* Refers to a control method in which the controller output is proportional to the error, its time history, and the rate at which it is changing. The error is the difference between the observed and desired values of the variable that is under control action. *3.* Proportional plus integral plus derivative control, used in processes where the controlled variable is affected by long lag times.

proportional limit The maximum stress at which strain remains proportional to stress and there is no plastic deformation.

proportional pitch In computer printers, a typeface in which each character is a different width, such as a W versus an I.

proportioning probe A probe used in leak testing in which the ratio of air to tracer gas can be changed without changing the amount of flow transmitted to the detector.

protected material Material used in inner lenses for optical devices, where such lenses are protected from exposure to the sun by an outer lens made of materials meeting the requirements for exposed plastics.

protocol A formal set of conventions or rules for the exchange of information, including the procedures for establishing and controlling transmissions on the multiplex signal bus (message administration) and the organization, meaning, and timing associated with the bits of data (message transfer).

protocol data unit (PDU) The form in which each of the seven OSI layers passes data to the layer below, after accepting data (SDUs) from the layer above, and adding its own header (PCI). Also, the form in which each of the layers passes from the layer above, after accepting data from the layer below, and stripping off its header.

protons Positively charged subatomic particles having a mass of 1.67252×10^{-24} gram, slightly less than that of an electron.

prototype A preproduction model suitable for evaluating the design, functionality, operability, and form of a product, but not necessarily its durability and reliability.

PS *See* polystyrene.

PSD *See* power-spectral density.

pseudo instruction *1.* A symbolic representation in a compiler or interpreter. *2.* A group of characters having the same general form as a computer instruction, but never executed by the computer as an actual instruction. Synonymous with quasi instruction.

pseudo-operations A group of instructions that, although part of a program, do not perform any application-related function. They generally provide information to the assembler.

pseudoplastic A material that exhibits flow (permanent deformation) at all values of shear stress, although in most cases the flow that occurs below some specific value (an apparent yield stress) is low and increases negligibly with increasing stress.

pseudopotentials Factors in an approximate method for calculation of energy bands in solids by the use of an approximation that includes the many body effect.

pseudo-random The property of satisfying one or more of the standard criteria for statistical randomness, but being produced by a definite calculation process.

PSK *See* phase shift keying.

PSU *See* polysulfone.

psychrometer A device consisting of two thermometers, one of which is covered with a water-saturated wick, used for determining relative humidity; for a given set of wet-bulb and dry-bulb temperature readings, relative humidity is read from a chart. Also known as wet-and-dry-bulb thermometer.

PTFE *See* polytetrafluoroethylene.

P-type material A semiconductor material that has been doped with an excess of acceptor impurity atoms, so that free holes are produced in the material.

puckers Areas on a prepreg material where the material has blistered.

pulled surface In laminar plastics, a breaking or lifting of the surface from the body.

pull test *1.* A test to determine the bond strength of a lead to an interconnecting surface, usually perpendicular to the surface, by pulling to failure. *2. See* tensile test.

pulp *See* slime.

pultrusion Process of pulling continuous lengths of resin-impregnated fiber through a shaped, heated die to produce lengths of reinforced plastic.

pumice A light-colored, vesicular, glassy rock commonly having the composition of a rhyolite.

punching A method of cold extruding, cold heading, hot forging, or stamping in a machine whereby the mating die sections control the shape or contour of the part.

PUR *See* polyurethanes.

purge *1.* To cause a liquid or gas to flow from an independent source into the impulse line(s). *2.* To introduce air into the furnace and the boiler flue passages in such volume and manner as to completely replace the air or gas-air mixture contained therein.

purging Elimination of an undesirable gas or material from an enclosure by means of displacing the undesirable material with an acceptable gas or material.

purity The degree to which a substance is free of foreign materials.

purple plague One of several gold-aluminum compounds formed when bonding gold to aluminum and activated by exposure to moisture and high temperature, resulting in brittle, time-based bond failure.

pusher furnace A type of continuous furnace in which parts to be heated are periodically charged into the furnace in containers, which are pushed along the hearth against a line of previously charged containers, thus advancing the containers toward the discharge end of the furnace, where they are removed.

PVAC *See* polyvinyl acetate.

PVAL *See* polyvinyl alcohol.

PVB *See* polyvinyl butyryl.

PVC *See* polyvinyl chloride.

PVDF *See* polyvinylidene fluoride.

PVF *See* polyvinyl fluoride.

pycnometer A container of precisely known volume that is used to determine the density of a liquid by weighting the filled container and dividing the weight by the known volume. Also spelled pyknometer.

pyknometer *See* pycnometer.

pyrazines Compounds that contain a six-membered heterocyclic ring containing nitrogen atoms in the 1 and 4 positions.

pyridines Compounds that contain a six-membered heterocyclic ring containing one nitrogen atom.

pyrimidines Compounds that contain a six-membered heterocyclic ring containing nitrogen atoms in the 1 and 3 positions.

pyroelectric detectors Detectors of visible, infrared, and ultraviolet radiation which rely on the absorption of radiation by pyroelectric materials.

pyroelectric effect Polarization that occurs in dielectric materials with a change in temperature.

pyrolysis *1.* Chemical decomposition by the action of heat. *2.* The formation of a material by decomposition of a compound at high temperature.

pyrometer Any of a broad class of temperature-measuring instruments or devices. The term was originally applied only to devices for measuring temperatures well above room temperature, but now it applies to devices for measuring temperatures in almost any range. Some typical pyrometers include thermocouples, radiation pyrometers, and thermistors, but usually not thermometers. *See also* optical pyrometer.

pyrometry The study of elevated temperature measurement.

pyrophoricity The property of a material to burn when exposed to air.

pyrophoric powder A metal powder that burns when exposed to air.

pyrophyllite A white, greenish, gray, or brown phyllosilicate mineral that resembles talc.

pyrotechnic A mixture of chemicals designed to produce heat, light, noise, smoke, or gas.

pyroxenes A group of dark, rock-forming silicate minerals.

pyrrhotite A common reddish-brown to bronze hexagonal mineral.

pyrroles Compounds that contain a five-membered heterocyclic ring containing one nitrogen atom.

Q

q Symbol for volumetric flow rate.

QBC *See* queue control block.

QC *See* quality control.

QCB *See* queue control block.

QCD *See* quantum chromodynamics.

Q **factor** *1.* A rating factor for electronic components such as coils, capacitors, and resonant circuits; equals reactance divided by resistance. *2.* In a periodically repeating mechanical, electrical, or electromagnetic process, the ratio of energy stored to energy dissipated per cycle.

quadrant detectors Detectors that are divided up into four angularly symmetric sectors or quadrants. The amounts of radiation incident on each quadrant can be compared to one another for applications such as making sure that a beam is centered on the detector.

quadratic programming In operations research, a particular case of nonlinear programming in which the function to be maximized or minimized and the constraints are quadratic functions of the controllable variables. Contrast with convex programming, dynamic programming, linear programming, and mathematical programming.

quadrupole mass spectrometer A type of mass spectrometer employing a filter consisting of four conductive rods electrically connected in such a manner that, by varying the absolute potential applied to the rods, all ions except those possessing a specific mass-to-charge ratio are prevented from entering the detector.

qualification tests Tests that are run on one or more units that are representative of the production article, to test the performance, endurance, environmental, and special features of a design to demonstrate specification compliance and suitability of the unit for production and use in service.

qualified *1.* Competent, suited, or having met the requirements for a specific material, position, or task. *2.* To declare competent or capable.

qualitative Term used to describe inductive analytical approaches that are oriented toward relative, nonmeasurable, and subjective values.

qualitative analysis Analysis in which the elements of a sample are identified, but not necessarily the amount of each.

quality *1.* A measure of the degree to which a material or item conforms to applicable specification and workmanship standards. *2.* The composite of all characteristics or attributes, including performance of an item. *3.* A property that refers to the tendency of an item to be made to specific specifications or the customer's express needs, or both.

quality assurance A system of activities whose purpose it is to provide assurance that the overall quality control job is, in fact, being done effectively.

quality characteristics The properties of an item or process that can be measured, reviewed, or observed and that are identified in the drawings, specifications, or constructural requirements. Reliability becomes a quality characteristic when so defined.

quality conformance tests Tests specified by the procuring agency to monitor and maintain continued quality.

quality control (QC) The overall system of activities, whose purpose is to provide a quality of product or service that meets the needs of users; also the use of such systems.

quality factor For a material, the ratio of elastic modulus to loss modulus.

quantitative Term used to describe inductive or deductive analytical approaches that are oriented toward the use of numbers or symbols used to express a measurable quantity.

quantitative analysis Analysis in which the amount of one or more elements in a sample are measured.

quantity *1.* A specified or indefinite number or amount. *2.* A large amount or number.

quantity of light (luminous energy) *See* luminous energy.

quantization The subdivision of the range of values of a variable into a finite number of nonoverlapping, and not necessarily equal, subranges or intervals, each of which is represented by an assigned value within the subrange. For example, a person's age is quantized for most purposes with the quantum of one year. *See also* measurement.

quantization distortion Inherent distortion introduced when a range of values for a wave attribute is divided into a series of smaller subranges.

quantization level A particular subrange for a quantized wave, or its corresponding symbol.

quantize To convert information from an analog pulse (as from a multiplexer) into a digital representation of that pulse. *See* encoder.

quantum chromodynamics A gauge theory describing the interaction between quarks and gluons.

quantum efficiency A measure of the efficiency of conversion or utilization of light or some other form of energy.

quantum electronics The branch of electronics that essentially deals with lasers and laser devices which require quantum theory for their exact description.

quantum mechanics The theory of matter, electromagnetic radiation, and the interaction between matter and radiation.

quantum number A number required to characterize a quantum state.

quantum theory The theory first stated by Max Planck that all electromagnetic radiation is emitted and absorbed in quanta, each of magnitude hv, h being the Planck constant and v the frequency of the radiation.

quark parton model A theoretical model that summarizes our understanding of how protons and neutrons are made up of the fundamental subparticles called quarks.

quarter hard A temper of metal between dead soft and half hard.

quartz A natural transparent form of silica, which may be marketed in its natural crystalline state, or crushed and remelted to form fused quartz.

quasi instruction Same as pseudo instruction.

quasi-isotropic laminate A laminate that has nearly uniform properties in all directions of the plies.

quaternary alloy An alloy that contains four principal elements.

quench-age embrittlement Low-carbon steel loss of ductility upon aging at room temperature after rapid cooling.

quench aging Natural or artificial aging of a ferrous material caused by the precipitation of an iron carbide or an iron nitride or a complex of both in alpha iron supersaturated with these compounds. Supersaturation is achieved by rapidly cooling the heated material from Ar_1 (the temperature at which steel transforms from austenite to ferrite or to ferrite plus cementite is completed during cooling).

quench annealing Annealing an austenitic ferrous alloy by solution heat treatment.

quench cracking The tendency of hard metals to fracture when quenched.

quench hardening Hardening a suitable ferrous alloy by austenitizing and then cooling at a rate such that a substantial amount of austenite transforms to martensite. Depending on alloy content, this rate could be an air quench, an oil quench, a water quench, or a brine quench.

quench hardening copper alloys Hardening suitable alloys by betatizing and quenching to develop a martensite-like structure.

quenching Usually thought to mean cooling from an elevated temperature more rapidly than would occur in still air. When applicable, the following more specific terms should be used: direct quenching, fog quenching, hot quenching, interrupted quenching, oil quenching, selective quenching, spray quenching, time quenching, and water quenching.

query language A means of getting information from a database without the need to write a program.

queue *1.* Waiting line resulting from temporary delays in providing service. *2.* In data processing, a waiting list of programs to be run next. *3.* Any list of items; for example, items waiting to be scheduled or processed according to system- or user-assigned priorities.

queue control block (QCB) A control block that is used to regulate the sequential use of a programmer-defined facility among requesting tasks.

queued access method Any access method that automatically synchronizes the transfer of data between the program using the access method and input/output devices, thereby eliminating delays for input/output operations.

queuing An ordered progression of items into and through a system or process, especially when there is waiting time at the point of entry.

queuing discipline The rules or priorities for queue formation within a system of "customers" and "servers," as well as the rules for arrival time and service time.

queuing theory A form of probability theory useful in studying delays or line-ups at servicing points.

Quevenne scale A specific gravity scale used in determining the density of milk; a difference of 1° Quevenne is equivalent to a difference of 0.001 in specific gravity, and therefore 20° Quevenne expresses a specific gravity of 1.020.

quinoxalines A group of heterocyclic compounds consisting of a benzene ring condensed with a diazine ring.

R

R *See* Rankine.

r *See* röentgen.

Ra The chemical symbol for radium.

RA See reduction of area.

rack *1.* A standardized steel framework designed to hold 19-in.-wide panels of various heights that have units of electronic equipment mounted on them. Also known as a relay rack. *2.* A frame for holding or displaying articles.

rad *See* radian.

radial distribution function analysis A diffraction technique that provides the distribution of interatomic distances present in a test specimen.

radial marks *See* chevron pattern.

radian Metric unit for a plane angle.

radiance Radiant flux per unit solid angle per unit of projected source area.

radiant flux The rate of flow of radiant energy with respect to time.

radiant fluxmeter A device that measures the amount of radiant flux emitted or absorbed. The units typically would be watts per unit area.

radiant intensity The energy emitted per unit time per unit solid angle along a specific linear direction.

radiation A process of emitting energy electromagnetically. (Thermal radiation differs from other forms of heat transfer in that its speed of propagation equals that of light and no intervening medium is required for its transmission.) See also thermal radiation.

radiation chemistry The branch of chemistry concerned with the chemical effects, including decomposition, of energetic radiation of particles of matter.

radiation counter Instrument used for detecting or measuring moving subatomic particles by a counting process.

radiation damage A general term for the deleterious effects of radiant energy— either electromagnetic radiation or particulate radiation—on physical substances or biological tissues.

radiation energy The energy of a photon or particle in a radiation beam.

radiation error The difference between the gas temperature and the sensed temperature caused by radiant heat transfer between the sensing element and surrounding areas.

radiation intensity The amount of radiant energy passing through an area in a specific time.

radiation pressure Pressure exerted upon any material body by electromagnetic radiation incident upon it.

radiation pyrometer An instrument that uses the radiant power emitted by a hot object in determining its temperature.

radiation thermometer Any of several devices for determining the temperature of a body by measuring its emitted radiant energy, without physical contact between the sensor and the body.

radiation transport The study of radiation from emission to absorption.

radiation trapping Confinement of radiation with a magnetic field.

radiator *1.* Any source of radiant energy, especially electromagnetic radiation. *2.* Device that dissipates the heat from something as from water or oil, not necessarily by radiation only.

radical A very reactive chemical intermediate.

radio-activation analysis Minute quantitative and qualitative analysis derived from the characteristics of radioactive emissions.

radioactive half life The time it takes for the radioactivity of a specific nuclide to be reduced by one half.

radioactive tracer A radioactive nuclide used at small concentration to follow the progress of some physical, chemical, or biological process.

radioactivity Spontaneous disintegration of atomic nuclei with emission of corpuscular or electromagnetic radiations. The number of spontaneous disintegrations per unit mass and per unit time of a given unstable (radioactive) element, usually measured in curies.

radio-frequency heating *See* electronic heating.

radio-frequency spectrometer A device that measures the intensity of radiation absorbed or emitted by atoms or molecules.

radiography A form of nondestructive testing that involves the use of ionizing radiation (x-rays or gamma rays) to detect flaws, characterize internal structure, or measure thickness of metal

parts or the human body; it may involve determining the attenuation of radiation passing through a test object.

radioisotope *1.* A radioactive isotope of a chemical element. *2.* A nonpreferred synonym for radionuclide.

radioluminescence Emission of light due to radioactive decay of a nuclide.

radiometer Instrument for detecting and, usually, measuring radiant energy.

radiometric quantities Properties of light based on purely physical measurement without relation to eye response. Standard units are the watt and joule.

radiometric analysis A method of quantitative chemical analysis based on measuring the absolute disintegration rate of a radioactive substance of known specific activity.

radionuclide Any nuclide that undergoes spontaneous radioactive decay.

radix Also called the base number; the total number of distinct marks or symbols used in a numbering system. For example, since the decimal numbering system uses ten symbols (0, 1, 2, 3, 4, 5, 6, 7, 8, 9), the radix is ten. In the binary numbering system the radix is two, because there are only two marks or symbols (0, 1).

radix-50 A storage format in which three ASCII characters are packed into a sixteen-bit word.

radix complement A number obtained by subtracting each digit of the given number from one less than the radix, then adding to the least significant digit, executing all required carries; in radix two (binary), the radix complement of a number is used to represent the negative number of the same magnitude.

radix notation A positional representation in which the significances of any two adjacent digit positions have an integral ratio called the radix of the less significant of the two positions. Permissible values of the digit in any position range from zero to one less than the radix of that position.

radix number The quantity of characters for use in each of the digital positions of a numbering system. In the more common number systems the characters are some or all of the Arabic numerals. Unless otherwise indicated, the radix of any number is assumed to be 10.

radix point Also called base point, binary point, decimal point, and other names, depending on the numbering system; the index that separates the integral and fractional digits of the numbering system in which the quantity is represented.

Raman shifter A device that alters the wavelength of light by inducing Raman shifts in the light passing through it. Raman shifts are changes in photon energy caused by the transfer of vibrational energy to or from the molecule.

Raman spectrum The modified frequencies that occur when matter is irradiated by a beam of radiant energy.

ramp encoder An analog-to-digital conversion process whereby a binary counter is incremented during the generation of a ramp voltage; when the amplitude of the ramp voltage is equal to the amplitude of the voltage sample, the counter clock is inhibited. The counter contents therefore contain the binary equivalent of the sampled data.

random access device A device in which the access time is effectively independent of the location of the data. Synonymous with direct access device.

random error(s) *1.* The chance variation encountered in all experimental work despite the closest possible control of variables. It is characterized by the random occurrence of both positive and negative deviations from the mean value for the method, the algebraic average of which will approach zero in a long series of measurements. *2.* Errors that are not systematic, are not erratic, and are not mistakes.

random event The occurrence of an event affected by chance alone. For example, heads or tails on a flipped coin occurs at random.

random file A collection of records, stored on random access devices such as discs. Algorithms are used to define the relationship between the record key and the physical location of the record.

randomization Random assignment of specimens to treatments to minimize unknown environment effects, improve validity of experimental error estimates, make possible statistical tests of significance, and make possible the calculation of confidence intervals.

random noise Oscillations whose instantaneous amplitudes occur, as a function of time, according to a normal (Gaussian) curve.

random numbers *1.* Expressions formed by sets of digits selected from a sequence of digits in which each successive digit is equally likely to be any of the digits. *2.* A series of numbers obtained by chance. *3.* A series of numbers considered appropriate for satisfying certain statistical tests. *4.* A series of numbers believed to be free of conditions that might bias the result of a calculation.

random sample As commonly used in acceptance sampling theory, the process of selecting sample units in such a manner that all units under consideration have the same probability of being selected.

random uncertainty The uncertainty associated with a random error.

random variable (r.v.) *1.* A function defined on a sample space, or a transformation that associates a real number with each point in a sample space. *2.* Variables characterized by random behavior in assuming their different possible values. Mathematically, they are described by their probability distribution, which specifies the possible values of a random variable together with the probability associated (in an appropriate sense) with each value.

random walk The path followed by a particle that makes random scattering collisions with other particles in a gaseous or liquid medium.

range *1.* For instrumentation, the set of values over which measurements can be made without changing the sensitivity of the instrument. *2.* The extent of a measuring, indicating, or recording scale. *3.* An area within defined boundaries or landmarks used for testing vehicles, artillery, or missiles, or for other test purposes. *4.* The maximum distance a vehicle, aircraft, or ship can travel without refueling. *5.* The distance between a weapon and a target. *6.* The maximum distance a radio, radar, sonar or television transmitter can send a signal without excessive attenuation. *7.* The set of values that a quantity or function may assume. *8.* The difference between the highest and lowest value that a quantity or function may assume. *9.* The difference between the maximum

and minimum values of physical output over which an instrument is designed to operate normally. *See also* error range.

rangeability *1.* Describes the relationship between the range and minimum quantity that can be measured. *2.* The ratio of the maximum flow rate to the minimum flow rate of a meter. *3.* (installed rangeability) The ratio of maximum to minimum flow within which limits the deviation from a desired installed flow characteristic does not exceed some stated limits. *4.* (inherent rangeability) A property of the valve alone, may be defined as the ratio of maximum to minimum flow coefficients between which the gain of the valve does not deviate from a specified gain by some stated tolerance.

range check In data processing, a validation that data are within certain limits.

ranging The measurement of distance by timing how long it takes a light, radio-frequency sound, or ultrasound pulse to make a round trip from the source to a distant object.

rank *1.* To arrange in an ascending or descending series according to importance. *2.* Position in some ascending or descending series.

Rankine An absolute temperature scale in which the zero point is defined as absolute zero (the point where all spontaneous molecular motion ceases) and the scale divisions are equal to the scale divisions in the Fahrenheit system; 0°F equals approximately 459.69°R.

rapid burning Material that transmits a flame across either surface at a rate too fast to measure accurately; therefore, no calculation is required. Examples of the materials in this category are extremely thin films that burn rapidly, or napped surfaces that "flash."

rapid quenching Rapid cooling of molten metals or alloys to achieve maximum uniformity in the crystal structure. Applicable in producing metal powder.

rapid solidification *See* rapid quenching.

rare earth catalyst A catalyst in which the active material is a rare earth element such as lanthanum and cerium. Note: The rare earth elements range in atomic number from 57 to 71.

rare gases Gases such as helium, neon, argon, krypton, xenon, and radon, all of whose shells of planetary electrons contain stable numbers of electrons, making the atoms are almost completely chemically inactive.

rare gas-halide lasers A class of lasers in which the inert gases are used as the amplifying medium.

rate control *See* derivative control.

rated capacity The manufacturers stated capacity rating for mechanical equipment, for instance, the maximum continuous capacity in pounds of steam per hour for which a boiler is designed.

ratio of specific heats Specific heat at constant pressure divided by specific heat at constant volume.

ratio spectrofluorometer A type of instrument used in chemical assaying to determine proportions when two compounds are similar in bioassay and spectrophotometry, but differ markedly in fluorescence; quantitative assays can be made either by proportionality or by use of a linearity curve.

raw material Sheet, plate, bar, billet, forging, or castings. Usually identified by a heat or lot number and usually tested destructively for acceptance.

raw particle count The actual number of particles counted in each size of a specified size range or above a specified size or sizes.

Rayleigh-Benard convection The flow of a fluid contained between horizontal thermally conducting plates and heated from below. The Rayleigh number is proportional to the temperature difference between the plates.

Rayleigh disc A special form of acoustic radiometer used to measure particle velocities.

Rayleigh equation Ratio of red to green required by each observer to match spectral yellow.

Rayleigh scattering Any scattering process produced by spherical particles whose radii are smaller than about one tenth the wavelength of the scattered radiation.

Rayleigh's law of scattering Law stating that, when heterogeneities of a transmitting medium have average dimensions somewhat smaller than the wavelength of the incident energy, the fraction of the incident flux scattered is inversely proportional to the fourth power of the wavelength.

Rayleigh waves *1.* Two-dimensional barotropic disturbances in a fluid having one or more discontinuities in the vorticity profile. *2.* Surface waves associated with the free boundary of a solid, such that a surface particle describes an ellipse whose major axis is normal to the surface and whose center is at the undisturbed surface.

ray tracing A procedure used in the graphical determination of the path followed by a single ray of radiant energy as it travels through media of varying indices of refraction.

Rb The chemical symbol for rubidium.

RBSC *See* reaction bonded silicon carbide.

RBSN *See* reaction bonded silicon nitride.

Re Chemical symbol for rhenium. *See also* Reynolds number.

reaction A chemical transformation or change brought about by the interaction of two substances.

reaction bonded silicon carbide (RBSC) A ceramic cladding material used in very high-temperature applications.

reaction bonded silicon nitride (RBSN) A ceramic material used in very high-temperature applications.

reaction bonding Chemical combining of ingredients.

reaction cell A cell at which a chemical reaction is occurring.

reaction factor *See* load factor.

reaction products The substances formed in a chemical reaction—the desired items as well as the unwanted fumes, sludge, residues, etc.

reaction time The element of uptime needed to initiate a mission, measured from the time command is received.

reactive diluent Pertaining to epoxy formulations, a compound containing one or more epoxy groups which functions mainly to reduce the viscosity of the mixture.

reactor *1.* A circuit element that introduces capacitative or inductive reactance. *2.* A vessel in which a chemical reaction takes place. *3.* An enclosed vessel in which a nuclear chain reaction takes place.

readability On an instrument, the smallest fraction of the scale that can be easily read—either by estimation or by use of a vernier.

reading The indicated value determined from the scale of an indicating instrument, or from the position of the index on a recording instrument with respect to an appropriate indicating scale.

readout In data processing, the display of information on the CRT screen or other display unit.

read time *See* access time.

reagent A substance used to detect, measure, or react to other substances.

real number Any number that can be represented by a point on a number line.

real time *1.* In data processing, the actual time that is required to solve a problem. *2.* Pertaining to the actual time during which a physical process transpires. *3.* Pertaining to the performance of a computation during the actual time that the related physical process transpires so that results of the computation can be used in guiding the physical process.

real-time clock A clock that indicates the passage of actual time, in contrast to a fictitious time set up by the computer program; for example, elapsed time in the flight of a missile, whereby a 60-second trajectory is computed in 200 actual milliseconds, or a 0.1 second interval is integrated in 100 actual microseconds. *See* clock, real time.

real-time executive *See* executive software.

real-time input Input data inserted into a system at the time of generation by another system.

real-time interrupt process (RIP) Software within TELEVENT that responds to "events" (interrupts).

real-time language (RTL) A computer language designed to work on problems of a time-critical nature.

real-time output Output data removed from a system at time of need by another system.

real-time processing *1.* The processing of information or data in a sufficiently rapid manner so that the results of the processing are available in time to influence the process being monitored or controlled. *2.* Computation that is performed while a related or controlled physical activity is occurring so that the results of the computation can be used to guide the process.

real-time program A program that operates concurrently with an external process that it is monitoring or controlling, meeting the needs of that process with respect to time.

real-time system A system that responds within the time scale of the process being controlled, i.e., whose response time depends on the process dynamics or time constants.

reasonableness test A test that provides a means for detecting a gross error in calculation by comparing results against upper and lower limits representing an allowable reasonable range.

Réaumur scale A temperature scale having 0° as the ice point and 80° as the steam point; the scale is little used outside of the brewing, winemaking, and distilling industries.

REB *See* relativistic electron beams.

recalescence A phenomenon associated with the transformation of gamma iron to alpha iron on the cooling (supercooling) of iron or steel, revealed by the brightening (reglowing) of the metal surface owing to the sudden increase in temperature caused by the fast liberation of the latent heat of transformation. Contrast with decalescence.

recarburize *1.* To increase the carbon content of molten cast iron or steel by adding carbonaceous material, high-carbon pig iron, or a high-carbon alloy. *2.* To carburize a metal part to return surface carbon lost in processing. Also known as carbon restoration.

receiver gage A gage that is calibrated in engineering units, and receives the output of a pneumatic transmitter.

receiving gage A fixed gage designed to inspect several dimensions on a part, and also inspect the relationships between dimensions.

recipe The complete set of data and procedure that defines the control requirements of a particular product manufactured by a batch process. A recipe consists of a header, equipment requirements, procedure, and formula.

recombination The reaction between an ion and one or more electrons that returns the ionized element or molecule to the neutral state.

recombination coefficient A measure of the specific rate at which oppositely charged ions join to form neutral particles (a measure of ion recombination).

record *1.* A segment of a file consisting of an arbitrary number of words or characters. *2.* In data processing, a group of data that contains all the information about a single item.

record (verb) To store on some permanent (generally magnetic) medium, as on a magnetic tape or disk.

recorded value The value of a measured variable as determined from the position of a trace or mark on chart paper, or as determined from a permanent or semipermanent effect on an alternative recording medium.

recorder *1.* An instrument that makes and displays a continuous graphic, acoustic, or magnetic record of a measured variable. *2.* A measuring instrument in which the values of the measured variable are recorded.

recording thermometer *See* thermograph.

recovery ratio The ratio of the indicated temperature to the total temperature.

recovery time *1.* The time interval, after a specified event (for example, overload, excitation transients, output short-circuiting) after which a transducer again performs within its specified tolerances. *2.* In a radiation counter tube, the time that must elapse after detection of a photon of radiation until the instrument can deliver another pulse of substantially full response level upon interaction with another photon of ionizing radiation. *3.* Regarding a heat-treatment furnace, the time required to return to the set temperature after a cold charge has been placed in the furnace.

recrystallization *1.* The change from one crystal structure to another, as occurs on heating or cooling through a transformation temperature. *2.* The formation of a new, strain-free grain structure from that existing in cold-worked metal, usually accomplished by heating.

recrystallization annealing Annealing cold-worked metal to produce a new grain structure without phase change.

recrystallization temperature The approximate minimum temperature at which complete recrystallization of a cold-worked metal occurs within a specified time.

recrystallized beta grain Strain-free grains of beta phase that precipitate relatively little or no alpha phase during a decoration aging treatment.

rectangular coordinates *See* Cartesian coordinates.

recursive Pertaining to a process that is inherently repetitive. The result of each repetition is usually dependent on the result of the previous repetition.

red plague A powdery brown-red cuprous oxide deposit sometimes found on silver-coated copper conductors and shield braids. It is fungus-like in appearance and will appear in random spots along the length of a conductor or shield.

reducing atmosphere An atmosphere that tends to: (a) promote the removal of oxygen from a chemical compound or alloy; or (b) promote the reduction of immersed materials.

reduction *1.* Gain of electrons by a constituent of a chemical reaction. *2.* Removal of oxygen from a chemical compound or alloy. *3.* The process of preparing subsamples from the original sample.

reduction catalyst A catalyst that promotes the chemical reduction of nitrogen oxides (NO_x) by reaction with carbon monoxide (CO), free hydrogen (H_2), or hydrocarbon (HC). The desired products of the chemical reaction are nitrogen gas, carbon dioxide, and water.

reduction of area (RA) The area difference between the initial cross section of a test specimen and the resulting cross section after applied tension and fracture within the gage length.

redundancy Added features in a communication and/or data processing system that are not essential for the specified operation, but that allow the detection of errors or continuation of operation in case of defects.

redundancy check An automatic or programmed check based on the systematic insertion of components or characters, used especially for checking purposes.

redundant character A character specifically added to a group of characters to ensure conformity with certain rules which can be used to detect computer malfunction.

Redwood scale A time-based viscosity scale used predominantly in Great Britain; it is similar in concept to the Saybolt scale used in the United States.

re-enterable load module A load module that can be used concurrently by two or more tasks.

re-entrancy The property whereby a callable program can be called and executed before it has completed the execution from a previous call. The results of the previous call are not affected.

reentrant Describes a program that can be interrupted at any point by another program, and then resume from the point at which it was interrupted.

re-entrant program A program that can be used for various tasks.

re-entry point The instruction at which a program is re-entered from a subroutine.

referee test In an instance when more than one test procedure is permitted to identify a particular characteristic, the test that is to be used in case of dispute. For example, depth of decarburization can be measured via several procedures, and some specifications will identify the referee test to be used, for example, via microhardness traverse.

refereed test A predetermined destructive or nondestructive test made by a regulatory body or a disinterested organization, often to fulfill a regulatory requirement; in some cases, the test may be done by the regulated organization and merely witnessed by an agent of the regulatory body.

reference conditions Standard ambient conditions of 70°F and 29.00 in. Hg. The reference temperature lies halfway between test temperature limits of 60 and 80°F.

reference electrode A reversible electrode used for measuring the potentials of other electrodes.

reference level *See* datum plane.

reference plane *See* datum plane.

reference plastic A clear polystyrene plastic standard selected for exposure as a check on a test apparatus and operating conditions.

refined metal Metal derived from raw materials in the form of virgin elements, master alloy, and/or revert melted in any combination, replenished as needed, fluxed for non-metallics, and degassed as necessary.

refinery gas The commercially noncondensible gas resulting from fractional distillation of crude oil, or the cracking of crude oil or petroleum distillates. Refinery gas is either burned at the refineries or supplied for mixing with city gas.

reflectance The ratio of the radiant flux reflected by a body to that incident upon it.

reflected power The power traveling toward the amplifier (or generator) reflected by the load caused by impedance mismatch between the transmission line and the load.

reflected waves Shock waves, expansion waves, or compression waves reflected by another wave incident upon a wall or other boundary. In electronics, radio waves reflected from a surface or object.

reflection The process whereby a surface discontinuity turns back a portion of the incident radiation into the medium through which the radiation approached.

reflection factor The ratio of the reflected luminous flux to the incident luminous flux. It is usually expressed as a percentage.

reflective *1.* Throwing back rays or images; reflecting. *2.* Of or produced by reflection.

reflectivity The quality or condition of being reflective.

reflectometer A photoelectric instrument for measuring the proportion of light reflected from a given surface.

reflux The recycle stream that is returned to the top of the column. This stream supplies a liquid flow for the rectifying section that enriches the vapor stream moving up the column. Material in the stream is condensate from the overhead condenser. Reflux closes the energy balance by removing heat introduced at the reboiler.

refracted wave(s) The resultant wave train produced when an incident wave crosses the boundary between its original medium and a second medium; in many cases, only a portion of the wave crosses the boundary, with the remainder being reflected from the boundary.

refraction Deflection of radiant energy from a straight path in passing from one medium to another.

refraction loss Reduction in amplitude or some other wave characteristic due to the refraction occurring in a nonuniform medium.

refractive index The ratio of phase velocity of a wave in free space to phase velocity of the same wave in the specific medium. *See also* refractivity.

refractivity The algebraic difference between an index of refraction and unity.

refractometer Instrument for measuring the index of refraction of a liquid, gas, or solid.

refractory *1.* A material of very high melting point with properties that make it suitable for such uses as furnace linings and kiln construction. *2.* The quality of resisting heat.

refractory coating Pyrolytic material used for coating other materials exposed to high temperatures.

refractory metal A metal having an extremely high melting point, for example, tungsten, molybdenum, or tantalum. In a broad sense, it refers to metals having melting points above the range of iron, cobalt, and nickel.

regenerator(s) *1.* Devices used in a thermodynamic process for capturing and returning to the process heat that would otherwise be lost. *2.* A repeater, that is, a device that detects a weak signal in a fiber-optic communication system, amplifies it, cleans it up, and retransmits it in optical form.

register(s) *1.* Accurate matching or superimposition of two or more images. *2.* Alignment with respect to a reference position or set of coordinates. *3.* A subassembly of the burner on a furnace or oven which directs airflow into the combustion chamber. *4.* The component of a meter that counts the revolutions of a rotor or individual pulses of energy and indicates the number of counts detected. *5.* In data processing, the specific location of data in memory.

regression analysis The statistical counterpart or analog of the functional expression, in ordinary mathematics, of one variable in terms of others.

regrowth alpha Alpha that grows on pre-existing (primary) alpha during cooling from some temperature high in the alpha-beta field.

regulate The act of maintaining a controlled variable at or near its setpoint in the face of load disturbances.

regulator A device for controlling pressure or flow in a process. *See also* controller, self-operated (regulator).

reheating *1.* An imprecise term denoting an additional heating applied between different mechanical operations or successive steps of the same operation. *2.* For hardening after carburizing, i.e., not using a direct quench from the carburizing temperature.

reheating aluminum and aluminum alloys Heating to hot-working temperature. Improvement of chemical or structural uniformity is incidental.

reinforced plastics Resins with reinforcing fibers or other woven material with resulting strength properties superior to the basic resin.

reinforcement A strong material bonded into a matrix to improve its mechanical properties.

reinforcing materials Fibers, filaments, fabrics, and other substances used for strengthening of matrices in composite materials.

relational data base In data processing, an information base that can draw data from another information base outside the original information base.

relational operator In data processing, a symbol used to determine a relationship to be tested, such as "greater than."

relative density bottle *See* specific gravity bottle.

relative humidity The ratio, expressed as a percentage, of the amount of water present in a given volume of air at a given temperature to the amount required to saturate the air at that temperature.

relative luminosity The ratio of measured luminosity at a particular wavelength to measured luminosity at the wavelength of maximum luminosity.

relative rigidity In mechanical measurement, the ratio of the modulus at some arbitrary frequency, temperature, or time to the modulus at some reference frequency, temperature, or time.

relative viscosity The ratio of the absolute viscosities of the solution and the solvent of the solution.

relativistic electron beam Beam of electrons traveling at approximately the speed of light.

relaxation The relief of stress by creep. Some types of tests are designed to provide diminution of stress by relaxation at constant strain, as frequently occurs in service.

relaxation method (mathematics) An iterative numerical method for solving elliptic partial differential equations, for example, a Poisson equation.

relaxation time In general, the time required for a system, object, or fluid to recover to a specified condition or value after disturbance.

relaxed stress The initial stress less the remaining stress.

relay rack *See* rack.

release agent *See* parting agent.

release film An impermeable layer of film that does not bond to the resin being cured.

reliability The probability that a device will function without failure over a specified time period or amount of usage at stated conditions.

reliability analysis *See* analysis, reliability.

reliability, assessed The reliability of an item determined within stated confidence limits from tests or failure data on nominally identical items. Results can only be accumulated (combined) when all of the conditions are similar.

reliability assurance The exercise of positive and deliberate measures to provide confidence that a specified reliability will be obtained.

reliability, basic The ability of an item to perform its required functions without failure for the duration of a specified mission or life profile.

reliability, inherent The potential reliability of an item present in its design.

relocatable A program that can be moved about and located in any part of a system memory without affecting its execution.

relocatable coding Absolute coding containing relative addresses, which when derelativized, may be loaded into any portion of a computer's programmable memory and will execute the given action properly. The loader program normally performs the derelativization.

relocate In programming, to move a routine from one portion of storage to another and to adjust the necessary address references so that that routine, in its new location, can be executed.

relocation dictionary The part of an object or load module that identifies all relocatable address constants in the module.

reluctance Resistance of a substance to the passage of magnetic lines of force; it is the reciprocal of magnetic permeability. Also known as magnetic resistance.

remainder In the presentation of chemical composition, identifies the basis element from which the alloy is made and is assumed to be present in an amount approximately equal to the difference between 100% and the sum percentage of the alloying elements and listed impurities.

remanence The magnetic flux density that remains in a magnetic circuit after the removal of an applied magnetomotive force.

remote In data processing, a term used to refer to any devices that are not located near the main computer.

remote access Pertaining to communication with a data-processing facility by one or more stations that are distant from that facility.

remote sensing Detecting, measuring, indicating, or recording information without actual contact between an instrument and the point of observation, for example, as in optical pyrometry.

repeatability *See* precision.

replaced test (no-test) A test whose results are considered to be untrue because of identified causes other than properties of the material being tested (for example, errors in specimen machining or testing). *See also* invalid test value.

replacement test A test made as a result of a replaced test.

replication Repetition of observation or measurement to increase precision and to provide a means for measuring dispersion.

reprocessed wool The resulting fiber when wool has been woven or felted into a wool product which, without ever having been utilized in any way by the ultimate consumer, subsequently has been made into a fibrous state.

reproducibility The ability of an instrument to duplicate, with exactness, measurements of a given value. Usually expressed as a percent of span of the instrument. *See* precision.

repulsion *1.* The act of repelling, or the state of being repelled. *2.* In physics, the mutual action by which bodies or particles tend to repel each other. *See* force.

reserved variable Any variable available only to specific programs in the system. Contrast with global variable.

reset control *See* integral control.

residual element *1.* Element present in an alloy in small quantity, but not added intentionally. *2.* An unspecified element, originating in raw materials, melting equipment (for example, furnace refractory lining), or melting fluxes-slags-atmospheres used in the melting practice.

residual error The error remaining after attempts at correction.

residual gas analysis (RGA) The analysis of residual gases in vacuum systems by mass spectrometry.

residual method In magnetic particle inspection, the application of particles after the magnetizing force is removed.

residual standard deviation *See* standard error of estimate.

residual strain Deformation associated with residual stress.

residual stress In structures, any stress in an unloaded body. These stresses arise from local yielding of the material due to machining, welding, quenching, or cold working.

residue check *1.* Any modulo n check. *2.* A check of numerical data or arithmetic operations in which the number A is divided by n and the remainder B accompanies A as a check digit.

resilience *1.* The ratio of energy output to energy input in a rapid (or instantaneous) full recovery of a deformed specimen. *2.* In elastomeric or rubberlike materials subjected to and relieved of stress, the ratio of energy given up on recovery from the deformation to the energy required to produce the deformation. Resilience of an elastomer is usually expressed in percent.

resin *1.* An organic substance of natural or synthetic origin characterized by being polymeric in structure. Most resins are of high molecular weight and consist of long-chain or network molecular structure. *2.* A polymer or insulating compound used for extrusions or molding.

resin content The amount of resin in a laminate expressed as either a percentage of total weight or total volume.

resin/fiber dust Nuisance dust composed of a mixture of resin and fiber formed from solid material by crushing, grinding, drilling, etc., of nonmetallic composites.

resin, liquid An organic polymeric liquid that becomes a solid when converted into its final state for use.

resin matrix composites Composite materials utilizing a matrix of filaments and/or fibers of glass, metal, or other material bound with a polymer or resin.

resinography The study of the properties and behavior of resins, polymers, and plastics.

resinoid Any of the class of thermosetting synthetic resins, either in their initial temporarily fusible state or in their final infusible state.

resin pocket An apparent accumulation of excess resin in a small, localized section visible on cut edges of molded surfaces, or internal to the structure and nonvisible.

resin-rich area Localized area filled with resin and lacking reinforcing material.

resin-starved area Localized area of insufficient resin, usually identified by low gloss, dry spots, or fiber showing on the surface.

resistance *1.* The material property that causes dissipation of energy when an electric current flows through a substance (unit of measure is the ohm (Ω)). When 1 ampere of current flows through 1 ohm of resistance, 1 watt of power is dissipated. *2.* Restraining forces contributed to the load. *3.* In electricity, the factor by which the square of the instantaneous conduction current must be multiplied to obtain the power lost by heat dissipation or other permanent radiation of energy away from the electrical current. *4.* In mechanics, the opposition by frictional effects to forces tending to produce motion.

resistance brazing Brazing by resistance heating, the joint being part of the electrical circuit.

resistance coefficients *See* resistance.

resistance drop The voltage drop in phase with the current.

resistance strain gage A fine wire or similar device whose electrical resistance changes in direct proportion to the amount of elastic strain it is subjected to.

resistance thermometer A temperature-measuring device in which the sensing element is a resistor that has a known variation in electrical resistance with temperature.

resistance welding A group of welding processes in which heat is obtained from resistance of the work to electrical current in a circuit of which the work is one part. Spot welding is an example of resistance welding.

resistivity A measure of the resistance of a material to electric current either through its volume or on a surface.

resistor An electrically conductive material shaped and constructed so that it offers a known resistance to the flow of electricity.

resistor-transistor logic (RTL) A form of logic circuit that uses resistors and transistors and performs not or nor logic.

resite *See* C-stage.

resitol *See* B-stage.

resol *See* A-stage.

resolution *1.* The act of deriving from a sound, scene, or other form of intelligence, a series of discrete elements from which the original may subsequently be synthesized. *2.* The degree to which nearly equal values of a quantity can be discriminated. *3.* The fineness of detail in a reproduced spatial pattern. *4.* The degree to which a system or a device distinguishes fineness of detail in a spatial pattern.

resolution sensitivity The smallest change in an input that produces a discernible response.

resolver Any means for determining the mutually perpendicular components of a vector quantity.

resolving power *1.* A measure of the ability to respond to small changes in input. *2.* The ability of an optical device to separate the images of two objects very close together. *3.* The ability of a mono-chromator to separate two lines in a multi-line spectrum. *See also* resolution.

resolving time The minimum separation time between events that will enable a counting device to detect and respond to both events.

resonance *1.* The reinforced vibration of a body exposed to the vibration at the frequency of another body. *2.* An oscillation of large amplitude in a mechanical or electrical system caused by a relatively small periodic stimulus of the same or nearly the same period as the natural oscillation period of the system.

resonance lines Spectral lines that occur either as absorption or emission lines.

resonant frequency *1.* The frequency at which a given system or object will respond with a maximum amplitude when driven by an external sinusoidal force of constant amplitude. *2.* The frequency at which maximum amplitude occurs for a given input force vibration system.

resonator(s) Generally, a pair of mirrors located at either end of a laser medium, which cause light to bounce back and forth between them while passing through the laser medium.

resource Any facility of the computing system or operating system required by a job or task, and including main storage, input/output devices, the central processing unit, data sets, and control processing programs.

resource manager A general term for any control program function responsible for the allocation of a resource.

responder *See* transponder.

response *1.* Experimental specimen measurement or result, due to its exposure to a treatment combination. *2.* The motion or output resulting from an excitation or input under specified conditions.

response curve A line drawn through at least six points established by calibration gases, which determines the sensitivity of the analytical instrument to unknown concentrations.

response time *1.* The time between the initiation of an operation from a computer terminal and the receipt of results at the terminal. *2.* The time it takes for a controlled variable to react to a change in input.

responsiveness The ability of an instrument or control device to follow wide or rapid changes in the value of a dynamic measured variable.

restart In electronic computing, the process of recommencing a computing function from a known point in a program following computer system failure or other unusual event during execution of a task.

restart address The address at which a program can be restarted; normally, the address of the code required to initialize variables, counters, and the like.

resultants The sums of two or more vectors.

retarder *See* inhibitor.

retentivity The ability of a material to retain magnetism after a magnetic force is removed.

retest A repeat of a test by the same laboratory, using the same method, equipment (of equivalent accuracy or better), and sample. Usually performed in response to suspect or non-conforming results from the original test(s).

reticulated foam A three-dimensional, net-like, open-cell material, made of a flexible polyurethane compound.

retrofit *1.* Modification and upgrading of older control systems. *2.* Parts, assembly, or kit that will replace similar components originally installed on equipment.

retroreflection The process by which illumination is returned by an object directly or generally back to the source of that illumination; reflection characterized by the flux in an incident beam being returned in directions close to the direction from which it came, this effect occurring over a wide range of incidence angles.

reused wool The resulting fiber when wool or reprocessed wool has been spun, woven, knitted, or felted into a wool product which, after having been used in any way by the ultimate consumer, subsequently has been made into a fibrous state.

reverse drawing Drawing, especially a deep-drawn part, a second time in a direction opposite to the original draw direction.

reverse impact test A test method in which one side of a test specimen is impacted, and the reverse side inspected for damage.

reverse osmosis The application of pressure to stop or reverse the transport of solvent through a semipermeable membrane separating two solutions of different solute concentration.

reverse phase chromatography (RPC) Chromatography with a nonpolar stationary phase coupled with a polar mobile phase.

reverse polarity *1.* An electrical circuit in which the positive and negative electrodes have been interchanged. *2.* An arc-welding circuit in which the electrode is electrically positive and the workpiece electrically negative.

reverse time *See* reaction time.

reverse video A cathode ray tube (CRT) screen display of dark characters on a light background—the opposite of the usual CRT screen display.

reversing switch An electrical switch whose function is to reverse connections, on demand, of one part of the circuit.

revolutions per minute (rpm) A standard unit of measure for rotational speed.

rework Restoring an item to a condition exactly conforming to original design specifications.

REX "A" hardness The hardness of a sealant as measured by a REX "A" hardness gage.

Reynolds number A dimensionless criterion of the nature of flow in pipes. It is proportional to the ratio of dynamic forces to viscous forces: the product of diameter, velocity, and density, divided by absolute viscosity.

Reynolds stress In the mathematical treatment of a viscous, incompressible, homogeneous fluid in turbulent motion, the stress that represents the transfer of momentum due to turbulent fluctuations.

RFI protector A device that protects a computer from strong radio or television transmissions.

RGB output An output model for specifying a color, or a gray-scale, as a combination of the three primary colors of light (red, green, and blue) in particular concentrations. The intensity of each primary color is specified by a number in the range of 0 to 1. If all three colors have equal intensity, the perceived result is a pure gray on the scale from black to white.

RGA *See* residual gas analysis.

Rh Chemical symbol for rhodium.

rheocasting Use of partially solidified metal alloys (fractions solids) fed directly into a casting machine for forming into machine parts.

rheology The study of the deformation and flow of matter.

rheopectic substance A fluid whose apparent viscosity increases with time at any constant shear rate.

rheostat An adjustable variable resistor.

Richardson-Dushman equation *See* thermionic emission.

Richardson number A nondimensional number arising in the study of shearing flows of a stratified fluid.

ridge A longitudinal line where the thickness of the metal is slightly greater than the thickness adjacent.

rigid foams Foams used in applications of sound absorption and vibration dampening.

rigidity Resistance of a body to instantaneous change of shape.

rigid plastics For purposes of general classification, a plastic that has a modulus of elasticity either in flexure or in tension greater than 690 MPa (100 ksi) at 23°C (70°F) and 50% relative humidity.

rigid resin A resin having a modulus high enough to be of practical importance.

rimmed steel A type of steel characterized by gaseous effervescence when cooling in the mold. This results in metal relatively free of voids. Sheet and strip products made from rimmed steel exhibit good surface quality.

Ringelmann chart A series of four rectangular grids of black lines of varying widths printed on a white background, and used as a criterion of blackness for determining smoke density.

ringing time In ultrasonic testing, the length of time that a piezoelectric crystal continues to vibrate after the ultrasonic pulse has been generated.

RIP *See* real-time interrupt process.

ripple Regular or irregular variations in voltage around the nominal d-c voltage level during steady-state operation of the system.

riser *See* feedhead.

rms *See* root mean square.

RMS Root mean square magnitude of a signal.

Rn Chemical symbol for radon.

Rockwell hardness number The number resulting from a Rockwell hardness test. The number will relate to one of several available scales, for example A, B, C, 15-N, 30-N, etc.

Rockwell hardness test An indentation hardness test based on the depth of penetration of a specified penetrator into the specimen under certain arbitrarily fixed conditions.

röentgen A quantity of x-ray or gamma-ray radiation that produces, in air, ions carrying one electrostatic unit of electrical charge of either sign per 0.001293 gram of air.

Roentgen rays An alternative term for x-rays.

roller leveler breaks Obvious transverse breaks usually 1/8 to 1/4 inch apart caused by the sheet fluting during roller leveling. These will not be removed by stretching.

roller leveler lines Lines running transverse to the direction of leveling. These may be seen upon stoning or light sanding after leveling and before drawing. Moderate stretching will usually remove them.

roller leveling Leveling by passing flat stock through a machine having a series of small-diameter staggered rolls.

rolling Reducing the cross-sectional area of metal stock, or otherwise shaping metal products, through the use of rotating rolls.

roll-over error For an analog-to-digital converter with bipolar input range, the output difference for inputs of equal magnitude but opposite polarity. Specified in counts or LSBs.

roll threading Making threads by rolling the workpiece between two grooved die plates, one of which is in motion, or between rotating grooved circular rolls.

Ronchi grating *See* optical grating.

Ronchi test An improvement on the Foucault knife-edge test for curved mirrors, in which the knife edge is replaced with a transmission grating with 15 to 80 lines per centimeter, and the pinhole source is replaced with a slit or a section of the same grating.

room temperature A temperature in the range of 20 to 30°C (68 to 86°F).

room-temperature curing adhesive An adhesive that sets within an hour at room temperature and later reaches full strength without heating.

room temperature vulcanizing Curing at room temperature by the use of chemical reactions versus heat and pressure.

root crack A crack at the root or base of a weldment.

root mean square (rms) *1.* A measure of surface roughness. *2.* A means of expressing a-c voltage in terms of the d-c voltage; equal to peak a-c voltage divided by the square root of two.

root-mean-square errors In statistics, the square root of the arithmetic mean of the squares of the deviations of the various items from the arithmetic mean of the whole.

rosebuds Concentric rings of distorted coating, giving the effect of an opened rosebud.

rosette-type strain gage *1.* A type of resistance strain gage having three individual gage elements arranged to measure strain in three different directions simultaneously; typical arrangement has two elements oriented 90° to each other and the third at 45° to the first two. *2. See* strain rosette.

rotary furnace A circular furnace constructed so that the hearth and workpieces rotate around the axis of the furnace during heating.

rotational flow *See* vortices.

rotational transition A change in the rotational state of a molecule. Rotational transitions involve less energy than either electronic or vibrational transitions, and typically correspond to wavelengths in the far infrared, longer than about 20 micrometers.

rough developed blank A blank that will require trimming after being formed.

roughness Relatively finely spaced surface irregularities, the height, width, and direction of which establish the predominant surface pattern.

roughness height rating Quantitative expression of the roughness of a surface.

roughness width The distance in inches between successive ridges that constitute the predominant pattern of the surface roughness.

round *See* round-off.

rounding error The error resulting from rounding off a quantity by deleting the less significant digits and applying some rule of correction to the part retained; for example, 0.2751 can be rounded to 0.275 with a rounding error of .0001. Synonymous with round-off error. Contrast with truncation error.

round-off Synonymous with round. *See also* rounding error and half-adjust.

round-off error Same as rounding error.

round robin testing Testing of specimens from the same sample by different laboratories and/or by different test methods or equipment. Such testing is usually motivated by the fact that two laboratories (vendor and purchaser) develop different data from the same lot(s) of material. It can also involve multiple laboratories to develop consensus data to represent a standard.

routine A subdivision of a program consisting of two or more instructions that are functionally related; therefore, a program. Clarified by subroutine and related to program.

roving A number of yarns, strands, tows, or ends collected into a parallel bundle with little or no twist.

roving ball The supply package offered to the winder, consisting of a number of ends or strands wound to a given outside diameter onto a length of cardboard tube.

roving cloth A textile fabric, coarse in nature, woven from rovings.

RPC *See* reverse phase chromatography.

rpm *See* revolutions per minute.

RQL *See* rejectable quality level.

RTL *1. See* real-time language. *2. See* resistor-transistor logic.

Ru Chemical symbol for ruthenium.

rubber *1.* A term that includes both natural and synthetic types. *2.* A material that is capable of recovering from large deformations quickly and forcibly in the vulcanized state.

rubber modulus The tensile stress at a specified elongation. A measure of resistance to deformation.

run *1.* In data processing, to start a program on the computer. *2.* The TELEVENT executive command to initiate a background program.

rung Group of program elements in a ladder diagram. The group controls a single output element (coil or function).

Runge-Kutta method A method for the numerical solution of an ordinary differential equation.

running A movement or flowing of wet topcoat over the sealer caused by a paint-sealer incompatibility or an excessive application of topcoat.

runout table A roll table used to receive a rolled or extruded section.

run time The length of time between the beginning and the end of a program execution.

run-time error In data processing, an error that occurs during a program operation which may or may not cause the program to stop.

rupture In breaking strength or tensile strength tests, the point at which a material physically separates, as opposed to yield strength.

rupture strength *1.* The stress in a material at failure based on the ruptured cross-sectional area itself. *2. See* rupture stress.

rupture stress The amount of stress at breaking or fracture. Same as rupture strength.

rust Corrosion product consisting primarily of hydrated iron oxide—a term properly applied only to iron and ferrous alloys. The result of oxidation.

rust proofing The application of coatings intended to prevent or greatly reduce the formation of rust on steel parts.

Rutherford *1.* 10^{-6} radioactive disintegration per second. *2.* A quantity of a nuclide having an activity equal to one Rutherford.

r.v. *See* random variable.

S

s *See* second; has largely replaced sec as the preferred abbreviation.

S *1. See* siemens. *2. See* insulation resistance. *3. See* solubility.

S₁ *See* dead load stress.

Sabin A unit of measure of sound absorption equivalent to one square foot of a perfectly absorptive surface.

sacrificial protection Reduction of corrosion of a metal in an electrolyte by galvanically coupling it to a more anodic metal. A form of cathodic protection.

saddling Forming a seamless ring by forging a pierced disk over a mandrel.

safety hardener A curing agent that causes only a minimum toxic effect on the human body, either on contact with the skin or as concentrated vapor in the air.

safe working pressure *See* design pressure.

sag The "droop" of a sealant after it is applied to an overhead or vertical surface.

sagging Run-off or flow-off of adhesive from an adherend surface due to application of excess or low-viscosity material.

salt beds Deposits of sodium chloride and other salts resulting from the evaporation and/or precipitation of ancient oceans.

salt spray test A test in which a specimen is sprayed with a fine mist of a saline solution to accelerate corrosion.

SAM *See* scanning auger microscopy.

sample *1.* A random selection of units from a lot, usually made for the purpose of evaluating the characteristics of the lot. *2.* The material or part that is evaluated for color and appearance match to the master standard.

sample cell The portion of the analytical instrument through which the sample gas being analyzed passes.

sample mean The arithmetic average of the measurements in a sample. The sample mean is an estimator of the population mean.

sample median The value of the middle observation if the sample size is ordered from smallest to largest.

sampling *1.* Obtaining the values of a function for discrete, regularly, or irregularly spaced values of the independent variable. *2.* Selecting only part of a production lot or population for inspection, measurement, or testing. *3.* The removal of a portion of a material for examination or analysis. *4.* In statistics, obtaining a sample from a population.

SAN A copolymer of styrene and acrylonitrile with good toughness, strength, and transparency.

sandblasting Grit blasting, especially when the abrasive is ordinary sand. *See* grit blasting.

sanding Smoothing a surface with abrasive cloth or paper; usually implies use of paper covered with adhesive-bonded flint or quartz fragments.

sandwich braze A joining technique for reducing thermal stress in a brazed joint, in which a shim is placed between the opposing surfaces to act as a transition layer.

sandwich construction A technique of producing composite materials which consists of gluing hard outer sheets onto a center layer, usually a foamed or honeycomb material.

SAP *See* sintered aluminum powder.

saponification corrosion Formation of a soap by the reaction of corrosion products with some organic coatings.

satin finish A type of metal finish produced by scratch brushing a polished metal surface to produce a soft sheen.

saturant A substance that saturates.

saturate *1.* To cause to become completely penetrated, impregnated, or soaked. *2.* To cause to be so completely filled, charged, or treated with something else that no more can be taken in. *3.* In chemistry, to cause a substance to combine to the full extent of its combining capacity with another.

saturated air Air that contains the maximum amount of water vapor that it can hold at its temperature and pressure.

saturated steam Steam at the temperature corresponding to its pressure.

saturated temperature The temperature at which evaporation occurs at a particular pressure.

saturated water Water at its boiling point.

saturation A circuit condition in which an increase in the driving or input signal no longer produces a change in the output.

saturation current *1.* In ionic conduction, the current obtained when the applied voltage is sufficient to collect all of the ions present. *2.* In an electromagnet, the excitation current required to produce magnetic saturation.

sawing Cutting a workpiece with a band, blade, or circular disk having teeth.

sawtooth edge *See* checked edges.

Saybolt color scale A standardized color scale used primarily in the petroleum and pharmaceutical industries to grade the yellowness of pale products.

Saybolt Furol viscosimeter An instrument similar to a Saybolt Universal viscosimeter, but with a larger diameter tube for measuring the viscosity of very thick oils.

Saybolt Universal viscosimeter An instrument for determining viscosity by measuring the time it takes an oil or other fluid to flow through a calibrated tube.

sb *See* stilb.

S_b *See* modulus of rupture in bending.

S-basis The minimum material property value specified by standards-writing organizations.

SBC *See* single board computer.

SBR A copolymer of styrene and butadiene that is a major commercial elastomer used in the rubber industry.

SBS *See* short beam shear.

Sc Chemical symbol for scandium.

scab corrosion Cosmetic corrosion caused by breakdown of the surface protection system, often proceeded by blisters.

scabs Elongated patches of loosened metal which have been rolled into the surface of the sheet or strip.

scale *1.* A thick metallic oxide, usually formed by heating metals in air. *2.* A graduated series of markings, usually used in conjunction with a pointer to indicate a measured value. *3.* A graduated measuring stick, such as a ruler. *4.* A device for weighing objects. *5.* A hard coating or layer of materials on surfaces of boiler pressure parts.

scale effect Any variation in the nature of the flow and in the force coefficients associated with a change in value of the Reynolds number, i.e., caused by change in size without change in shape.

scale factor *1.* The coefficients used to multiply or divide quantities in a problem in order to convert them so as to have them lie in a given range of magnitude, for example, plus one to minus one. *2.* A constant multiplier that converts an instrument reading in scale divisions to a measured value in standard units. *3.* In analog computing, a proportionality factor that relates the value of a specified variable to the circuit characteristic that represents it in the computer. *4.* In digital computing, an arbitrary factor applied to some of the numerical quantities in the computer to adjust the position of the radix point so that the significant digits occupy specific positions.

scale length The distance that the pointer of an indicating instrument, or the marking device of a recording instrument, travels in moving from one end of the instrument scale to the other, measured along the baseline of the scale divisions.

scale units The units of measure stated on an instrument scale.

scaling *1.* The formation at high temperature of thick corrosion product layers on a metal surface. *2.* The deposition of water-insoluble constituents on a metal surface.

scalloping Uneven wipe at the outer periphery of a pattern.

scalloping (earing) The formation of scallops (ears or marked unevenness) around the top edge of a drawn cup, caused by differences in the directional properties of the sheet metal used.

scalp To remove the surface layer of a billet, slab, or ingot, thereby removing surface defects that might persist through later operations.

scan *1.* Collection of data from process sensors by a computer for use in calculations, usually obtained through a multiplexer. *2.* Sequential interrogation of devices or lists of information under program control. *3.* A single sweep of PC applications program operation. The scan operates the program logic based on I/O status, and then updates outputs and input status. The time required for this is called the scan time. *4.* To examine an area, volume, or portion of the electromagnetic spectrum, point by point, in an ordered manner.

scanning acoustic microscopy An inspection method to examine the microstructure of a material or detect areas of debonding in composite materials.

scanning auger microscopy (SAM) Determining the distribution of elements on the surface of a material by measuring the intensity of auger electrons.

scanning electron microscope (SEM) An instrument that provides a visual image of the surface of an item. It scans an electron beam over the surface of a sample held in a vacuum and measures any of several resultant particle counts or energies. Provides depth of field and resolution significantly exceeding light counts microscopy and may be used at magnifications exceeding 50,000 times.

scanning electron microscopy (SEM) Determining the surface characteristics of a material.

scanning transmission electron microscopy (STEM) Determining the distribution of elements in a material by measuring electrons transmitted through a sample.

scan rate *1.* A single sweep of PC applications program operation. The scan operates the program logic based on I/O status, and then updates output and inputs status. The time required for this is called scan time. 2. Sample rate, in a predetermined manner, each of a number of variables intermittently.

scarfing Cutting surface areas of metal objects, ordinarily by using a gas torch. The operation permits surface defects to be cut from ingots, billets, etc.

scarf joint A joint in which the mating surfaces are at an angle to the axis of the mating pieces.

scattering A collision or other interaction that causes a moving particle or photon of electromagnetic energy to change direction.

scattering coefficients Measures of the attenuation due to scattering of radiation as it traverses a medium containing scattering particles.

scattering cross sections The hypothetical areas normal to the incident radiation that would geometrically intercept the total amount of radiation actually scattered by a scattering particle. They are also defined, equivalently, as the cross-sectional areas of isotropic scat-

terers (spheres) that would scatter the same amount of radiation as the actual amount.

scattering loss A reduction in the intensity of transmitted radiation due to internal scattering in the transmission medium or to roughness of a reflecting or transmitting surface.

scavenger *1.* A reactive metal added to molten metal to combine with and remove dissolved gases or other impurities. *2.* A chemical added to boiler water to remove oxygen.

Schering bridge A type of a-c bridge circuit particularly useful for measuring the combined capacitive and resistive qualities of insulating materials and high-quality capacitors.

schlieren Areas of a transparent material with varying refraction.

Schottky barrier A simple metal-to-semiconductor interface that exhibits a nonlinear impedance.

scintillation spectrometer A scintillation system so designed that it can separate and determine the energy distribution in heterogeneous radiation.

sclerometer An instrument that determines hardness of a material by measuring the force needed to scratch or indent the surface with a diamond point.

scleroscope An instrument that determines hardness of a material by measuring the height to which a standard steel ball rebounds when dropped from a standard height.

scleroscope hardness number A number derived from the rebound height of the hammer dropped on a test specimen.

scleroscope test *See* scleroscope.

scoop trim A trimmed surface that is concave.

scoring *1.* The method by which fiber-board may be depressed, or partially cut in basically linear configurations in any direction, which will later facilitate bending along the depressed or cut scores into various three-dimensional shapes. *2.* Marring or scratching of a formed part by metal pickup on the punch or die.

scouring *1.* Physical or chemical attack on internal surfaces of process equipment. *2.* Mechanical finishing or cleaning using a mild abrasive and low pressure.

scrap *1.* Solid material, including inspection rejects, suitable for recycling as feedstock in a primary operation such as plastic molding, alloy production, or glass remelting. *2.* Narrowly, any unusable reject at final inspection.

scratch An elongated surface discontinuity which is infinitely small in width compared to length.

scratch hardness *1.* A measure of the resistance of minerals or metals to scratching; for minerals it is defined by comparison with 10 selected minerals comprising the Mohs scale. *2.* A method of measuring metal hardness in which a cutting point is drawn across a metal surface under a specified pressure, and hardness is determined by the width of the resulting scratch.

scratch pad An intermediate work file that stores the location of an interrupted program, and retrieves the program when the interruption is complete.

scratchpad memory A high-speed, limited-capacity computer information store that interfaces directly with the central processing unit; it is used to supply the central processor with data for immediate computation, thus avoiding delays that would be encountered by interfacing with main memory.

scratch register Addresses of scratch pad storage locations that can be referenced by the use of only one character.

screen *1.* In data processing, the plane surface of a CR (cathode ray tube) that is visible to the user. *2.* One of a set of sieves used to classify granules or particles by size. *See also* shield.

screen analysis A method of finding the particle size distribution of any loose, flowing aggregate by sifting it through a series of standard screens with holes of various sizes and determining the proportion that passes each screen.

screening *1.* The process of performing 100% inspection, or exposure to stress, on product lots and removing the defective units from the lots. *2.* Separation of an aggregate mixture into two or more portions according to particle size, by passing the mixture through one or more standard screens. *3.* Removing solids from a liquid-solid mixture by means of a screen.

scribe Intentional paint damage typically used for material evaluation during corrosion testing.

scrim A low-cost reinforcing fabric made from continuous-filament yarn in an open-mesh construction.

scuffing *1.* Metal surface degradation resulting from adhesive wear. *2.* A dull mark or blemish, sometimes due to abrasion, on a smooth or polished surface.

scum A film of impurities on the surface of a liquid or a solid.

Se Chemical symbol for selenium.

seal *1.* To close a fuel tank or vessel to make it leakproof by the application of

351

sealant to fasteners, seams, and any other possible leak path. *2.* Any device or system that creates a nonleaking union between two mechanical components.

sealant A material applied to a joint in paste or liquid form which hardens or cures in place, forming a seal against gas or liquid entry.

seal coat *1.* A layer of bituminous material flowed onto macadam or concrete to prevent moisture from penetrating the surface. *2.* A preliminary coating to seal the pores in a material such as wood or unglazed ceramic.

sealers Products applied to joints or seams to prevent the entry of moisture or contaminants. Paint coatings applied to prevent the undesirable interaction of a subsequent coating with a previous coating or to enhance adhesion or corrosion protection.

sealing *1.* Impregnating castings with resins to fill regions of porosity. *2.* Immersing anodized aluminum parts in boiling water to reduce porosity in the anodic oxide film.

seal tests Tests that are devised to assess the sealing characteristics for high- and low-temperature sealing application, fluid compatibility, and shear resistance under laboratory simulated service conditions.

seal weld A weld used primarily to obtain tightness and prevent leakage.

seam *1.* An extended-length weld. *2.* A mechanical joint, especially one made by folding edges of sheet metal together so that they interlock. *3.* A mark on ceramic or glass parts corresponding to the mold parting line. *4.* On the surface of metal, narrow openings that subse-quently have been closed but not welded during hot working; they appear as straight-line cracks, usually resulting from a defect developed in casting or in working.

seamless ring rolling Hot-rolling of a circular blank, with a hole in the center, to form a weldless circular ring by continuous compressive forces exerted by a main roll on the outer diameter against a pin on the inner diameter. Shaped cross sections may be obtained by appropriate contouring of the pin and roll. The height of the ring is controlled by auxiliary rolls.

seamless tubing Tubular products made by piercing and drawing a billet, or by extrusion.

seam lines A continuous line of small beads.

seam welding A process for making a weld between metal sheets which consists of a series of overlapping spot welds; it is usually done by resistance welding, but may be done by arc welding.

season cracking A historical term usually applied to stress-corrosion cracking of brass.

sec *See* second.

SEC *See* size-exclusion chromatography.

second A unit of time in the metric and English systems.

secondary bonding Bonding of two cured components where the only chemical or thermal reaction is that of the adhesive.

secondary electron An energetic electron set in motion by the transfer of momentum from primary electromagnetic or particulate radiation.

secondary emission Emission of subatomic particles of photons stimulated by primary radiation; for example, cosmic rays impinging on other particles and causing them, by disruption of their electron configurations or even of their nuclei, to emit particles or photons, or both, in turn.

secondary hardening The hardening phenomenon that occurs during relatively high-temperature tempering of certain steels containing one or more carbide-forming alloying elements (for example, the high-speed steel family). Up to an optimum combination of tempering time and temperature, the reaction results either in the retention of hardness or an actual increase in hardness.

secondary ion mass spectroscopy (SIMS) Determining the distribution of elements on the surface of a material by measuring the mass of ions emitted from the surface.

secondary storage The storage facilities not an integral part of the computer but directly connected to and controlled by the computer, for example, magnetic drum and magnetic tapes.

section modulus (frame section modulus) The engineering concept that relates shape to section and stiffness. It takes into account frame depth, flange width, and material thickness. All other things being equal, the frame with the largest section modulus will have the greatest strength and stiffness, i.e., the ability to more effectively resist deflection under load.

sediment *1.* Matter in water that can be removed from suspension by gravity or mechanical means. *2.* A non-combustible solid matter that settles out at the bottom of a liquid; a small percentage is present in residual fuel oils.

sedimentation *1.* Classification of metal powders according to the rate at which they settle out of a fluid suspension. *2.* Removal of suspended matter either by quiescent settling or by continuous flow at high velocity and extended retention time to allow the matter to deposit out.

Seebeck coefficient *See* Seebeck effect.

Seebeck effect The establishment of an electric potential difference tending to produce a flow of current in a circuit of two dissimilar metals, the junctions of which are at different temperatures.

seed(s) A small, single crystal of semiconductor material used to start the growth of a single large crystal from which semiconductor wafers are cut.

seedy appearance A paint-sealer incompatibility causing pigment flocculation beading to a surface that is spotted with raised "grainy" looking particles.

seek To position the access mechanism of a direct-access storage device at a specified location.

seek time The time taken to execute operation.

segment In computer software programming, the division of a routine.

segregation *1.* Nonuniform distribution of alloying elements and impurities in a cast or wrought metal macrostructure or microstructure. *2.* The variation in chemical composition resulting from natural phenomena in the solidification of a steel ingot. The various elements of the steel having lowest freezing points are concentrated in parts of the ingot last to solidify. *3.* Keeping process streams apart. *4.* A series of close, parallel, narrow, and sharply defined wavy lines of color on the surface of a molded plastics part which differ in shade from surrounding areas and make it appear as if the components have separated. *5.* The tendency of refuse of varying compositions to deposit selectively in different parts of the unit.

selection Addressing a terminal and/or a component on a selective calling circuit. *See also* lockout and polling.

selective carburizing Carburizing only selected surfaces of a workpiece by preventing absorption of carbon by all other surfaces.

selective case hardening Case hardening only selected surfaces of a workpiece.

selective corrosion The selective corrosion of certain alloying constituents from an alloy (as dezincification), or in an alloy (as internal oxidation).

selective heating Intentional heating of only certain portions of a workpiece.

selective-ion electrode A type of pH electrode that involves use of a metal-metal-salt combination as the measuring electrode, which makes the electrode particularly sensitive to solution activities of the anion in the metal salt.

selective plating Any of several methods of electrochemically depositing a metallic surface layer at only localized areas of a base metal, the remaining unplated areas being masked with a nonconductive material during the plating step.

selective quenching Quenching only certain portions of a workpiece.

selective reinforcement The addition of advanced composite materials to selected areas for local augmentation of strength or stiffness.

selective surfaces Surfaces, often coated, for which the spectral optical properties, such as reflectance, absorptance, emittance, or transmittance vary significantly with wavelength. Such properties are of interest in solar energy applications.

selector A device for choosing objects or materials according to predetermined attributes.

self absorption Attenuation of radiation due to absorption within the substance that emits the radiation.

self-calibrating test method Testing that is essentially controlled by the equipment, with results reported by the equipment, independent of operator technique (other than specimen loading); equipment is calibrated by testing certified standard specimens prior to and after testing required specimens.

self-checking code Same as error detecting code.

self diffusion (solid state) The spontaneous movement of an atom to a new site in a crystal of its own species.

self-excited vibration Vibrations in which the vibratory motion produces cyclic forces that sustain the vibration.

self-extinguishing The quality of a material to stop burning once the source of a flame is removed. For example, the material ignites on either surface, but the flame extinguishes itself before reaching the first scribed line.

self-extinguishing/no burn rate Refers, for example, to a material that stops burning before it has burned for 60 seconds from the start of timing, and has not burned more than 50.8 mm (2 in.) from the point at which the timing was started.

self-extinguishing resin A resins formulation that will burn in the presence of a flame, but will extinguish itself within a specified time after the flame is removed.

self-hardening steel *See* air-hardening steel.

self-skinning foam Foam that develops a resilient outer surface when curing.

SEM *See* scanning electron microscope.

semiconductor *1.* An electronic conductor with resistivity in the range between metals and insulators, in which the electric-charge-carrier concentration increases with increasing temperature over some temperature range. Note: Certain semiconductors possess two types of carriers, namely, negative electrons and positive holes. *2.* A material whose resistivity lies in the broad range between conductors and insulators.

semiconductor controlled rectifier An alternate name used for the reverse-blocking triode-thyristor. Note: The name of the actual semiconductor material (selenium, silicon, etc.) may be substituted in place of the word semiconductor in the naming of the components.

semiconductor devices Electron devices in which the characteristic distinguishing electronic conduction takes place within semiconductors.

semiconductor diodes Two-electrode semiconductor devices utilizing the rectifying properties of junctions or point contacts.

semiconductors, II-VI Semiconductors composed of elements from groups II and VI of the periodic table—sometimes extended to cover elements with valances of 2 and 6. Typical II-VI compounds are cadmium telluride and cadmium selenide.

semiconductors, III-V Semiconductors composed of atoms from groups III and V of the periodic table, such as gallium (III) and arsenic (V), which form gallium arsenide.

semiconductor strain gage A type of strain-measuring device particularly well suited to use in miniature transducer elements; it consists of a piezoresistive element that is either bonded to a force-collecting diaphragm or beam or diffused into its surface.

semicrystalline In plastics, materials that exhibit localized crystallinity.

semikilled steels Steels that have characteristics intermediate between those of killed and rimmed steels. During the solidification of semikilled steel, some gas is evolved and entrapped within the body of the ingot. This tends to compensate for the shrinkage that accompanies solidification.

semi-metallic friction material A friction material having organic binders formulated with metallic fibers and/or metal powders.

semirigid plastic Any plastics material having an apparent modulus of elasticity of 10,000 to 100,000 psi.

Sendzimir mill A cold rolling mill having two work rolls of 1 to 2-1/2 inch diameter each, backed up to by two rolls twice that diameter, and each of these backed up by bearings on a shaft mounted eccentrically so that rotating it increases the pressure between bearings and backup rolls.

sense *1.* To examine, particularly relative to a criterion. *2.* To determine the present arrangement of some element of hardware, especially a manually set switch. *3.* To read punched holes or other marks.

sensibility reciprocal A balance characteristic equal to the change in load required to vary the equilibrium position by one scale division at any load.

sensible heat Heat that causes a temperature change.

sensitivity *1.* Measure of the ability of a device or circuit to react to a change in some input. Also, the minimum or required level of an input necessary to obtain rated output. *2.* The ratio of the change in transducer output to a change in the value of the measurand.

sensitivity analysis Repetition of an analysis with different quantitative values for cost or operational assumptions or estimates such as hit-kill probabilities, activity rates, or research and development costs, to determine their effects for the purposes of comparison with the results of basic analysis.

sensitivity coefficient The shape of the straight line representing the best fit to the calibration values determined by the method of least squares within the channel amplitude class.

sensitization In stainless steels, a phenomenon whereby chromium at the grain boundaries is reduced by carbide precipitation to below the passive level. Sensitization makes the alloy susceptible to intergranular corrosion and can be caused by improper welding, furnace brazing, milling, or heat treatment.

sensitized stainless steel Any austenitic stainless steel having chromium carbide deposited at the grain boundaries. This deprives the base alloy of chromium, thereby resulting in more rapid corrosion in aggressive media.

sensitizing heat treatment A heat treatment, whether accidental, intentional, or incidental (as during welding), that causes precipitation of constituents at grain boundaries, often causing the alloy to become susceptible to the intergranu-

lar corrosion or intergranular stress corrosion cracking.

sensitometer An instrument for determining the sensitivity of light-sensitive materials.

sensor *1.* A device responsive to the value of the measured quantity. *2.* A transducer that converts a parameter (at a test point) to a form suitable for measurement (by the test equipment).

separation *1.* The point at which the bonded substrates disconnect from one another at the adhesive. *2.* An action that disunites a mixture of two phases into the individual phases. *3.* Partition of aggregates into two or more portions of different particle size, as by screening. *4.* The removal of dust from a gas stream.

separator *1.* Any machine for dividing a mixture of materials according to some attribute such as size, density, or magnetic properties. *2.* A device for separating materials of different specific gravity using water or air. *3.* A cage in a ball-bearing or roller-bearing assembly. *See* cage.

septum Adhesive and prepreg (or a metal sheet) cured between two pieces of core.

sequence checking routine A routine that checks every instruction executed and prints out certain data; for example, the routine may print out the coded instructions with addresses, and the contents of each of several registers, or it may be designed to print out only selected data, such as transfer instructions and the quantity actually transferred.

sequence monitor Computer monitoring of the step-by-step actions that should be taken by the operator during a startup and/or shutdown of a power unit. As a minimum, the computer would check that certain milestones have been reached in the operation of the unit. The maximum coverage would have the computer check that each required step is performed, that the correct sequence is followed, and that every checked point falls within its prescribed limits. Should an incorrect action or result occur, the computer would record the fault and notify the operator.

sequential control A class of industrial process-control functions in which the objective of the control system is to sequence the process units through a series of distinct states (as distinct from continuous control). *See* sequence monitor.

sequential sampling A method of inspection that involves testing an undetermined number of samples, one by one, until enough test results have been accumulated to allow an accept/reject decision to be made.

serial *1.* In reference to digital data, the presentation of data as a time-sequential bit stream, one bit after another. *2.* In PCM telemetry, the transfer of information on a bit by bit basis. *3.* In data transfer operations, a procedure that handles the data one bit at a time in contrast to parallel operations.

serial accumulator A register that receives data bits in serial or sequence and temporarily holds the data for future use.

serial computer *1.* A computer having a single arithmetic and logic unit. *2.* A computer, some specified characteristic of which is serial; for example, a computer that manipulates all bits of a word serially. Contrast with parallel

computer.

serial I/O Method of data transmission in sequential mode, one bit at a time. Only one line is needed for the transmission. However, it takes longer to send/receive the data than parallel I/O.

serial operation The organization of data manipulation within circuitry whereby the digits of a word are transmitted one at a time along a single line. The serial mode of operation is slower than parallel operation, but utilizes less complex circuitry.

serial-parallel Pertaining to processing that includes both serial and parallel processing, such as the one that handles decimal digits serially, but handles the bits that comprise a digit in parallel.

service test A test of a material conducted under simulated or actual operational conditions to determine whether the specified requirements or characteristics are satisfied.

servo A transducer type in which the output of the transduction element is amplified and fed back so as to balance the forces applied to the sensing element or its displacements. The output is a function of the feedback signal.

session Layer 5 of the open system interconnection (OSI) model.

set *1.* Permanent strain in a metal or plastics material. *2.* A collection. *3.* To place a storage device into a specified state, usually other than that denoting zero or blank. *4.* To place a binary cell into the state denoting one. *5.* In simulation theory, sets consist of entities with at least one common attribute. Additionally, entities may own any number of sets. Sets may be arranged (topologically ordered) on a "first-in, first-out," "last-in, last-out," or ranked basis. *6.* A combination of units, assemblies, or parts connected together or used together to perform a single function, as in a television or radar set. *7.* A group of tools, often with at least some of the individual tools differing from others only in size. *8.* In plastics processing, conversion of a liquid resin or adhesive into a solid material. *9.* Hardening of cement, plaster, or concrete.

settling Partial or complete separation of heavy materials from lighter ones by gravity.

setup *1.* An arrangement of data or devices to solve a particular problem. *2.* In a computer that consists of an assembly of individual computing units, the arrangement of interconnections between the units, and the adjustments needed for the computer to solve a particular problem. *3.* Preliminary operations—such as control adjustments, installation of tooling or filling of process fluid reservoirs—that prepare a manufacturing facility or piece of equipment to perform specific work.

set-up driver A routine capable of accepting raw set-up and control information, converting this information to static stores, dynamic stores, or control words, and loading or transmitting the converted data to the associated module in order to achieve the desired effect.

sexadecimal number A number, usually of more than one figure, representing a sum in which the quantity represented by each figure is based on a radix of sixteen. Synonymous with hexadecimal number.

S-glass *1.* A magnesium aluminosilicate compound that produces high-tensile-strength glass filaments. *2. See* structural glass.

shaft encoder A device for indicating the angular position of a cylindrical member. *See also* gray code and cyclic code.

shake The intermediate-frequency (5–25 Hz) vibrations of the sprung mass as a flexible body.

shakeout Removing sand castings from their molds.

shake table *See* vibration machine.

shake-table test A durability test in which a component or assembly is clamped to a table or platen and subjected to vibrations of predetermined frequencies and amplitudes.

shape memory alloys Martensitic alloys (titanium-nickel) that exhibit shape recovery characteristics through stress-induced transformation and reorientation. Reverse transformation during heating restores the original grain structure of the high temperature phase.

shattercrack *See* flake.

shear *1.* Force that causes or tends to cause two contiguous parts of the same body to slide relative to each other in a direction parallel to their plane of contact. *2.* A tool that cuts metal or other material by the action of two opposing blades that move along a plane approximately at right angles to the surface of the material being cut.

shear crimping Buckling of the compressive facing due to low core shear modulus. Usually causes the core to fail in shear at the crimp.

shear disturbances *See* S waves.

shear edge The cutoff edge of the mold.

shear fracture Fracture in which crystals are separated by sliding or tearing under a shear stress.

shear lip A characteristic of ductile fractures in which the final portion of the fracture separation occurs along the direction of principal shear stress, as exhibited in the cup and cone fracture of a tensile-test specimen made of relatively ductile material.

shear modulus The ratio of shearing stress to shearing strain within the proportional limit of the material.

shearography Developed for strain measurements, the process provides a full-field video strain gauge, in real time, over large areas.

shear rate The rate of deformation (of a fluid) tangent to an applied stress.

shear rigidity The sandwich property that resists shear distortions.

shear spinning A metal forming process in which sheet metal or light plate is formed into a part having rotational symmetry by pressing a tool against a rotating blank and deforming the metal in shear until it comes in contact with a shaped mandrel. The resulting part has a wall thinner than the original blank thickness.

shear stability The ability of a fluid to withstand shearing without permanent loss of viscosity.

shear strain Deformation resulting from an applied shear stress.

shear strength The stress required to effect fracture in the plane of the material cross section.

shear stress Component of any stress lying in the plane of area where stress is measured. The existence of shear stress in fluid is evidence of viscosity.

shear test A test designed to apply a shear force upon a specimen, thereby identifying the material's shear strength under stated conditions.

shear wave(s) *1.* In ultrasonic testing, sound waves directed in such a way that defects not detectable by the longitudinal waves would be detected, for example, surface or near surface defects. *2.* A wave in an elastic medium in which any element of the medium along the wave changes its shape without changing its volume. *See also* S waves.

sheet Any flat material intermediate in thickness between film or foil and plate; specific thickness limits for sheet depend on the type of material involved, and sometimes also on other dimensions such as width. For example, for steel, the descriptive dimensions are 0.006–0.1875 inches thickness and 24 inches minimum width.

sheet metal A flat-rolled metal product generally thinner than about 0.25 inches.

sheet molding compound A mixture of resins, fibers, and fillers that is processed into sheets ready for molding.

shelf life The length of time that a material or product can be stored before losing required properties.

shellac A flammable resinous material, produced by a species of insect found in India, used to make a water-resistant coating for wood by dissolving the resin in alcohol.

shell hardening A surface hardening process whereby a suitable steel workpiece, when heated through and quench hardened, develops a martensitic layer or shell that closely follows the contour of the piece and surrounds a core of essentially pearlitic transformation product. This result is accomplished through a proper balance between section size, steel hardenability, and severity of quench. *See also* case hardening, definition 2.

shield *1.* A conductive layer of material (usually wire, foil, or tape) applied over a wire or cable to provide electrical isolation from external interference. Also known as a screen. *2.* Any barrier to the passage of interference-causing electrostatic or electromagnetic fields. An electrostatic shield is formed by a conductive layer, such as a foil, surrounding a cable core. *3.* An attenuating body that blocks radiation from reaching a specific location in space, or that allows only radiation of significantly reduced intensity to reach the specific location.

shielded enclosure A mesh or sheet metallic housing designed expressly for the purpose of separating electromagnetically the internal and external environment.

shielded metal arc welding (SMAW) Welding in which metals are heated with an arc between a covered metal electrode and the work. Shielding is obtained through decomposition of the electrode covering. Pressure is not used and filler metal is obtained from the electrode.

shim A thin piece of material, usually metal, that is placed between two surfaces to compensate for slight variations in dimensions between two mating parts, and to bring about a more proper alignment or fit.

shock resistance The ability to absorb mechanical shock without cracking, breaking, or excessively deforming. *See also* impact strength.

shore hardness *1.* The relative hardness of an elastomer obtained by use of a shore durometer instrument. *2.* An instru-

ment measure of the surface hardness of an insulation or jacket material.

shore hardness test *See* scleroscope.

short beam shear (SBS) A flex test of a material with a low span-to-thickness ratio.

short beam shear strength The interlaminar shear strength of a parallel-fiber-reinforced plastic material as determined by three-point flexural loading of a short segment cut from a ring specimen.

short fiber composites Composites in which the reinforcement fiber length is short compared to the length of the component, which results in reduced tensile strength and rigidity.

short finish Gray appearance of a polished glass surface resulting from this operation not being carried to the point at which all traces of the previous grinding or smoothing operation are removed.

shortness A form of brittleness in alloys, usually brought about by grain boundary segregation; it may be referred to as hot, red, or cold, depending on the temperature range in which it occurs.

short run Failure of molten metal to completely fill the mold cavity.

shorts *1.* Large particles remaining on a sieve after the finer portion of an aggregate has passed through the screen. *2.* A slang shop word regarding bar lengths; any bar shorter in length than allowed by specification is a short.

short-transverse *1.* In mechanical testing of rectangular bars, the shorter of the two transverse directions. Contrast with long transverse. *2.* Direction of maximum contraction of a metal during forging.

shot *1.* Small, spherical particles of a metal. *2.* Small, roughly spherical steel particles used in a blasting operation to remove scale from a metal surface. *3.* An explosive charge.

shot blasting Impingement of a surface with metal shot to remove deposits or scale.

shot peening Impingement of a surface with shot to improve surface properties.

shrinkage cavity A void left in cast metals as a result of shrinkage during solidification.

shrink fitting A process for forming metal assemblies which allows the outside member, when heated to a practical temperature, to assemble easily with the inside member.

shrink forming A process for forming metal parts in which the inner fibers of a cross section undergo a reduction in a localized area by the application of heat, cold upset, or mechanically induced pressures.

SI *1. See* Systeme International d'Unites and International System of Units. *2. See* silicone.

Si Chemical symbol for silicon.

sialon Any composition containing silicon, aluminum, oxygen, and nitrogen, and usually produced by the high-temperature reactions among the gradients.

side force Force that acts perpendicular to the nominal plane of symmetry of the vehicle.

side strain *See* edge strain.

siemens Metric unit of conductance.

sieve *1.* A meshed or perforated sheet, usually of metal, used for straining liquids, classifying particulate matter, or breaking up masses of loosely adherent or softly compacted solids. *2.* A meshed sheet with apertures of uniform standard size used as an element of a set of screens for determining particle size distribution of a loose aggregate.

sieve diameter (of a particle) (d) *See* particle diameter (sieve diameter).

sieve fraction The portion of a loose aggregate mass that passes through a standard sieve of given size number, but does not pass through the next finer standard sieve; usually expressed in weight percent.

sievert The SI unit for absorbed ionizing radiation.

sight glass A glass tube, or a glass-faced section of a process line, used for sighting liquid levels or taking manometer readings.

sighting tube A tube, usually made of a ceramic material, that is used primarily for directing the line of sight for an optical pyrometer into a hot chamber.

sigma phase An extremely brittle Fe-Cr phase in Fe-Ni-Cr alloys which can form at elevated temperatures.

signal *1.* A visual, audible, or other indication used to convey information. *2.* The intelligence, message, or effect to be conveyed over a communication system. *3.* A signal wave; the physical embodiment of a message. *4.* (computing systems) The event or phenomenon that conveys data from one point to another. *5.* (control, industrial control) Information about a variable that can be transmitted in a system.

signal, error *See* error signal.

signal, input *See* input signal.

signal-to-noise ratio The ratio of the value of a recorded signal channel to that of a simultaneously recorded shorted channel.

sign bit A single bit, usually the most significant bit in a word, which is used to designate the algebraic sign of the information contained in the remainder of the word.

sign-check indicator An error-checking device, indicating no sign or improper signing of a field used for arithmetic processes. The machine can, upon interrogation, be made to stop or enter into a correction routine.

sign digit In coded data, a digit incorporating 1 to 4 binary bits which is associated with an item of data to indicate its algebraic sign.

significance *1.* Results that show deviations between a hypothesis and the observations used as a test of the hypothesis greater than can be explained by random variation or chance alone are called statistically significant. *2.* In positional representation, the factor, dependent on the digit position, by which a digit is multiplied to obtain its additive contribution in the representation of a number.

significance level The probability of rejecting the null hypothesis when it is actually true.

significant digit(s) *1.* Any digit that is necessary to define a value or quantity. *2.* A set of digits, usually from consecutive columns beginning with the most significant digit different from zero and ending with the least significant digit whose value is known and assumed relevant; for example, 2300.0 has five significant digits, whereas 2300 probably has two significant digits; however, 2301 has four significant digits and 0.0023 has two significant digits.

significant out of tolerance condition A condition in the laboratory or inspection station that results in the change of the disposition of material (for example, from conforming to nonconforming).

significant surfaces Surfaces that are visible when the finished component is assembled onto the vehicle and is observed in normal viewing position. Nonsignificant surfaces that may cause corrosion products to run onto significant surfaces are significant. Also, nonvisible surfaces, when designated by the customer, are significant.

significant wave height The average of the highest one-third of the wave height population. Wave height is measured trough to crest.

silica Crystalline quartz used in the manufacture of refractories, insulators, and glass. *See also* silicon dioxide.

silica fibers Fibers with better strength and temperature resistance than glass fibers.

silica gel A colloidal, highly absorbent silica used as a dehumidifying and dehydrating agent, as a catalyst carrier, and sometimes as a catalyst.

silica glass A transparent or translucent material consisting almost entirely of fused silica (silicon dioxide). Also known as fused silica or vitreous silica.

silicon A basic material used to make semiconductors that has limited capacity for conductivity. It is also commonly added to steel.

silicon bronze A corrosion-resistant alloy of copper and 1 to 5% silicon that has good mechanical properties.

silicon carbide Whisker-like reinforcement fiber used in metal matrix composites.

silicon dioxide The chemically resistant dioxide of silicon.

silicone (SI) *1.* A type of elastomer having the basic polymer of dimethyl polysiloxane, with various attached vinyl or phenyl groups. *2.* A generic name for semiorganic polymers of certain organic radicals; they can exist as fluids, resins, or elastomers, and are used in diverse materials such as greases, rubbers, cosmetics, and adhesives.

siliconizing Producing a surface layer alloyed by diffusing silicon into the base metal at elevated temperature.

silicon nitride An engineering ceramic used in high-temperature and wear-resistant applications.

silicon-on-insulator semiconductors *See* SOI (semiconductors).

silicon-on-sapphire (SOS) Pertaining to the technology in which monocrystalline silicon films are epitaxially deposited onto a single-crystal sapphire substrate to form a structure for the fabrication of dielectrically isolated elements.

silky fracture A type of fracture surface appearance characterized by a fine texture, usually dull and nonreflective, typical of ductile fractures.

silver solder A brazing alloy composed of silver, copper, and zinc that melts at a temperature below that of silver but above that of lead-tin solder.

simple balance A weighing device consisting of a bar resting on a knife edge and two pans, one suspended from each end of the bar; to determine precise weight, an unknown weight on one of the pans is approximately balanced by known weights placed in the other pan, and a precise balance is obtained by sliding a very small weight along the bar until a pointer attached to the bar at the balance point indicates a null position.

simple harmonic vibration A vibration at a point in a system in which the displacement with respect to time is described by a simple sine function.

simplex method A finite iterative algorithm used in linear programming whereby successive solutions are obtained and tested for optimality.

SIMS *See* secondary ion mass spectroscopy.

simulate *1.* To represent certain features of the behavior of a physical or abstract system by the behavior of another system. *2.* To represent the functioning of a device, system, or computer program by another; for example, to represent the functioning of one computer by another, to represent the behavior of a physical system by the execution of a computer program, or to represent a biological system by a mathematical model.

simulation *1.* The representation of certain features of the behavior of a physical or abstract system by the behavior of another system. For example, the representation of physical phenomena by means of operations performed by a computer, or the representation of operations of a computer by those of another computer. *2.* Using computers, electronic circuitry, models, or other imitative devices to gain knowledge about operations and interactions that take place in real physical systems.

simulation framework A simulation system capable of integrating and running two or more simulation algorithms in a single simulation environment.

simulator *1.* A device, system, or computer program that represents certain features of the behavior of a physical or abstract system. *2.* A program that simulates the operation of another device or system. In the case of microprocessors, a simulator allows the execution of a microprocessor object program on a computer that is different from the microprocessor for which the program has been written. The simulator provides a range of debugging tools which allows the programmer to correct errors in the program.

single board computer (SBC) A complete computer, including memory, clock, and input/output ports assembled on a single board.

single crystal A crystalline material with a single continuous lattice.

single density A computer diskette that can store approximately 3,400 bits per inch.

single event upsets Radiation-induced errors in microelectronic circuits caused when charged particles (usually from the radiation belts or from cosmic rays) lose energy by ionizing the medium through which they pass, leaving behind a wave of electron-hole pairs.

single fracture In a composite material, the fracture of both the matrix and the fibers.

single lap specimen In adhesive testing, a specimen made by bonding the over-lapped edges of two sheets or strips of material, or by grooving a laminated assembly.

single pass test Filter performance tests in which containment that passes through a test filter is not allowed to recirculate back to the test filter.

single sampling A type of inspection in which an entire lot or production run (population) is accepted or rejected based on results of inspecting a single group of items (sample) selected from the population.

single-test specimen blank The volume of material from which only one test specimen can be machined.

sink A reservoir into which material or energy is rejected.

sinkhead *See* feedhead.

sinter To thermally seal or fuse a material.

sintered aluminum powder (SAP) A pure aluminum sintered compact with higher strength and creep resistance than pure aluminum produced by conventional methods.

sintered carbides Carbide particles compacted and sintered on cutting tool tips.

sintered silicon nitride (SSN) An engineering ceramic.

sintering Heating a powder metal compact at a temperature below the melting point to form diffusion bonds between the particles.

sintering temperature The temperature at which a given powdered compact will densify to a certain desired density, perhaps 90% of the theoretical density during a certain heating period.

size *1.* A specified value for some dimension that establishes an object's comparative bulk or magnitude. *2.* One of a set of standard dimensions used to select an object from among a group of similar objects to obtain a correct fit. *3.* In welding, the joint penetration of a groove weld, or the nugget diameter of a spot weld, or the length of the nominal legs of a fillet weld. *4.* A material such as casein, gum, starch, or wax used to treat the surface of leather, paper, or textiles.

size change With respect to movement as a result of heat treatment, refers to the predictable change, which is based on laboratory studies and experience. This does not include changes due to distortion.

size distribution The study of the size of objects or features and their distribution.

size-exclusion chromatography (SEC) A liquid chromatography technique to separate molecules by their size.

sizing *1.* Applying a material on a surface in order to fill pores and thus reduce the absorption of the subsequently applied coating. *2.* The process of applying a size. This term is sometimes incorrectly used when a finish to improve adhesion is being described.

skin *1.* A general term for a thin exterior covering; may be applied to the exterior walls of a building, the exterior covering of an airplane, a protective covering made of wood or plastics sheeting, or a thin layer on a mass of metal that differs in composition or some other attribute from the main mass of metal. *2.* (natural skin) The smooth surface of a latex foam rubber product, formed by contact with mold or cover plates. *3. See* facings.

skin lamination Subsurface separation that usually results in surface rupture.

skin passing Cold rolling of sheets to improve flatness, impart a desired surface finish, or reduce the tendency to stretch or strain and flute. This expression should not be confused with the term "cold reduction," which involves a substantial reduction in thickness. Sometimes called temper rolling.

skull A layer of solidified metal or dross on the walls of a pouring vessel after the metal has been poured.

slab *1.* A flat piece of concrete that spans beams, piers, columns, or walls to make a floor, roof, or platform. *2.* A relatively thick piece of metal whose width is at least twice its thickness—generally used to describe a mill product intermediate between an ingot and a flat rolled product such as sheet or plate.

slack quenching The incomplete hardening of steel due to quenching from the austenitizing temperature at a rate slower than the critical cooling rate for the particular steel, resulting in the formation of one or more transformation products in addition to martensite.

slag Molten or fused refuse.

sleek A scratch having boundaries that appear polished.

slides (microscopy) Rectangular pieces of glass on which objects are mounted for microscopic examination.

slime *1.* A soft, viscous, or semisolid surface layer—often resulting from corrosion or bacterial action, and often having a foul appearance or odor. *2.* A mudlike deposit in the bottom of a chemical process or electroplating tank. *3.* A thick slurry of very fine solids. Also known as mud, pulp, or sludge.

slip *1.* A deformation process involving shear motion of a specific set of crystallographic planes. *2.* Designates the difference between input and output rpm. It may also be expressed as a percent of input. *3.* The permissible length of axial travel. *4.* A term commonly used to express leakage in positive-displacement flowmeters. *5.* A suspension of ground flint or fine clay in water which is used in making porcelain or in decorating ceramic ware. Sometimes called a slurry.

slippage *1.* Fluid leakage along the clearance between a reciprocating-pump piston and its bore. Also known as slippage loss. *2.* Movement that unintentionally displaces two solid surfaces in contact with each other. *3.* Movement of a gas phase through or past a gas-liquid interface instead of driving the interface forward; especially applicable to certain phenomena in petroleum engineering.

slippage loss *See* slippage.

slip plane A crystallographic plane along which dislocations move under local shear stresses to produce permanent plastic strain in ductile metals.

slit *1.* A long, narrow opening—often used for directing and shaping streams of radiation, fluids, or suspended particulates. *2.* To cut sheet metal, rubber, plastics, or fabric into sheet or strip stock of precise width using rotary cutters, knives, or shears. *3.* In expressing steel quality, fine longitudinal lines seen on a machined surface. They may be actual voids caused by gas or rupturing, may contain inclusions, or may be due to differences in microstructure.

sliver *1.* Surface ruptures somewhat similar in appearance to skin laminations, but usually more prominent. *2.* A thin fragment attached at one end to the surface of flat-rolled metal and rolled into the surface during reduction.

slow neutron A free (uncombined) neutron having a kinetic energy of about 100 eV or less. *See also* thermal neutrons.

slow strain rate technique An experimental technique for evaluating susceptibility to stress corrosion cracking. It involves pulling the specimen to failure in uniaxial tension at a controlled slow strain rate while the specimen is in the test environment, and examining the specimen for evidence of stress corrosion cracking.

slub An abruptly thickened place in a yarn.

sludge *See* slime.

slugging Producing a substandard weld joint by adding a separate piece of material that is not completely fused into the joint.

sluice *1.* A waterway fitted with a vertical sliding gate for controlling the flow of water. *2.* A channel for draining away excess water.

slump *See* sag.

slump test A quality control test for determining the consistency of concrete; the amount of slump is expressed as the decrease in height that occurs when a conical mold filled with wet concrete is inverted over a flat plate and then removed, leaving the concrete behind.

slurry *1.* A suspension of fine solids in a liquid which can be pumped or can flow freely in a channel. *2. See* slip. *3.* A semi-liquid refractory material used to repair furnace linings. *4.* An emulsion of soluble oil and water used as a cutting fluid in certain machining operations.

slushing compound A non-drying oil, grease, or similar organic compound that, when coated over a metal, affords at least temporary protection against corrosion.

SMA *See* styrene-maleic anhydride.

smash A place in the fabric where a number of warp or filling yarns have been broken.

SMAW *See* shielded metal arc welding.

S_{min} *See* minimum stress.

smoke A dispersion of fine solid or liquid particles in a gas.

smudge A dark residue on the surface of sheet steel.

smut A reaction product sometimes left on the surface of the sheet after pickling or annealing.

snaky edges Carbonaceous deposits in a wavy pattern along the edges of the annealed strip.

snap temper A precautionary interim stress-relieving treatment applied to high hardenability steels immediately after quenching to prevent cracking because of delay in tempering them at the prescribed higher temperature.

Snell's law Law stating that the ratio of sines of angles of incidence and refraction is the reciprocal of the ratio of refractive indices of initial and final media.

S-N$_f$ diagram Related to fatigue testing, a plot of stress against the number of cycles to failure. The stress can be S_{max}, S_{min}, or S_a. The diagram indicates the S-N$_f$ relationship and a specified probability of survival. For N$_f$, a log scale is almost always used. For S, a log scale is used most often, but a linear scale is sometimes used.

snowflake *See* flake.

snubber A generic term that is commonly applied to mechanisms that employ dry friction to produce damping of suspension systems.

soaking Holding at a selected temperature to assure uniformity of microstructure throughout or simply to await the next operation.

soak test A test in which sealants are soaked in fluids as specified in specifications to determine resistance to these materials. The conditions are somewhat harsher than are found in actual use.

soap bubble test A leak test consisting of applying soap solution to the external surface or joints of a system under internal pressure and observing the location, if any, at which bubbles form indicating the existence of a gas leak.

soda-lime-silica glass A common glass used for commercial containers.

sodium A highly reactive alkali metal.

sodium sulfates Sodium compounds containing the $-SO_4$ group.

soft anneal copper and copper alloy An imprecise term used to indicate the formability of cold-rolled and annealed products. Its use is discouraged. The desired product is properly described as "fully recrystallized; grain size 0.025–0.090 mm."

soft conversion The process of changing a measurement from inch-pound to equivalent metric units within acceptable measurement tolerances without changing the physical configuration of the item. *See also* hard conversion.

softening The act of reducing scale-forming calcium and magnesium impurities from water.

softening agent A substance—often an organic chemical—that is added to another substance to soften it.

softening point The temperature at which an adhesive will soften to the point where it will be unable to sustain a predetermined load.

soft magnetic alloys Ferrous alloys that are easily magnetized, but lose magnetic properties when the magnetic field is removed.

soft nitriding A misnomer for ductile nitriding. *See* nitriding.

soft paint A paint film that has not achieved its specified hardness. This can be caused by the migration of plasticizers into the topcoat, from the sealer, or by the sealer chemically retarding the curing system of the topcoat.

software *1.* Computer programs, routines, programming languages, and systems. *2.* The collection of related utility, assembly, and other programs that are desirable for properly presenting a given machine to a user. *3.* Detailed procedures to be followed, whether expressed as programs for a computer or as procedures for an operator or other person. *4.* Documents, including hardware manuals and drawings, computer-program listings and diagrams, etc.

software priority interrupt The programmed implementation of priority interrupt functions. *See* priority interrupt.

software quality assurance A program to ensure that software used to control tests and/or generate data is not altered without adequate validation and documentation control.

soft water Water that contains little or no calcium or magnesium salts, or water from which scale-forming impurities have been removed or reduced.

SOI (semiconductors) Semiconductor devices consisting of a silicon layer coupled to an electrically insulating layer.

solar simulators Devices that produce thermal energy, equivalent in intensity and spectral distribution to that from the sun; used in testing materials.

solar thermal propulsion Proposed energy source for spacecraft propulsion, obtained by passing hydrogen through a heat exchanger placed at the focal point of a large parabolic dish solar concentrator mirror.

solder A joining alloy with a melting point below about 450°F, such as certain lead-base or tin-base alloys.

solder glass A special glass that softens below about 900°F, and that is used to join two pieces of higher-melting glass without deforming them.

soldering A joining process that uses an alloy, melting below 800°F, to form an electrical joint without alloying the base metals.

soldering embrittlement Penetration by molten solder along grain boundaries of a metal with resulting loss of mechanical properties.

solder short *See* bridging.

solid cryogen cooling Cooling with solidified cryogenic fluids.

solid damping *See* damping, structural.

solid electrolytes Single crystals, certain alloys, alkaline metals, and other compact compounds used in galvanic cells (batteries).

solidification The change in state from liquid to solid in a material as its temperature passes through its melting temperature or melting range on cooling.

solid laminate A structurally reinforced resin-impregnated composite cured to a solid state containing no sandwich layers of honeycomb, plastic foam, or other material.

solid phase chemical dosimeter A device for measuring radioactivity using specific plastics or glasses that change optical density when exposed to radiation.

solid solution A single, solid homogenous crystalline phase containing two or more chemical species.

solid-state Pertaining to circuits and components using semiconductors.

solid-state device Any element that can control current without moving parts, heated filaments, or vacuum gaps. All semiconductors are solid-state devices, although not all solid-state devices are semiconductors (for example, transformers).

solid-state relay A relay constructed exclusively of solid-state components.

solid-state welding Any welding process that produces a permanent bond without exceeding the melting point of the base materials and without using a filler metal.

solidus The temperature at which a metal, or phase, completes freezing on cooling or begins to melt on heating.

solubility coefficient The percentage of water or other fluid absorbed by a material at saturation at a given temperature.

solubility parameter Symbolized by delta, a measure of the energy required to separate the molecules of a liquid.

solubility (S) The air solubility in volume percent, measured at 32°F and one atmosphere pressure, which will dissolve in a petroleum liquid when the air in equilibrium with a liquid is at a partial pressure of 760 mm of Hg. In the basic V/L formula, "S" is expressed in units of volume of air, measured at 32°F and one atmosphere pressure, dissolved in 100 volumes of fuel at 60°F.

solute The element that is dissolved in a solvent.

solution A liquid containing dissolved substances.

solution heat treatment Heating an alloy to a suitable temperature, holding at that temperature long enough to cause one or more constituents to enter into solid solution, and then cooling rapidly enough to hold these constituents in solution.

solutionizing Another name for solution heat treatment, used principally in copper-beryllium technology.

solvation The process of swelling, gelling, or dissolving of a material by a solvent; for resins, the solvent can be plasticized.

solvent That portion of a liquid or solid solution that is present in the largest amount.

solvent activated adhesive A dry-film adhesive that is rendered tacky by the application of a solvent just prior to use.

solvent adhesive An adhesive with a volatile liquid as its vehicle.

solvent retention The occurrence of solvent residues in chemical or material end products or intermediates.

sonic testing An improper expression for ultrasonic testing.

sorbates Gas taken up by sorbents.

sorbents The materials that take up gas by sorption.

sorption The taking up of gas by absorption, adsorption, chemisorption, or any combination of these processes.

SOS *See* silicon-on-sapphire.

sound absorption The change of sound energy into some other form, usually heat, in passing through a medium or on striking a surface.

sound intensity In a specified direction at a point, the average rate of sound energy transmitted in the specified direction through a unit area normal to this direction at the point considered.

sound-level meter An electronic instrument for measuring noise or sound levels in either decibels or volume units.

sound pressure level (SPL) The intensity of a sound wave which, in decibels, equals $20 \log (P_s/P_r)$, where P_s is the pressure produced by the sound and P_r is a stated reference pressure.

source resistance The output resistance of the source.

SP *See* stack pointer.

spall To detach material from a surface in the form of thin chips whose major dimensions are in a plane approximately parallel to the surface.

spalling *1.* The spontaneous chipping, fragmentation, or separation of a surface or surface coating. *2.* The breaking off of the surface of refractory material as a result of internal stresses.

span *1.* The algebraic difference between the limits of the range. *2.* A structural dimension measured in a straight line between two specific extremities, such as the ends of a beam or two columnar supports. *3.* The dimension of an airfoil, such as the wings of an aircraft, from tip to tip, measured in a straight line. *4.* The difference between maximum and minimum calibrated measurement values. For example, an instrument having a calibrated range of 20–120 has a span of 100.

span adjustment *See* adjustment, span.

span error The difference between actual span and ideal span, usually expressed as a percent of ideal span.

span gas A single calibrating gas blend routinely used in calibration of an instrument such as those used for detecting hydrocarbons, carbon monoxide, and nitric acid. *See* normalizing gas.

spangle The characteristic crystalline form in which a hot-dipped zinc coating solidifies on steel strip.

spark recorder A type of recorder in which sparks passing between a metal pointer and an electrically grounded plate periodically burn small holes in recording paper as it moves slowly across the face of the plate; sparks are produced at regular intervals by a cir-

cuit powering an induction coil, and the varying lateral position of the moving pointer creates the trace.

spark source mass spectrometry A method to determine the composition and concentration of elements in a material where high-voltage spark in a vacuum is used to produce ions on a sample specimen.

spark test *1.* A test used to locate pinholes in wire or cable insulation by the application of an electrical potential across the insulation while it is passing through an ionized field. *2.* A test to identify general chemistry of a metal based on spark characteristics compared to that of a known sample; particularly used in separating mixed items.

spatial coherence The coherence of light over an area of the wavefront of a beam; where the beam hits the surface.

spatial isotropy *See* isotropy.

spatial resolution The precision with which an optical instrument can produce separable images of different points on an object.

spatter Particles of molten metal expelled during a welding operation which become adhered to an adjacent surface.

SPD *See* spectral power distribution.

specific acoustic impedance The complex ratio of sound pressure to particle velocity at a given point within the medium.

specific acoustic reactance The imaginary component of specific acoustic impedance.

specific acoustic resistance The real component of specific acoustic impedance.

specific adhesion Adhesion between materials that are held together by valence forces.

specification *1.* A description of an item or system that defines in detail its functional performance capabilities, and the installation, environmental, and operational requirements or limitations. *2.* A list of requirements that must be met when making a material, part, component, or assembly; installing it in a system; or testing its attributes or functions. *3.* A set of standard requirements applicable to any product or process within the jurisdiction of a given standards-making organization; an industry consensus standard.

specific gravity (sp gr) The ratio of the density of a material to the density of the water at the same conditions. The density of any material divided by that of water at a standard temperature.

specific gravity bottle A small flask used to determine density; its precise weight is determined when empty, when filled with a reference liquid such as water, and when filled with a liquid of unknown density. Also known as a density bottle or relative density bottle.

specific gravity, gas The density of a gas compared to the density of air.

specific gravity, liquid The density of a liquid compared to the density of water.

specific heat (sp ht) *1.* The quantity of heat, expressed in Btu, required to raise the temperature of 1 lb of a substance 1°F. *2.* The ratio of the thermal capacity of a substance to that of water. The specific heat at constant pressure of a gas is designated c_p. The specific heat at constant volume of a gas is designated c_v. The ratio of the two (c_p/c_v), is called the ratio of specific heats, k.

specific heat ratio The ratio of the specific heat of a material at constant pressure and the specific heat at constant volume.

specific humidity The weight of water vapor in a gas-water vapor mixture per unit weight of dry gas.

specific properties Material properties divided by the material density.

specified data That basic information furnished in the request for a test.

specimen A specific portion of a material or laboratory sample upon which a test is performed or which is selected for that purpose. Typically a sample or coupon is machined or ground into a specimen.

specimen blank The volume of material that encompasses one or more single-test specimens. Also referred to as a coupon or sample.

specimen temperature The temperature detected by placing or locating a temperature-sensing device in or on the specimen. In most cases, temperature gradients that develop within flexing rubber specimens make it necessary to define the precise points and techniques used to measure temperature.

speckle holography An imaging technique in which a speckle pattern results from laser illumination of a diffusely reflecting surface when interference occurs between the fields passing through the various portions of a lens aperture. Information about the motion of an object can then be obtained from the imaged fringes resulting from the translation of two speckle patterns.

speckle interferometry An imaging process in which the pattern on the image plane of an interferometer is the result of interference between two mutually coherent, but randomly speckled, fields of two lens-formed images from laser illuminated, diffusely reflecting surfaces.

spectral absorption *See* absorption spectra.

spectral analysis A frequency decomposition of the analog input signals. Identification of the frequency spectrum.

spectral density The amount of a signal level at each frequency or portion of the spectrum.

spectral distribution Relative distribution of radiant energy or flux in the spectrum.

spectral emissivity The ratio, at a specified wavelength, of thermal radiation emitted from a non-blackbody to that emitted from a blackbody at the same temperature.

spectral lines *See* line spectra.

spectral power distribution (SPD) The relative power emitted by a source as a function of wavelength.

spectral reflectance Spectral reflectance is the reflectance as a function of the wavelength of the incident light.

spectral response *See* spectral sensitivity.

spectral response curve A curve giving the relative response of a system (such as a photodetector) as a function of the measuring wavelength.

spectral sensitivity In electronics, radiant sensitivity considered as a function of wavelength. In physics, the response of a device or material to monochromatic light as a function of wavelength; also known as spectral response.

spectrochemical analysis A method to determine the elements and quantity of each in a sample by measuring the wavelengths and intensities of spectral lines.

spectrofluorometer An instrument for determining chemical concentration by fluorometric analysis using two mono-

chromators—one to analyze the wavelength of strongest emission and the other to select the wavelength of best excitation in the sample.

spectrograph An instrument that uses photography to record spectral ranges.

spectrometer A spectroscope that includes an angular scale for measurement of the angular deviation and wavelengths of the components of the spectrum.

spectrophotometer Device for measuring spectral transmittance, spectral reflectance, or relative spectral emittance.

spectrophotometric titration Instrumented titration in which the end point is determined by measuring a change in absorbed radiation with a spectrophotometer.

spectrophotovoltaics The enhancement of solar cell productivity by concentrating and subdividing the sunlight spectrum and focusing on specific spectrum efficient solar cells.

spectropolarimeter *See* polarimeter.

spectroradiometer An instrument that measures power as a function of wavelength.

spectroreflectometer A device that measures the reflectance of a surface as a function of wavelength.

spectroscope A device that spreads out the spectrum for analysis. The simplest type is a prism or diffraction grating that spreads out the spectrum on a piece of paper or ground glass.

spectroscopic analysis Identification of chemical elements by characteristic emission and absorption of light rays.

spectrum *1.* Spatial arrangement of components of radiant energy in order of their wavelengths. *2.* Magnitude versus frequency display.

spectrum analyzer *1.* An instrument for measuring the distribution of energy among the frequencies emitted by a pulse magnetron. *2.* An electronic instrument for analyzing the output, amplitude, and frequency of audio- or radio-frequency generators or amplifiers under normal or abnormal operating conditions.

spectrum display unit *1.* An adjunct to a radio receiver that displays the radio spectrum in and on each side of the carrier being received. *2.* On a telemetry receiver, a device that displays the spectrum at and on both sides of the frequency to which the receiver is tuned.

specular reflection Reflection in which the reflected radiation is not diffused; reflection as from a mirror.

specular transmission density The value of photographic density obtained when only the normal component of transmitted flux is measured for source illumination whose rays are perpendicular to the plane of the film.

speed *See* velocity.

speed of response *See* response time and time constant.

SPF *See* superplastic forming.

sp gr *See* specific gravity.

sphalerite *See* zincblende.

spherical aberration A lens defect that makes rays from the peripheral part of the lens focus at a different point than do rays from the central portion of the lens, producing an image lacking in contrast.

spherical powder Metal powders with round shapes.

spheroidal powders Metal powders with oval shapes.

spheroidite Spherical shaped iron or alloy carbides distributed through a ferrite matrix.

spheroidizing In processing steel, heating and cooling to create a microstructure consisting of ferrite with carbide formed into balls or spheroids. Such would be a preferred annealed microstructure. *See* annealing.

spherulite Radiating lamellar crystals that are present in most crystalline plastics.

spherulitic-graphite cast iron *See* ductile iron.

sp ht *See* specific heat.

spike A transient that exceeds peak ripple for a period less than 150 microseconds. Spikes are sometimes high-frequency oscillations resulting from sudden load variations.

spill The accidental release of a hazardous chemical or radioactive liquid from a process system or a container.

spin glass A magnetic alloy in which the concentration of magnetic atoms is such that below a certain temperature their magnetic moments are no longer able to fluctuate thermally in time, but are still directed at random in loose analogy to the atoms of ordinary glass.

spinning *1.* Production of plastics filament by extrusion through a spinneret. *2.* Forming sheet metal into rotationally symmetrical shapes such as bowls or cones by pressing a round-ended tool against the flat stock and forcing it to conform to the shape of a rotating mandrel.

spiral-flow test Determining the flow characteristics of a resin where the resin flows down a spiral tube. The length and weight of the resin that flows into the tube provides a relative measurement of its flow properties.

spiral trim A trimmed surface that has a spiral pattern.

SPL *See* sound pressure level.

splat powder Metal powder with a flat, thin shape.

splice *1.* To connect two pieces, forming a single longer piece, as in connecting the ends of wire, rope, or tubing; the connection may be made by any of several methods including weaving and welding, and may be made with or without a connector. *2.* A permanent junction between two optical fiber ends. It can be a mechanical splice, formed by gluing or otherwise attaching the ends together mechanically, or a fusion splice, formed by melting the ends together.

splintering A combination of cracking and delamination of the outer skin.

split-beam colorimeter An instrument for determining the difference in radiation absorption by the sample at two wavelengths in the visible or ultraviolet region.

split-beam ultraviolet analyzer An instrument for monitoring the concentration of a specific chemical substance in a process stream or coating by measuring the amount of ultraviolet light absorbed at one wavelength and comparing it to the amount at a reference wavelength that is only weakly absorbed by the sample. Also known as a dual beam analyzer.

sponge *See* sponge metal.

sponge metal Any metal mass produced by decomposition or chemical reduction of a compound at a temperature below the melting temperature of the metal. The expression, commonly referred to as simply "sponge," is applied to forms of iron, the platinum-group metals, titanium, and zirconium.

sponge rubber Cellular rubber consisting predominantly of open interconnecting cells made from a solid rubber compound.

spontaneous combustion Ignition of combustible materials following slow oxidation without the application of high temperature from an external source.

spool *1.* The drum of a hoist. *2.* The movable member of a slide-type hydraulic valve. *3.* A reel or drum for winding up thread or wire. *4.* A relatively short transition member (also known as a spool piece) for making a welded connection between two lengths of pipe.

spool piece *See* spool.

sporification corrosion Formation of a soap by the reaction of corrosion products with some organic coatings.

spot check A type of random inspection in which only a very small percentage of total production is checked to verify that a process remains within its control limits.

spot welding A form of resistance welding in which a weld nugget is produced along the interface between two pieces of metal, usually sheet metal, by passing electric current across the joint, which is clamped between two small-diameter electrodes or between an electrode and an anvil or plate.

spray A mechanically produced dispersion of liquid drops in a gas stream; the larger the drops, the greater must be the gas velocity to keep the drops from separating out by gravity.

spray angle The angle included between the sides of the cone formed by liquid fuel discharged from mechanical, rotary atomizers and by some forms of steam or air atomizers.

spray painting A process in which compressed air atomizes paint and carries the resulting spray to the surface to be painted.

spray quenching Quenching in a spray of liquid.

spray tower A duct through which liquid particles descend countercurrent to a column of gas; a fine spray is used when the object is to concentrate the liquid, a coarse spray when the object is to clean the gas by entrainment of the solid particles in the liquid droplets.

spread The weight of an adhesive per a thousand square feet applied to a joining area.

spread reflection Reflection of electromagnetic radiation from a rough surface with large irregularities.

springback *1*. Movement of a part in the direction of recovering original size or shape upon release of elastic stress. *2*. The amount of elastic deflection that occurs in cold-formed material upon release of the forming force; movement is in a direction opposite to the direction of plastic flow. *3*. In flash, upset, or pressure welding, the amount of deflection in the welding machine due to the upsetting pressure.

spring brass (70% copper-30% zinc) A material that is most often supplied as strip or wire and is recommended for light-duty springs.

spring rate *1*. The change of load of a spring per unit deflection, taken as a mean between loading and unloading at a specified load. *2*. The amount of force or torque produced by a spring as a function of deflection or position relative to a reference.

spring steel Carbon or low-alloy steel that is cold-worked or heat-treated to give it the high yield strength normally required in springs; if it is a heat-treatable composition, the springs may be formed prior to heat treatment (hardening).

spring temper A level of hardness and strength for nonferrous alloys and some ferrous alloys corresponding approximately to a cold-worked state two-thirds of the way from full hard to extra spring temper.

sprung mass Considered to be a rigid body having equal mass, the same center of gravity, and the same moments of inertia about identical axes as the total sprung weight.

sprung weight All weight that is supported by the suspension, including portions of the weight of the suspension members.

sputtering Dislocation of surface atoms of a material from bombardment of high-energy atomic particles.

sputter-ion pump *See* getter-ion pump.

square mesh A weave in wire cloth or textile fabric in which the number of wires or threads per inch is the same both with the weave and in the cross-weave direction.

square wells The impurity potential areas that bound an electron or hole in semiconducting crystals such as silicon.

squeeze films Thin viscoelastic fluid films squeezed between two usually planar structures to serve as sealants, load dampers, lubricants, etc.

squeeze roll One of two opposing rollers designed to exert pressure on a material passing between them.

squeeze time In resistance welding, the time from initial application of pressure until welding current begins to flow.

S$_{ra}$ *See* column average stress.

S$_{rm}$ *See* column maximum stress.

SSC *See* sulfide stress cracking.

SSN *See* sintered silicon nitride.

stability *1*. The ability of a component or device to maintain its nominal operating characteristics after being subjected to changes in temperature, environment, current, and time. *2.* The ability of a transducer to retain its performance characteristics for a relatively long period of time. *3.* Freedom from undesirable deviation. *4.* A measure of the controllability of a process. *5.* The relative ability of a substance to retain its mechanical, physical, and chemical properties during service. *6.* The relative ability of a chemical to resist decomposition during storage. *7.* The relative ability of a waterborne vessel to remain upright in a moving sea. *8.* The state of a system if the magnitude of the response produced by an input variable, either constant or varied in time, is limited and related to the magnitude of the input variable.

stability of a linear system Property of a linear system whereby, having been displaced from its steady state by an external disturbance, it comes back to that steady state when the disturbance has ceased.

stabilization In carbon fiber forming, the process used to render the carbon fiber precursor infusible prior to carbonization.

stabilization annealing A heat treatment used with some non-age-hardening alloys to cause a precipitation of carbides of a form and composition that do not sensitize the alloy to intergranular corrosion and to stabilize the alloy against becoming sensitized during subsequent elevated temperature exposures.

stabilization heat treatment An intermediate temperature precipitation heat treatment used with many age-hardening nickel alloys to cause a precipitation of discontinuous chromium carbides at grain boundaries prior to a lower temperature aging heat treatment, which will cause a fine gamma-prime precipitation within the grains. This two-step precipitation results in an optimization of tensile and creep-rupture properties.

stabilized core Honeycomb cores in which the cells have been filled with a specified reinforcing material for the purpose of supporting the cell walls during machining.

stabilized honeycomb compressive strengths The compressive strength of honeycomb materials for which the plane surface of the test specimen has been stabilized with either a plastic resin or by the attachment of facings.

stabilizer(s) *1.* An airfoil or combination of airfoils, considered a single unit, with the principal function of maintaining stable flight for an aircraft or missile. *2.* Any chemical added to a formulation for the chief purpose of maintaining mechanical or chemical stability throughout the useful life of the substance.

stabilizing treatment Any of various treatments—mechanical or thermal—intended to promote dimensional or microstructural stability in a metal or alloy.

stabilizing treatment, aluminum and aluminum alloys An imprecise term used to denote a treatment above room temperature, but below the recrystallization temperature applied: (a) to cold-worked materials of some non-heat-treatable alloy systems to reduce the tendency to age soften; (b) to some types of solution-treated artificial aging alloys in order to improve stability of mechanical properties and of dimensions (*see* overaging); (c) to other types of solution-treated artificial aging alloys to control the size and distribution of the precipitate to improve resistance to intergranular corrosion or exfoliation corrosion and to stress corrosion cracking; (d) to still other types of age-hardening alloys to reduce the tendency to age naturally.

stabilizing treatments, ferrous A treatment applied for the purpose of stabilizing the dimensions of a workpiece or the structure of a material such as (a) before finishing to final dimensions, heating a workpiece to or somewhat beyond its operating temperature and then cooling to room temperature a sufficient number of times to ensure stability of dimensions in service; (b) transforming retained austenite in those materials that retain substantial amounts when quench hardened (*see* cold treatment); (c) heating a solution treated austenitic stainless steel that contains controlled amounts of titanium or columbium plus tantalum to a temperature below the solution heat-treating temperature to cause precipitation of finely divided, uniformly distributed carbides of those elements, thereby substantially reducing the amount of carbon available for the formation of chromium carbides in the grain boundaries upon subsequent exposure to temperatures in the sensitizing range.

stack *1.* The portion of a chimney above roof level. *2.* Any structure that contains flues for discharging waste gases to the atmosphere. *3.* A vertical conduit that, due to the difference in density between internal and external gases, creates a draft at its base. *4.* An area of memory set aside for temporary storage, or for procedure and interrupt linkages. A stack uses the last-in, first-out (LIFO) concept. As items are added to ("pushed on") the stack, the stack pointer decrements; as items are retrieved from ("popped off") the stack, the stack pointer increments.

stacking A lamination sequence in which the warp surface of one ply is laid against the fill surface of the preceding ply.

stacking sequence A description of a laminate that details the ply orientations and their sequence in the laminate.

stack pointer (SP) Contains the address of the top (lowest) address of the processor-defined stack.

stage In electronics, the portion of a circuit between the control tap of one tube or transistor and the control tap of another.

staggered-intermittent fillet welding Welding a T joint on both sides of the tee in such a manner that the weld bead is segmented, with the segments on either side being opposite gaps between segments on the opposing side.

stain *1.* A nonprotective liquid coloring agent used to bring out the grain in decorative woods. *2.* A permanent or semipermanent discoloration on wood, metal, fabric, or plastic caused by a foreign substance. *3.* Any colored organic compound used to prepare biological specimens for microscopic examination.

staining A discoloring of the topcoat due to a sealer-topcoat interaction.

stainless alloy Any member of a large and complex group of alloys containing iron, at least 5% (usually at least 12%)

chromium, and often other alloying elements, and whose principal characteristic is resistance to atmospheric corrosion or rusting. Also known as stainless steel.

stainless steel *See* stainless alloy.

stamping Virtually any metal-forming operation carried out in a press.

standard addition Procedure whereby mall amounts of a substance being measured are added to the sample to establish a response function.

standard air Dry air weighing 0.075 pounds per cubic foot at sea level (29.92 inches barometric pressure) and 70°F.

standard atmospheric pressure A standard unit of atmospheric pressure defined as the pressure exerted by a 760-mm column of mercury at standard gravity (980.665 centimeters per second or 9.8066 cm/sec^2) at temperature zero degree Celsius. One standard atmosphere = 760 mm of mercury or 29.9213 in. of mercury or 1013.25 millibars.

standard cell A reference cell for electromotive force.

standard conditions Temperature of 59°F and pressure of 14.7 psi.

standard deviation The square root of the variance.

standard electrode potential The reversible potential for an electrode process when all products and reactants are at unit activity on a scale in which the potential for the standard hydrogen half-cell is zero.

standard error of estimate The standard deviation of the residual scatter of points about a curve fit. It is calculated by statistical methods from the analysis of data.

standard flue gas Gas weighing 0.078 pounds per cubic foot at sea level (29.92″ barometric pressure) and 70°F.

standard gage *1.* A highly accurate gage used only as a reference standard for checking or calibrating working gages. *2.* A set span across tracks of a railroad that measures 4 feet 8-1/2 inches (1.44 meters).

standardization *1.* The act or process of reducing something to, or comparing it with, a standard. *2.* A measure of uniformity. *3.* A special case of calibration in which a known input is applied to a device or system for the purpose of verifying the output or adjusting the output to a desired level or scale factor.

standardize *See* normalize.

standard leak A controlled finite amount of tracer gas allowed to enter a leak detector during adjustment and calibration.

standard sphere gap The maximum distance between the surfaces of two metal spheres, measured along a line connecting their centers, at which spark-over occurs when a dynamically variable voltage is applied across the spheres under standard atmospheric conditions; this value is a measure of the crest value of an alternating-current voltage.

standard volume Of a gas, defined as the volume at 25°C (77°F) and 1001.32 kPa (29.921 in. Hg abs.).

standing wave A wave in which, for any of the dependent wave functions, the ratio of its instantaneous value at one point on the wave to its instantaneous value at any other point does not vary with time.

standing wave meter An instrument for measuring the standing-wave ratio in a radio-frequency transmission line.

starved area An area in a plastic part that lacks sufficient resin to bond with a reinforcement fiber.

starved joint A portion of a joint that lacks sufficient adhesive to form a complete bond.

state *1*. Condition of a circuit, system, etc., such as the condition at the output of a circuit that represents logic 0 or logic 1. *2*. A description of the process in terms of its measured variables, or a description of the condition of a circuit or device as in "logic state 1."

statement A software instruction to a computer telling it to perform some sequence of operations.

state variables The output(s) of the memory element(s) of a sequential circuit.

static *1*. At rest, or at rest relative to a solid surface. *2*. Describes a structural test with application of a single increasing load. *3*. Communication interference due to discharge of static electricity.

static accuracy The degree to which the controlled temperature coincides with the specified or selected temperature after all transients have decayed. Static accuracy is usually specified as a deviation from nominal.

static fatigue Material that fails under a constant load, often the result of aging expedited by stress.

static hot pressing Applying a static load during hot pressing.

static information Information that remains constant during the time of relevant task performance.

static load The load that is imposed on a member when in a static state or 1 g condition.

static modulus The ratio of stress to strain when under static conditions.

static port An opening used as a source of ambient (static) pressure in the pitot-static system. One static port can generally be found on each side of an aircraft, in an area where there is usually no dynamic (positive or negative) pressure due to the motion of the aircraft through the air. Static air pressure is used to determine altitude and vertical velocity of an aircraft, and, when compared with dynamic or impact pressure from the pitot tube, it is used in determining airspeed. *See also* pitot-static tube and static tube.

static pressure *1*. The pressure of a fluid that is independent of the kinetic energy of the fluid. *2*. Pressure exerted by a gas at rest, or pressure measured when the relative velocity between a moving stream and a pressure-measuring device is zero.

static pressure gage An indicating instrument for measuring pressure.

static pressure tube *See* static tube.

static RAM Random access memory (RAM) that requires continuous power, but does not need to be refreshed as with dynamic RAM. Memory density is not as high as for dynamic RAM.

static rate With regard to an elastic member, the rate measured between successive stationary positions at which the member has settled to substantially equilibrium condition.

static rated load The maximum load that can be lifted under normal land conditions, without exceeding allowable strength limits.

static register A computer register that retains information in static form.

static stability The property of a physical system whereby constancy in its static and dynamic responses is maintained despite changes in its internal conditions and variations in its environment. Compare with dynamic stability.

static stress Constant stress resulting in failure without shock.

static temperature The temperature of a fluid as measured under conditions of zero relative velocity between the fluid and the temperature-sensitive element, or as measured under conditions that compensate for any relative motion.

static test *1.* Any measurement taken in a normally dynamic system under static conditions, for instance, a pressure test of a hydraulic system under no-flow conditions. *2.* Specifically, a test to verify structural characteristics of a rocket, or to determine rocket-engine thrust, while a rocket is in a stationary or hold-down position.

static tube A device used to measure static pressure in a stream of fluid. Normally, a static tube consists of a perforated, tapered tube with a branch tube for connecting it to a manometer; a related device called a static pressure tube consists of a smooth tube with a rounded nose that has radial holes in the tube behind the nose.

static unbalance That condition of unbalance for which the central principal axis is displaced only parallel to the shaft axis.

static weighing A method in which the net mass of liquid collected is deduced from tare (empty tank) and gross (full tank) weighings respectively made before the flow is diverted into the weighing tank and after it is diverted to the bypass.

statistical control Control of a process by statistical methods. A process is said

to be in a state of statistical control if the variations among the sampling results from it can be attributed to a stable pattern of chance causes.

statistical error *1.* Generally, any error in measurement resulting from statistically predictable variations in measurement system response. *2.* Specifically, an error in radiation-counter response resulting from the random time distribution of photon-detection events.

statistical model A probability distribution as a representation of time to failure.

statistical quality control Any method for controlling the attributes of a product or controlling the characteristics of a process that is based on statistical methods of inspection.

statuary bronze Any of several copper alloys used chiefly for casting ornamental objects such as statues; a typical composition is 90% Cu-6% Sn-3% Zn-1% Pb.

steady flow A flow in which the flow rate in a measuring section does not vary significantly with time.

steady state *1.* A condition in which circuit values remain essentially constant, occurring after all initial transients or fluctuating conditions have settled down. *2.* The condition of a substance or system whose local physical and chemical properties do not vary with time.

steady-state deviation The system deviation after transients have expired. *See* offset and deviation, steady-state.

steady-state flow *See* equilibrium flow.

steady-state motion Motion that continues unchanged over successive intervals of time. A stationary process or equilibrium condition. It can involve constant motion or cyclic motion in a repetitive pattern (for example, oscillatory).

steady-state response gain The ratio of change in the steady-state response of any motion variable with respect to change in input at a given time.

steady-state vibration Vibration in which the displacement at each point recurs for equal increments of time.

steam *1.* The vapor phase of water substantially unmixed with other gases. *2. See* water vapor.

steam attemperation Reducing the temperature of superheated steam by injecting water into the flow or passing the steam through a submerged pipe.

steam cure To hasten the curing cycle of concrete or mortar by the use of heated water vapor, at either atmospheric or higher pressure.

steam dryer A device for removing water droplets from steam.

steam-free water Water containing no steam bubbles.

steam purity The degree of contamination, with contamination usually expressed in ppm.

steam quality The percent by weight of vapor in a steam and water mixture.

steam scrubber A series of screens, wires, or plates through which steam is passed to remove entrained moisture.

steatite A ceramic produced for electrical applications.

steel Any alloy of iron with up to 2% carbon which may or may not contain other alloying elements to enhance strength or other properties.

Stefan-Boltzmann law One of the radiation laws which states that the amount of energy radiated per unit time from a unit surface area of an ideal black body is proportional to the fourth power of the absolute temperature of the black body.

stellite Any of a family of cobalt-containing alloys known for their wear resistance, corrosion resistance, and resistance to softening at high temperature.

STEM *See* scanning transmission electron microscopy.

step aging Aging a material at two or more temperatures.

step brazing Making a series of brazed joints in a single assembly by sequentially making up individual joints and heating each one at a lower temperature than the previous joint to maintain joint integrity of earlier joints; the process requires a lower-melting brazing alloy for each successive joint in the assembly.

step change The change from one value to another in a single increment in negligible time.

step soldering Making a series of joints by soldering them sequentially at successively lower temperatures.

step stress test A test consisting of several stress levels applied sequentially for periods of equal duration to a sample. During each period, a stated stress level is applied, and the stress level is increased from one step to the next.

step trim A trimmed surface having a discontinuity perpendicular to the contact line.

steradian Unit of solid angle, subtended at the center of a sphere 1 m radius by 1 m^2 area on that sphere.

stereochemistry Chemistry dealing with the arrangement of atoms and molecules in three dimensions.

stereoisomer An isomer in which atoms are linked in the same order, but differ in arrangement.

stereospecific plastic The definite order or arrangement of molecules in a polymer which allows close packing of the molecules.

sticker breaks Arc-shaped breaks usually located near the middle of a sheet.

stiffness *1.* The ratio of change of force (or torque) to the corresponding change in translational (or rotational) displacement of an elastic element. *2.* The ability of a metal or shape to resist elastic deflection.

stilb (sb) The CGS unit of luminance equal to one candela per square centimeter.

stilling basin An area ahead of the weir plate large enough to pond the liquid so that the liquid approaches the weir plate at low velocity. Also called a weir pond.

stimulate To cause an occurrence or action artificially, rather than waiting for it to occur naturally, for example, to stimulate an event.

stimulus That which stimulates, or excites to action or increased action. A stimulant.

stishovite A mineral consisting essentially of silicon trioxide.

stitch bonding A method of making wire connections on an integrated circuit board using impulse welding or heat and pressure to bond a connecting wire at two or more points while feeding the wire through a hole in the welding electrode.

stitching *1.* Making a seam in fabric using a sewing machine. *2.* Progressive welding of thermoplastics by successively pressing two small induction-heated electrodes against the material along a seam in a manner resembling the action of a sewing machine.

stitch welding Making a welded seam using a series of spot welds that do not overlap.

stochastic Pertaining to direct solution by trial and error, usually without a step-by-step approach, and involving analysis and evaluation of progress made, as in a heuristic approach to trial-and-error methods. In a stochastic approach to a problem solution, intuitive conjecture or speculation is used to select a possible solution, which is then tested against known evidence, observations, or measurements. Intervening or intermediate steps toward a solution are omitted. Contrast with heuristic.

stochastic processes Ordered sets of observations in one or more dimensions, each being considered as a sample of one item from a probability distribution.

stoichiometric For a specific chemical reaction, the exact proportions of substances that will combine with no excess of any reactant.

stoichiometric conditions In chemical reactions, the point at which equilibrium is reached, as calculated from the atomic weights of the elements taking part in the reaction; stoichiometric equilibrium is rarely achieved in real chemical systems, but rather, empirically reproducible equivalence points are used to closely approximate stoichiometric conditions.

Stokes A unit of kinematic viscosity (dynamic viscosity divided by sample density); the centistoke is more commonly used.

Stoke's diameter (of a particle) (d$_{st}$) *See* particle diameter (Stoke's diameter).

stop aging—aluminum alloys Employment of two different aging treatments to control the type of precipitate formed from a supersaturated alloy matrix in order to obtain the desired properties. The first aging treatment, sometimes referred to as intermediate or stabilizing, is usually carried out at a higher temperature than the second.

stop bit The last bit in an asynchronous serial transmission. Like the start bit, it is used for timing control and carries none of the message information.

storage Pertaining to a device in which data can be stored and from which it can be obtained at a later time. The means of storing data may be chemical, electrical, or mechanical.

storage buffer *1.* A synchronizing element between two different forms of storage, usually between internal and external storage. *2.* An input device in which information is assembled from external or secondary storage and stored, ready for transfer to internal storage. *3.* An output device into which information is copied from internal storage and held for transfer to secondary or external storage. Computation continues while transfers between buffer storage and secondary or internal storage or vice versa take place.

storage life The length of time that an item can be stored under specified conditions and still meet specified requirements.

storage protection An arrangement for preventing access to storage for either reading or writing, or both. *See* memory protect.

stored program *See* stored routine.

stored routine A series of instructions in storage to direct the step-by-step operation of the machine.

STP Abbreviation for standard temperature and pressure.

STPD Abbreviation for standard temperature and pressure, dry. Conditions comprising a temperature of 0°C (32°F) and one atmosphere absolute pressure (101.3 kPa or 760 mm of Hg), dry basis (partial pressure of water vapor equals zero).

straight polarity Arc welding in which the electrode is connected to the negative terminal of the power supply.

strain Deformation of material at any given point with respect to a specific plane passing through that point, expressed as change in length per unit length.

strain aging Natural aging of a ferrous material following cold plastic strain. When tested in tension, strain-aged low-carbon sheet exhibits discontinuous yielding, a decrease in ductility, and an increase in yield strength and hardness without substantial change in tensile strength as compared with unaged sheet. Appropriate restraining (temper rolling) temporarily restores continuous yielding.

strain fatigue *See* fatigue.

strain foil A type of strain gage made by photoetching a resistance element out of thin foil.

strain gage(s) *1.* A device that can be attached to a surface, usually with an adhesive, and that indicates strain magnitude in a given direction by changes in electrical resistance of fine wire; it may be used to measure strain due to static or dynamic applied loading, in tension or compression, or both, depending on design of the gage, bonding technique, and type of instrumentation used to determine resistance changes in the strain element. *2.* Converting a change of measurand into a change of resistance due to strain. *3.* A high-resistance, fine-wire or thin-foil grid for use in a measuring bridge circuit. When the grid is securely bonded to a specimen, it will change its resistance as the specimen is stressed. These devices are used in many forms of transducers.

strain hardening An increase in hardness and strength caused by plastic deformation at temperatures lower than the recrystallization range. Examples are cold rolling and cold drawing.

strain ratio Expressed as an 'r' value. The ratio of width to thickness strain determined in the uniform elongation portion of a tension test. It is a good measure of the crystallographic directionality of the material. It is also a good measure of deep drawability. The higher the 'r' value, the better the deep drawability.

strain rosette An assembly of two or more strain gages used for determining biaxial stress patterns. Also known as a rosette-type strain gage.

strain sensitivity A characteristic of a conductor that describes its resistance change in relation to a corresponding length change; it can be calculated as $\Delta R/R$ divided by $\Delta L/L$. When referring to a specific strain-gage material, strain sensitivity is commonly known as the gage factor.

strand A bundle or assembly of continuous fibers.

strand casting *See* continuous casting.

strand tensile test A tensile test of a single resin-impregnated strand of any fiber.

strategic materials Critical raw materials whose foreign source of supply is uncertain and subject to potential cut-

off. Examples of such materials are chromium, cobalt, manganese, and platinum group metals.

stratification Non-homogeneity existing transversely in a gas stream.

strats Portions of a sample lot that vary from the properties under study.

Straus test A test to determine the susceptibility of stainless steel to weld failure.

stray current corrosion Galvanic corrosion of a metal or alloy induced by electrical leakage currents passing between a structure and its service environment.

streamline flow *1.* A type of fluid flow in which flow lines within the bulk of the fluid remain relatively constant with time. *See also* laminar flow. *2. See* leakage.

strength The maximum stress that a material is capable of sustaining before rupture. Also refers to an ultimate (tensile or compression) strength.

strength weld A weld capable of withstanding a design stress.

stress *1.* The force per unit area of a body that tends to produce a deformation. *2.* The effect of a physiological, psychological, or mental load on a biological organism which causes fatigue and tends to degrade performance.

stress amplitude One-half the algebraic difference between the maximum stress and minimum stress in one cycle of repeated variable loading.

stress analysis In the design of a system or equipment, the evaluation of stress conditions existing in parts when loads are applied.

stress concentration *1.* A modification of the simple stress distribution and occurrence of localized high stresses due to the presence of shoulders, grooves, holes, keyways, threads, etc. *2.* In structures, a localized area of high stress.

stress concentration factor (K_t) The ratio of the greatest stress in the region of a notch or other stress raiser, as determined by advanced theory, photoelasticity or direct measurement of elastic strain, to the corresponding nominal stress. The higher the value of K_t, the greater the stress concentration.

stress corrosion Preferential attack of areas under stress in a corrosive environment, where such an environment alone would not have caused corrosion.

stress-corrosion cracking Cracking in a metal part due to the synergistic action of tensile stress and a corrosive environment, causing failure in less time than could be predicted by simply adding the effects of stress and the corrosive environment together. The tensile stress may be a residual or applied stress, and the corrosive environment need not be severe, but only must contain a specific ion that the material is sensitive to.

stress crack External or internal cracks in a plastic caused by tensile stresses less than that of its short-time mechanical strength, frequently accelerated by the environment to which the plastic is exposed.

stress cycles A variation of stress with time, repeated periodically and identically.

stress intensity factor Load-induced variable in tension, compression, and/or shear that is conducive to crack initiation and propagation and fatigue fracture in materials.

stress raiser A discontinuity or change in contour that induces a local increase in stress in a structural member.

stress ratio In fatigue testing, the ratio of the minimum stress to the maximum stress in one cycle, considering tensile stresses as positive, compressive stresses as negative.

stress relaxation *1.* The decrease in stress after a given exposure time to a constant strain. *2.* A characteristic of an elastomer whereby a gradual increase in deformation is experienced under constant load, after the initial deformation.

stress relief A design means to minimize the effects of stress, for example, a cable clamp.

stress relieving Heating to a suitable temperature, holding long enough to reduce residual stresses, and then cooling slowly enough to minimize the development of new residual stresses. Note: Stress relief may be accomplished by the application of other forms of energy, principally mechanical, either alone or in combination with thermal energy.

stress, residual The stress present in the body that is free of external forces or thermal gradients, usually the result of fabrication methods. In fastener fabrication, the stresses resulting from heading, forming, machining, grinding, and rolling are residual stresses and may be desirable or not depending on their location and type.

stress rupture test The tensile type test—both smooth bar and notched—conducted at a specified stress and temperature popular as an acceptance test; a minimum time to failure is specified. Also known as a creep rupture test.

stress-strain curve Readings of load and deformation for stress and strain plotted to obtain a stress-strain diagram.

stress-strain diagram *See* stress-strain curve.

stress-strain relationship Relationship between the stress or load on a structure, structural member, or specimen, and the strain or deformation that follows.

stress tensors Complete sets of stress components in a solid or fluid medium.

stress whitening An effect noted in nylon (crystalline materials) under stress loading, occurring as a result of molecular orientation, visible as a white area due to the change in the refractive index of the material.

stretchability The ability of a metal to be stretched over a punch without splitting.

stretcher leveling Removing warp and distortion in a piece of metal by gripping it at both ends and subjecting it to tension loading at stresses higher than the yield strength.

stretcher strain (Lüder's lines) Irregular surface patterns of ridges and valleys that develop during forming of annealed last or temper rolled, aged steel.

stretch forming Shaping a piece of sheet metal or plastics sheet by applying tension and then wrapping the sheet around a die form; the process may be performed cold or the sheet may be heated first. Also known as wrap forming.

stretching An operation in which the blank is stretched around the punch with no metal flow over the draw ring. The metal thickness is reduced.

striation In fatigue tests, the location of the crack after each cycle of stress.

strike *1.* A thin electroplated film to be followed by other plated coatings. *2.* A plating solution of high covering power and low efficiency used for electroplating very thin metallic films. *3.* A local crater or remelted zone caused by accidental contact between a welding electrode and the surface of a metal object. Also known as arc strike.

string *1.* In data processing, a group of consecutive characters. *2.* A linear sequence of entities, such as characters or physical elements.

stringer(s) A solid nonmetallic impurity in the parent metal, often the result of an inclusion that has been elongated during a rolling process. Usually a mixture of magnesium and aluminum oxides that results from inadequate molten metal treatment.

string-shadow instrument An indicating instrument in which the measured value is indicated by means of the shadow of a filamentary conductor whose position in an electric or magnetic field depends on the magnitude of the quantity being measured.

stringy alpha Platelet alpha that has been elongated and distorted by non-directional metal working, but not broken up or recrystallized. Also called "wormy alpha."

strip *1.* A flat-rolled metal product of approximately the same thickness range as sheet, but having a width range narrower than sheet. *2.* To mine stone, coal, or ore without tunneling, but rather by removing broad areas of the earth's surface to relatively shallow depths.

stripper A liquid chemical that can remove another material.

strobe pulse A pulse of light whose duration is less than the period of a recurring event or periodic function, and which can be used to render a specific event or characteristic visible so it can be closely observed.

stroboscope Device for presenting a rapid series of exposures of a related sequence of visual stimuli. An illusion of continuous motion may be produced. Also used to stop apparent motion of a moving object.

strongly coupled plasmas Highly compressed and collisional plasmas with electron densities on the order of 10^{24} per cubic centimeter or more. The mean kinetic and potential energies of particles in the plasma are typically of the same order of magnitude.

strontium An alkaline earth metal used as a deoxidizing agent for copper.

Strouhal number A nondimensional parameter defined as: $S = fh/V$, where f is frequency, V is velocity, and h is reference length.

structural adhesive Adhesive used for transferring required loads between adherents exposed to service environments typical for the structure involved.

structural analysis Determination of the stresses and strains in a structural member due to combined gravitational and applied service loading.

structural assembly One or more structural elements that together provide a basic structural function.

structural damage Damage that may affect the structural integrity.

structural fatigue *See* fatigue.

structural glass (S-glass) A magnesia/alumina/silicate glass reinforcement providing high strength.

structural sandwich construction A laminar construction comprising a combination of alternating dissimilar simple or composite materials assembled and intimately fixed in relation to each other so as to use the properties of each to attain specific structural advantages for the whole assembly.

structural steel Hot-rolled steel produced in standard sizes and shapes for use in constructing load-bearing structures, supports, and frameworks; some of the standard shapes are angles, channels, I-beams, H-beams, and Z-sections.

stud arc welding (SW) Metals are heated with an arc between a metal stud, or similar part, and the work. Once the surfaces to be joined are properly heated, they are brought together under pressure.

stud welding Producing a joint between the end of a rod-shaped fastener and a metal surface, usually by drawing an arc briefly between the two members, then forcing the end of the fastener into a small weld puddle produced on a metal surface.

styrene-acrylonitrile *See* SAN.

styrene copolymers *See* ABS and SAN.

styrene-maleic anhydride (SMA) A copolymer of styrene and maleic anhydride with greater heat resistance than typical ABS (acrylonitrile butadiene styrene) compounds.

subcooling The process of cooling a liquid below its condensing temperature for the particular saturated liquid pressure.

subcritical annealing A process anneal performed at a temperature below Ac_1. *See* transformation temperature.

subgrain A part of a crystal with an orientation that is somewhat different than other portions of the same crystal.

sublayer A subdivision of an open system interconnection (OSI) layer; for example, the IEEE 802 Standard divides the link layer into the logical link control (LLC) and media access control (MAC) sublayers.

sublimation The transition of a substance directly from the solid state to the vapor state, or vice versa, without passing through the intermediate liquid state.

submerged-arc welding An electric-arc welding process in which the arc between a bare-wire welding electrode and workpiece is completely covered by granular flux during welding.

subroutine *1.* A routine that is arranged so that control may be transferred to it from a master routine, and so that, at the conclusion of the subroutine, control reverts to the master routine. Such a subroutine is usually called a closed subroutine. *2.* A single subroutine with respect to another routine and a master routine with respect to a third.

subscale Subsurface oxides formed by reaction of a metal with oxygen that diffuses into the interior of the section rather than combining with metal in the surface layer.

subsieve analysis Determination of particle-size distribution in a powdered material, none of which is retained on a standard 44-micrometer sieve.

subsieve fraction The portion of a powdered material that passes through a standard 44-micrometer sieve.

substitutional element An alloying element with an atom size and other features similar to the titanium atom, which can replace or substitute for the titanium atoms in the lattice and form a significant region of solid solution in the phase diagram.

substrate *1.* A thermally stable material, usually catalytically inert, to which the active catalyst is affixed, in which it is imbedded, or in some other way joined. Pellets and monolith represent two physical forms of substrate. *2.* A surface underlying a coating such as paint, porcelain enamel, or electroplate.

subsurface corrosion Pockets of corrosion that develop beneath the surface of a material.

sub-tier laboratory A laboratory that does not belong to a direct material supplier. Systems for such laboratories must qualify as "independent."

suffuse To pour beneath, or diffuse beneath or upon. To overspread, as with a liquid, light, color to fill, or cover.

suffusion The act of suffusing or overspreading, as with a liquid. *See also* perfusion and diffusion.

sulfates Sulfuric acid or one of its salts.

sulfide stress cracking (SSC) Brittle failure by cracking under the combined action of tensile stress and corrosion in the presence of water and hydrogen sulfide.

sulfonated oil Mineral or vegetable oil treated with sulfuric acid to make an emulsifiable form of oil.

sulfurized oil Any of various oils containing active sulfur to increase film strength and load-carrying ability.

sulfate-carbonate ratio The proportion of sulfates to carbonates, or alkalinity expressed as carbonates, in boiler water. The proper maintenance of this ratio has been advocated as a means of inhibiting caustic embrittlement.

sulfur A yellow, crystalline non-metallic element.

sum The quantity resulting from the addition of an addend to an augend.

summation action A type of control-system action in which the actuating signal is the algebraic sum of two or more controller output signals, or in which it depends on a feedback signal that is the algebraic sum of two or more controller output signals.

superalloys *See* heat-resistant alloys.

supercalendered finish A shiny, smooth finish on paper obtained by subjecting the material to steam and pressure while passing it between alternating fiber-filled and steel rolls.

supercompressibility The extent to which behavior of a gas departs from Boyle's law.

superconductivity Near-zero electrical resistance exhibited by some metals. With such metals, extremely powerful currents and magnetic fields are possible.

superconductor A compound capable of exhibiting superconductivity—that is, an abrupt and large increase in electrical conductivity as the temperature of the material approaches absolute zero.

supercooling Cooling below the temperature at which a transformation will take place without developing the transformation.

superficial Rockwell hardness test A Rockwell test using very light indenter loads for testing thin sections of a material.

superfines The portion of a metal powder whose particle size is less than 10 micrometers.

superfinishing Producing a finely honed surface by rubbing a metal with abrasive stones.

superheat To raise the temperature of steam above its saturation temperature. The temperature in excess of its saturation temperature.

superheated steam Steam at a higher temperature than its saturation temperature.

superheating Heating above the temperature at which a transformation will take place without developing the transformation.

superhybrid materials Composites of polymers, boron-aluminum, and titanium.

superlattices Crystals grown by depositing semiconductors in layers whose thickness is measured in atoms.

supernatant liquor The liquid above settled solids, as in a gravity separator.

superplastic forming (SPF) A metal forming process that uses materials with high elongation-to-failure characteristics.

superplasticity The unusual ability of some metals and alloys to elongate uniformly by several thousand percent at elevated temperatures without separating.

supported adhesive film An adhesive supplied in a sheet or in a film form with an incorporated carrier that remains in the bond when the adhesive is supplied and used.

surface The exterior skin of a solid body, considered to have zero thickness.

surface-active agents *See* surfactant.

surface analyzer An instrument that measures irregularities in the surface of a body by moving a stylus across the surface in a predetermined pattern and producing a trace showing minute differences in height above a reference plane magnified as much as 50,000 times.

surface area *1.* The total amount of exterior area on a solid body. *2.* The sum of the individual surface areas of all the particles in a mass of particulate matter.

surface blowoff *1.* Removal of water, foam, etc. from the surface at the water level in a boiler. *2.* The equipment for such removal.

surface combustion The non-luminous burning of a combustible gaseous mixture close to the surface of a hot, porous refractory material through which it has passed.

surface condenser Any of several designs for inducing a change of state from gas to liquid by allowing the gas phase to come in contact with a surface such as a plate or tube which is cooled on the opposite side, usually by being in direct contact with flowing cooled water.

surface contamination Foreign matter on the surface of a material.

surface density Any amount distributed over a surface, expressed as amount per unit area of surface.

surface diameter (of a particle) (d_s) The diameter of a sphere having the same surface area as the particle.

surface drag Skidding of plastic resin along the surface of a mold due to improper mold temperatures, injection pressure, or injection speed.

surface finish The roughness of a surface after finishing, measured either by comparing its appearance with a set of standards of different patterns and lusters or by measuring the height of surface irregularities with a profilometer or surface analyzer.

surface flaws Irregularities of any sort which occur at only one place or at relatively infrequent and widely varying random intervals on a surface. A flaw may be a scratch, ridge, hole, peak, crack, check, etc. Unless otherwise specified, the effect of flaws shall not be included in the roughness height measurement.

surface gage *1.* A scribing tool in an adjustable stand that is used to check or lay out heights above a reference plane. *2.* A gage for measuring height above a reference plane.

surface hardening A generic term covering several processes applicable to a suitable ferrous alloy that produce by quench hardening only, a surface layer that is harder or more wear resistant than the core. There is no significant alteration of the chemical composition of the surface layer. The processes commonly used are induction hardening, flame hardening, and shell hardening. Use of the applicable specific process name is preferred. *See also* case hardening, definition 2.

surface irregularities Deviations from the nominal surface, such as (a) roughness: relatively finely spaced irregularities, the height, width, shape, and direction of which establish the predominant surface pattern; (b) waviness: irregularities of the nominal surface evidenced by recurrent forms of waves, where the waviness may be caused by factors such as machining deflections, vibrations, heat treatment, or warping strains; and (c) flaws: irregularities of any sort which occur at only one place or at relatively infrequent and widely varying random intervals on a surface. A flaw may be a scratch, ridge, hole, peak, crack, check, etc. Unless otherwise specified, the effect of flaws shall not be included in the roughness height measurement.

surface preparation Physical and/or chemical preparation of an adherent to make it suitable for adhesive bonding.

surface pressure *See* pressure.

surface resistivity The resistance between opposite sides of a unit square on the surface of a material.

surface roughness Minute pits, projections, scratches, grooves, and the like which represent deviations from a true planar or contoured surface on solid material.

surface tension The contractive force in the surface film of a liquid which tends to make the liquid occupy the least possible volume.

surface texture The finish of the surface of sheet steel presently described by the roughness (peak) height in micro inches and the peaks per inch.

surface treating Any of several processes for altering properties of a metal surface, making it more receptive to ink, paint, electroplating, adhesives, or other coatings, or making it more resistant to weathering or chemical attack.

surfacing Depositing filler metal on the surface of a part by welding or thermal spraying.

surfactant A chemical compound that reduces surface tension.

surge A non-oscillatory transient that exceeds peak ripple, is infrequent, and has a duration equal to or greater than 150 microseconds.

survivor curve A type of reliability curve that shows the average percent of total production of a given model or type of machine still in service after various lengths of service life in hours.

susceptibility The characteristic of an object that results in undesirable responses when subjected to electromagnetic energy.

susceptibility meter An instrument for determining magnetic susceptibility at low magnitudes.

susceptometer A device for measuring the magnetic susceptibility of ferromagnetic, paramagnetic, or diamagnetic materials.

suspended solids Undissolved solids in boiler water.

suspension *1.* A fine wire or coil spring that supports the moving element of a meter or other instrument. *2.* A system of springs, shock absorbers, and other devices that support the chassis of a motor vehicle on its running gear. *3.* A system of springs or other devices that supports an instrument or sensitive electronic equipment on a frame and reduces the intensity of mechanical shock or vibration transmitted through the frame to the instrument. *4.* A mixture of finely divided insoluble particles of solid or liquid in a carrier fluid (liquid or gas).

SW *See* stud arc welding.

swaging (swedging) Any of several methods of tapering or reducing the diameter of a rod or tube, most commonly involving hammering, forging, or squeezing between simple concave dies.

swapping The process of copying areas of memory to mass storage, and back, in order to use the memory for two or more purposes. Data are swapped out when a copy of the data in memory is placed on a mass storage device; data are swapped in when a copy on a mass storage device is loaded in memory.

swapping device A mass storage device that is especially suited for swapping because of its fast transfer rate.

S waves Waves in an elastic media that cause an element of the medium to change its shape without a change in volume. Mathematically, S waves are those whose velocity field has zero divergence.

sweat The condensation of moisture from a warm saturated atmosphere on a cooler surface. A slight weep in a boiler joint, but not in sufficient amount to form drops.

sweat cooling A process by which a body having a porous surface is cooled by forced flow of coolant through the surface from the interior.

swedging *See* swaging.

sweet crude Crude petroleum containing very little sulfur.

sweet gas Natural gas containing no hydrogen sulfide or mercaptans.

swell A sudden increase in the volume of steam in a water steam mixture below the water level.

symbolic logic The discipline that treats formal logic by means of a formalized artificial language or symbolic calculus whose purpose is to avoid the ambiguities and logical inadequacies of natural languages.

symmetrical laminate A composite laminate in which the sequence of plies below the laminate midplane is a mirror image of the stacking sequence above the midplane.

syndiotactic stereoisomerism A molecule in which side atoms or groups alternate in a regular pattern on opposite sides of a chain.

syneresis Exuding small quantities of liquid by a standing gel.

synergism An action whereby the total effect of two components or agents is greater than the individual effects of the components when simply added together; for instance, in stress-corrosion cracking, cracks form and propagate deep into a material in a much shorter time and at a much lower stress than could be predicted from known effects of stress and the corrosive environment.

syntactic foams Lightweight composites made by mixing hollow spheres of glass and polymers with a curable liquid resin.

syntax *1.* The structure of expressions in a language. *2.* The rules governing the structure of a language. *3.* In data processing, grammatical rules for software programming that specify how instructions can be written.

syntectic alloys Metallic composite materials characterized by a reversible convertibility of their solid phases into two liquid phases by the application of heat.

synthesis (chemistry) The application of chemical reactions to obtain desired chemical products.

synthetic cold rolled A hot-rolled pickled sheet given a sufficient final temper pass to impart a surface approximating that of cold-rolled steel.

synthetic lubricant Any of a group of lubricating substances that can perform better than straight petroleum products in the presence of heat, chemicals, or other severe environmental conditions.

synthetic metals Materials that do not occur in nature, but have the appearance and physical properties of true metals.

synthetic quench A water solution of polyalkylene glycol or other synthetic material used when minimum distortion or low residual stresses are desired.

synthetic rubber Synthetic elastomers made by polymerization of one or more monomers.

system An assembly of procedures, processes, methods, routines, or techniques united by some form of regulated interaction to form an organized whole.

systematic error *1.* An error in a set of measurements or control that can be predicted from scientific principles; individual errors from the same cause bias the value of the mean because they all act in the same direction (sense); the amount of error in each individual value may or may not have a direct mathematical relationship to the true value of the quantity. *2.* Error that cannot be reduced by increasing the number of measurements if the equipment and conditions remain unchanged. *3.* Any constant or reproducible error introduced into a measured or controlled value due to failure to control or compensate for a specific side effect.

systematic uncertainty The uncertainty associated with a systematic error.

system board The control center of a computer.

system check A check on the overall performance of the system, usually not made by built-in computer check circuits, for example, control totals, hash totals, and record counts.

system control *See* control system.

Systeme Internationale d'Unites (SI) The current International System of Units.

system error In a control system, the difference between the value of the ultimately controlled variable and its ideal value.

systems analysis The examination of an activity, procedure, method, technique, or a business to determine: (a) behavioral relationships or, (b) what must be accomplished and how.

systems engineering Designing, installing, and operating a system in a manner intended to achieve optimum output while conserving manpower, materials, and other resources.

systems integration The combining of subsystems each with numerous interfaces for the input and output of data, and each with specified functions vital to the planned success of the main system.

systems simulation The simulation of any dynamic system.

system test Determining performance characteristics of an integrated, interconnected assemblage of equipment under conditions that evaluate its ability to perform as intended and that verify suitability of its interconnections.

T

t *See* temperature.

T *1. See* tesla. *2. See* temperature. *3. See* transmittance.

Ta The chemical symbol for tantalum.

table look up A procedure for obtaining the function value corresponding to an argument from a table of function values.

tableting A method of compacting powdered or granular solids using a punch and die; used to make certain food products, dyes, and pharmaceuticals.

tack *1.* Stickiness of an adhesive or filament-reinforced resin prepreg material. *2.* The property of an adhesive that enables it to form a bond of measurable strength immediately after the adhesive and the adherent are brought into contact under low pressure.

tack-free A condition in which a plastic material can be dented with an inert object without sticking to it.

tack-free time The time required at standard conditions for a curing sealant to lose its surface tackiness as determined by placing a small piece of polyethylene film on its surface and then peeling the film away. The sealant is tack free when no sealant is removed by the film.

tackiness agent An additive that imparts adhesive qualities to a nonadhesive material.

tack weld *1.* Any small, isolated arc weld, especially one that does not bear load, but rather merely holds two pieces in a fixed relationship. *2.* A weld joint made by arc welding at small, isolated points along a seam.

tacky topcoat A more severe form of soft paint in which the sealer has interfered with the topcoat curing mechanism to the degree that fingerprints can be left on the surface of the paint.

tacticity The stereo-regular pattern of a long-chain polymer.

tag A unit of information whose composition differs from that of other members of the set so that it can be used as a marker or label. Also called a flat or sentinel.

tag (from data compressor) A unique sixteen-bit word, preselected by the operator, that precedes each data output word and identifies it.

Tag-Robinson colorimeter A laboratory device used to compare shades of color in oil products by varying the thickness of a column of the oil until its color matches that of a standard.

talcum powder A powder used in determining leak exit location; the powder turns bright red in fuel.

tangent galvanometer A galvanometer consisting of a small compass mounted horizontally in the center of a large vertical coil of wire; the current through the coil is proportional to the tangent of the angle the compass needle makes with its rest (no current) position.

tank test *See* wet dielectric test.

tantalum A metallic element used largely as an alloying element.

tantalum carbide A hard carbide used in cemented carbide tools.

tape A ribbon made of plastic, metal, paper, or other flexible material suitable for data recording by means of electromagnetic imprinting, punching or embossing patterns, or printing.

tape wrap A spirally or longitudinally applied insulating tape, wrapped around a conductor or cable, either insulated or uninsulated, and used as an insulation or mechanical barrier.

tapping Removing molten metal from a furnace.

tap test The practice of tapping a part with a solid object and listening for acoustic changes that may indicate flaws.

tare pressure The pressure loss between the pressure tapping points, as generated by the test equipment exclusive of the test valve.

tare weight In any weighing operation, the residual weight of any containers, scale components, or residue that is included in total indicated weight and must be subtracted to determine the weight of the live load.

target *1.* A goal or standard against which some quantity such as productivity is compared. *2.* A point of aim or object to be observed by visual means, electromagnetic imaging, radar, sonar, or similar noncontact method. *3.* In MS-DOS, the location where data are to be copied and stored.

target language The language into which some other language is to be translated.

target program An object program that has been assembled or compiled by a host processor for a target computer.

tarnish Surface discoloration of a metal resulting from formation of a thin film of corrosion product.

task *1.* *See* program. *2.* A specific "run time" execution of a program and its subprograms. *3.* In the MULTICS sense, a virtual processor. (A single processor may be concurrently simulating many virtual processors.) *4.* The execution of a segment on a virtual processor. *See* virtual processor. *5.* In RSX-11 (Real-Time Resource Sharing Executive) terminology, a load module with special characteristics; in general, any discrete operation performed by a program.

Tc Chemical symbol for technetium.

Te Chemical symbol for tellurium.

tear The removal or separation of a portion of the material.

tear resistance The property of an elastomeric material that enables it to resist tearing forces.

tear strength Force required to initiate or continue a tear in a material.

technetium A metallic element made by the neutron bombardment of molybdenum.

technical cohesive strength The measured stress at fracture in a notched tensile test.

technical evaluation An investigation to determine the suitability of materials, equipment, or systems to perform a specific function.

technical specifications A detailed description of the technical characteristics of a material, item, or system in sufficient detail to form the basis for design, development, production, and, in some cases, operation.

TED *See* transferred electron devices.

teeming Pouring molten metal into an ingot mold; most often used with reference to steel production.

tektites Small glassy bodies containing no crystals, composed of at least 65 percent silicon dioxide, bearing no relation to the geological formations in which they occur, and believed to be of extraterrestrial origin.

telescoping gage An adjustable gage for measuring inside dimensions such as hole diameters.

TELEVENT *See* high-level language(s).

tellurium A metallic element used as an alloying element.

telomer A polymer with molecules that are incapable of reaction with other monomers.

TEM *See* transmission electron microscopy.

temper *1.* In the heat treatment of ferrous alloys, to reheat after hardening for the purpose of decreasing hardness and increasing toughness without undergoing a eutectoid phase change. *2.* The relative hardness and strength of flat-rolled steel or stainless steel that cannot be further hardened by heat treatment. *3.* The relative hardness and strength of nonferrous alloys, produced by mechanical or thermal treatment, or both, and characterized by a specific structure, range of mechanical properties, or reduction of area during cold working. *4.* In glass manufacture, to anneal or toughen by heating below the softening temperature. *5.* In the production of casting molds, to moisten mold sand with water. *6.* To moisten and mix clay, mortar, or plaster to a consistency suitable for use.

temperature *1.* In general, the intensity of heat as measured on some definite temperature scale by means of any of various types of last sensors. *2.* In statistical mechanics, a measure of translational molecular kinetic energy (with three degrees of freedom). *3.* In thermodynamics, the integrating factor of the differential equation referred to as the first law of thermodynamics.

temperature, ambient The static temperature of the medium surrounding any unit under consideration.

temperature coefficient The rate of change of some physical property—electrical resistivity, for instance—with temperature; the coefficient may be constant or nearly constant, or it may vary itself with temperature.

temperature coefficient of resistivity The amount of resistance change of a material per degree of temperature rise.

temperature compensation Any construction or arrangement that makes a measurement device or system substantially unaffected by changes in ambient temperature.

temperature cycle A stress test in which the temperature of the medium (usually air) surrounding the test items is varied in a predetermined manner over the temperature range in such a way that the internal item temperature is kept at a fixed minimal increment from the medium temperature.

temperature dependence The characteristic of a material that is dependent on changes in the ambient temperature.

temperature, dew point The saturation temperature for a particular partial pressure of water vapor.

temperature, differential The difference between the internal temperature in a duct and a reference temperature; or, in a flowing system, the difference in temperature between two points due to temperature drop.

temperature, dry bulb The static temperature of an air-water vapor mixture.

temperature, dynamic The difference between the total temperature and the static temperature of a stream of compressible fluid.

temperature error The maximum change in output, at any measurand value within the specified range, when the transducer temperature is changed from room temperature to specified temperature extremes.

temperature error band The error band applicable over stated environmental temperature limits.

temperature gradient error The transient deviation in output of a transducer at a given measurand value when the ambient temperature or the measurand fluid temperature changes at a specified rate between specified magnitudes.

temperature load tests The conditioning of an item (that is, inflators, modules) to high/low temperature cycles.

temperature shock tests The conditioning of an item (that is, inflators, modules) to multiple, rapid thermal cycles.

temperature, static The temperature of a stream of compressible fluid, as measured by a temperature element that is moving with the stream so that there is no relative velocity between the measuring element and the fluid, and measured in the absence of radiation.

temperature, total The temperature of a stream of compressible fluid, as measured by a temperature element whose entire surface is subjected to the full dynamic pressure of the fluid stream and measured in the absence of radiation.

temperature, wet bulb The temperature at which liquid or solid water, by evaporating into air, can bring the air to saturation adiabatically. Wet bulb temperature (without qualification) is the temperature indicated by a wet bulb psychrometer constructed and used according to specifications.

temper brittleness Brittleness that results when certain steels are held within, or are cooled slowly through, a certain range of temperature below the transformation range. The brittleness is revealed by notched bar impact tests at or below room temperature.

temper color A thin, tightly adhering oxide skin only a few molecules thick, but with an identifying color that forms when steel is exposed to relatively low temperature in air or a mildly oxidizing atmosphere. The color, which ranges from straw to blue depending on the thickness of the oxide skin, varies with the exposure temperature and to a lesser magnitude with the exposure time. Failure analysis studies will make use of relevant observations regarding temper color.

tempered glass A single piece of specially treated sheet, plate, or float glass possessing mechanical strength substantially higher than annealed glass. When broken at any point, the entire piece breaks into small pieces that have relatively dull edges as compared to those of broken pieces of ordinary annealed glass. *Note*: Other terms such as "heat treated glass," "heat toughened glass," "case hardened glass," and "chemically tempered glass" also are used.

tempering *1.* Reheating a quench-hardened or normalized ferrous alloy to a temperature below the transformation range (Ac_1) and then cooling at any desired rate. *2.* Adding moisture to molding sand, clay, mortar, or plaster. *3.* Heating glass below its softening temperature.

tempering air Air at a lower temperature added to a stream of preheated air to modify its temperature.

tempering copper alloys Heating quench-hardened material to a temperature below the solution treatment temperature to produce desired changes in properties.

tempering ferrous Heating a quench-hardened or normalized ferrous alloy to a temperature below the transformation range to produce desired changes in properties.

temper rolling *1.* Cold rolling of sheets to improve flatness, impart a desired surface finish, or reduce the tendency to stretcher strain and flute. Temper rolling should not be confused with "cold reduction," which involves a substantial reduction in thickness. *2. See* skin passing.

tenacity The strength of a yarn or filament of a given dimension.

tensile bar *See* tensile specimen.

tensile specimen A bar, rod, or wire of several specified dimensions used in a tensile test. Also known as a tensile bar or tensile test specimen. Most are wrought, but some are cast.

tensile strength The unit stress at the highest load reached during the tension test. More significantly, the maximum force per unit of the original cross-sectional area of the sample which results in the rupture of the sample. It is calculated by dividing the maximum force by the original cross-sectional area.

tensile stress Force per unit cross-sectional area applied in tension.

tensile test A method of determining mechanical properties of a material by loading a machined or cast specimen of specified cross-sectional dimensions in uniaxial tension until it breaks; the test is used principally to determine tensile strength, yield strength, ductility (elongation and reduction of area), and modulus of elasticity. Also known as a pull test.

tensile yield strength The stress to cause a specified amount of inelastic strain, usually 0.2%. It is usually determined by constructing a line of slope E through 0.2% strain and zero stress (E = modulus of elasticity). The stress at which the constructed line intercepts the stress-strain curve is taken as the yield strength.

tension A force tending to produce elongation or extension.

tensors Arrays of functions that obey certain laws of transformation. A one-row or one-column tensor array is a vector.

terminal velocity The maximum velocity attainable, especially by a free-falling body under given conditions.

terne An alloy of lead used for coating steel.

terpolymer A polymer with three monomers.

tesla Metric unit for magnetic flux density.

test A procedure or action taken to measure, under real or simulated conditions, parameters of a system or component.

testability A design characteristic that allows the status of a unit to be confidently determined in a timely fashion.

testability measurement A measurement of the ease and adequacy of testing derived as a result of demonstration or failed history analysis.

testability protection A prediction of the ease and adequacy of testing developed through use of models or schematics.

test, acceptance *See* acceptance test.

test block A material that is representative of the standard for calibration.

test chamber An environmental chamber capable of controlling the test ambient temperatures within the parameters of the test document.

test, design verification Test performed to verify that known contributors to excessive downtime have been corrected or neutralized by appropriate design action.

test, field *See* field test.

test flow Any steady-state flow rate required to conduct a test.

test gage A pressure gage specially built for test service or other types of work requiring a high degree of accuracy and repeatability.

testing *See* inspection.

testing, nondestructive Testing of a nature that does not impair the usability of the item.

testing, quantitative Testing that monitors or measures the specific quantity, level, or amplitude of a characteristic to evaluate the operation of an item.

testing without replacement In life test sampling, a life test procedure in which failed units are not replaced.

test measurement and diagnostic equipment Any system or device used to evaluate the condition of an item to identify or isolate any actual or potential failures.

test, near surface With regard to ultrasonic testing, the test part surface through which the ultrasonic energy used for inspection initially enters the test part.

test program A set of directed commands, instructions, and data designed to automatically accomplish a test function.

test, qualification A test conducted under specified conditions, by, or on behalf of, the purchaser, using items representative of the production configuration, in order to determine compliance with item design requirements as a basis for production approval.

test records Records maintained by the laboratory; to be available for review, but not required to be furnished with the certification unless specified.

test specification Document describing the method(s) and procedure(s) by which material is to be tested.

test specimen The configuration of test material in which testing to determine properties is performed.

test to failure The practice of inducing increased electrical and mechanical stresses in order to determine the maximum capability of a material or device so that conservative use in subsequent application will, thereby, increase its life through derating determined by these tests.

tetrahydrofuran In organic chemistry, an intermediate and a solvent for polyvinyl chloride.

tetrol *See* furan.

tex A measurement unit for linear density of fibers, filaments, or yarns; the weight in grams per 1000 meters of a strand.

TGA *See* thermogravimetric analysis.

Th Chemical symbol for thorium.

thallium A metallic element resembling lead.

theoretical air The amount of air required to completely burn a given amount of a combustible material.

therm A unit of heat applied especially to gas. One therm equals 100,000 Btu.

thermal accommodation coefficients *See* accommodation coefficient.

thermal aging Exposure to a given thermal condition or a series of conditions for prescribed periods of time.

thermal analysis Determining transformation temperatures and other characteristics of materials or physical systems by making detailed observations of time-temperature curves obtained during controlled heating and cooling.

thermal-arrest calorimeter A device for measuring heats of fusion in which a sample is frozen under vacuum at sub-zero temperatures and thermal measurements are taken as the calorimeter warms to room temperature.

thermal barrier The zone of speed at which friction heat generated by rapid passage of an object through the atmosphere exceeds endurance compatible with the function of the object.

thermal bulb A device for measuring temperature in which the liquid in a bulb expands and contracts with changes in temperature, causing a Bourdon-tube element to elastically deform, thereby moving a pointer in direct relation to the temperature at the bulb.

thermal comfort The condition that expresses satisfaction with the thermal environment; measured by such factors as air temperature, relative humidity, air velocity, etc.

thermal conductance (C-factor) The rate of heat flow under steady-state conditions between two definite surfaces at uniform separation, divided by the difference of their average temperatures and by the area of one surface. The average temperature is one that adequately approximates that obtained by integrating the temperatures of the entire surface.

thermal conductivity (k-factor) The rate of heat flow through a homogeneous material under steady-state conditions, through unit area, per unit temperature gradient in the direction perpendicular to an isothermal surface.

thermal conductivity gage A device for measuring pressure in a high-vacuum system by observing changes in thermal conductivity of an electrically heated wire that is exposed to the low-pressure gas in the system.

thermal contraction Contraction caused by a decrease in temperature.

thermal decomposition The braking apart of complex molecules into simpler units by the application of heat.

thermal degradation Impairment of properties caused by exposure to heat.

thermal detector *See* bolometer.

thermal diffusion Spontaneous movement of solvent atoms or molecules to establish a concentration gradient as a direct result of the influence of a temperature gradient.

thermal efficiency *See* thermodynamic efficiency.

thermal electromotive force The electromotive force developed across the free ends of a bimetallic couple when heat is applied to a physical junction between the opposite ends of the couple. Also known as thermal emf.

thermal emf *1.* The electrical potential generated in a conductor or circuit due to thermal effects, usually differences in temperature between one part of the circuit and another. *2. See* thermal electromotive force.

thermal emission The process by which a body emits electromagnetic radiation as a consequence of its temperature only.

thermal endurance The time at selected temperature for a material or system of materials to deteriorate to some predetermined level of electrical, mechanical, or chemical performance under prescribed conditions of test.

thermal energy Energy that flows between bodies because of a difference in temperature.

thermal expansion A physical phenomenon whereby raising the temperature of a body causes it to change dimensions (usually increasing) in a manner characteristic of the material of construction.

thermal fatigue *1.* The failure of materials subjected to alternating heating and cooling. *2.* Failure induced by clinical varying temperature gradients that create cyclic thermal stress and strain in the material.

thermal instability The conditions of temperature gradient, thermal conductivity, and viscosity that lead to the onset of convection in a fluid.

thermal instrument Any instrument that measures a physical quantity by relating it to the heating effect of an electric current, such as in a hot-wire instrument.

thermal junction *See* thermocouple.

thermal neutron(s) *1.* Neutrons in thermal equilibrium with a medium in which they exist. *2.* A free (uncombined) neutron having a kinetic energy approximately equivalent to the kinetic energy of its surroundings.

thermal pollution Environmental temperature rise due to waste heat disposal.

thermal printer A printer that prints characters on paper using a high-speed heating element activating chemicals in the paper to form an image.

thermal processing Any process in which metals are exposed to controlled heating, soaking, or cooling.

thermal radiation *1.* Electromagnetic radiation that transfers heat out of a heated mass. *2.* Electromagnetic radiation that is absorbed by a grey or black body from a source at a higher temperature than the absorbing body.

thermal rating The maximum or minimum temperature at which a material will perform its design function without undue degradation.

thermal resistor An electronic device that makes use of the change in resistivity of a semiconductor with changes in temperature.

thermal shock A stress test in which the temperature of the medium surrounding the test items is varied as rapidly as possible in order to create large cyclic temperature gradients in the test items.

thermal spraying A method of coating a substrate by introducing finely divided refractory powder or droplets of atomized metal wire into a high-temperature plasma stream from a special torch, which propels the coating material against the substrate.

thermal stress *1.* Stress caused by the different expansion in different directions of plies in a laminate at different orientations. *2.* In metals or other materials, the stress that is developed if materials having different coefficients of thermal expansion are bolted or otherwise joined together and heated. *3.* The stress developed if a block of material is restrained from expanding while being heated or from contracting while being cooled.

thermal stress cracking Crazing or cracking of a material due to prolonged exposure to high temperatures.

thermal variable A characteristic of a material or system that depends on its thermal energy—temperature, thermal expansion, calorific value, specific heat, or enthalpy, for instance.

thermionic emission Direct ejection of electrons as the result of heating the material, which raises electron energy beyond the binding energy that holds the electron to the material.

thermionics The study of the emission of electrons as a result of heating.

thermistor *1.* A solid-state semiconducting device, the electrical resistance of which varies with the temperature. Its temperature coefficient of resistance is high, nonlinear, and negative. *2.* A specific resistor whose temperature coefficient of resistance is unusually high; it is also nonlinear and negative. It is often made by sintering a mixture of oxide powders of various metals, and is thus a solid-state semiconductor material in many types of shapes, such as disks, flakes, rods, etc., to which contact wires are attached. The resistance of the unit varies with temperature changes, and thus acts as a sensor for temperature change.

thermites Fire-hazardous mixtures of ferric oxide and powdered aluminum; upon ignition with a magnesium ribbon, the mixtures reach temperatures up to 4000°F (sufficient to soften steel).

thermit reactions Exothermic reactions between two materials that produce enough heat to weld metal.

thermit welding A fusion welding process in which a mixture of finely divided iron oxide and aluminum particles is ignited, reducing the iron oxide and producing a molten ferrous alloy which is then cast in a mold built up around the joint to be welded.

thermochemistry A branch of chemistry that treats the relations of heat and chemical changes.

thermocouple Also called a thermal junction. A device for measuring temperature in which two electrical conductors of dissimilar metals are joined at the point of heat application and a resulting voltage difference, directly proportional to the temperature, is developed across the free ends and is measured potentiometrically; a pair of dissimilar conductors so joined at two points that an electromotive force is developed by the thermoelectric effects when the junctions are at different temperatures.

thermocouple vacuum gage A device for measuring pressures in the range of about 0.005 to 0.02 torr by means of current generated by a thermocouple welded to the midpoint of a small heating element exposed to the vacuum chamber.

thermodynamic efficiency In thermodynamics, the ratio of the work done by a heat engine to the total heat supplied by the heat source.

thermodynamic equilibrium A very general result from statistical mechanics which states that if a system is in equilibrium, all processes that can exchange energy must be exactly balanced by the reverse process so that there is no net exchange of energy.

thermodynamics The study of the flow of heat.

thermoelasticity Dependence of the stress distribution of an elastic solid on its thermal state, or of its thermal conductivity on the stress distribution.

thermoelectric cooling A method of cooling a chamber based on the Peltier effect, in which an electric current is circulated in a thermocouple whose cold junction is coupled to the chamber; the hot junction dissipates heat to the environment. Also known as thermoelectric refrigeration.

thermoelectric emf The algebraic sum of the potential differences associated with (a) maintaining the junctions of two dissimilar conductors at different temperatures and (b) keeping the conductor in a thermal gradient.

thermoelectric heating A method of heating involving a device similar to one used for thermoelectric cooling, except that the direction of current is reversed in the circuit.

thermoelectric hygrometer A condensation-type hygrometer in which the mirror element is chilled thermoelectrically.

thermoelectric refrigeration *See* thermoelectric cooling.

thermoelectric series A tabulation of metals and alloys, arranged in order according to the magnitude and sign of their characteristic thermal emf.

thermoelectric thermometer A thermometer that uses a thermocouple or thermocouple array in direct contact with the body whose temperature is to be measured, and whose reading is given in relation to the reference junction whose temperature is known or automatically compensated for.

thermoforming A method of forming sheet plastic by heating it, then pulling it over a contoured mold surface.

thermogalvanic corrosion Corrosion resulting from a galvanic cell caused by a thermal gradient.

thermograph An instrument for recording air temperature. Also known as a recording thermometer.

thermographic inspection A method of optically viewing the temperature distribution on skins or components, through the measurement of the radiation produced at infrared wavelengths.

thermography Either of two methods—contact thermography or projection thermography—for measuring surface temperature using thermoluminescent materials.

thermogravimetric analysis (TGA) Analyzing the mass of various materials under different conditions of temperature and pressure.

thermomechanical treatment A general term covering a variety of processes combining controlled thermal and deformation treatments to obtain synergistic efforts such as improvement in strength without loss of toughness.

thermomechanics The combining of material forming processes with heat treatments in order to obtain specific material properties. *See* thermodynamics.

thermometer(s) An instrument for measuring temperature—usually involving a change in a physical property such as density or electrical resistance of a temperature-sensitive material.

thermomigration A technique for doping semiconductors in which exact amounts of known impurities are made to migrate from the cool side of a wafer of pure semiconductor material to the hotter side when the wafer is heated in an oven.

thermophoresis A process in which particles migrate in a gas under the influence of forces created by a temperature gradient.

thermophysics The science dealing with properties, changes, interactions, etc. of matter and energy as they are affected by heat. *See* thermodynamics.

thermopile(s) *1.* Transducers for converting thermal energy directly into electrical energy, composed of pairs of thermocouples that are connected either in series or in parallel. *2.* An array of thermocouples used for measuring temperature or radiant energy, or for converting radiant energy into power.

thermoplastic A classification of resin that can be readily softened and resoftened by heating.

thermoplastic films Materials with a linear macromolecular structure that will repeatedly soften when heated and harden when cooled.

thermoplastic polyesters (TPES) A group of thermoplastic polymers in which repeating molecules are joined by ester molecules.

thermoplastic polyimides (TPI) Polymers with good heat and chemical resistance.

thermoplastic polyurethanes (TPUR) Widely used molding polymers.

thermoplastic resin An organic solid that will repeatedly soften when heated and harden when cooled; examples include polystyrene, acrylics, polyethylene, vinyl polymers, and nylon.

thermoplastic rubber (TPR) Elastomers that can be molded at high rates of production.

thermoset *1.* A plastic that, when cured by application of heat or chemical means, changes into a substantially infusible and insoluble material. *2. See* thermosetting resin.

thermosetting A classification of resin that cures by chemical reaction when heated, and when cured cannot be softened by heating.

thermosetting resin (thermoset) An organic solid that sets up (solidifies) under heat and pressure, and cannot be softened and remolded readily; examples include phenolics, epoxies, melamine, and polyurea.

thermostat A temperature-sensitive element used as the prime actuator of a temperature regulator.

thermotropism The tendency to grow toward or away from a source of heat.

thickener Equipment for removing free liquid from a slurry or other liquid-solid mixture to give a solid or semisolid mass without using filtration or evaporation; usually, the process involves centrifuging or gravity settling.

thick film Pertaining to a film pattern usually made by applying conductive and insulating materials to a ceramic substrate by a silk-screen process. Thick films can be used to form conductors, resistors, and capacitors.

thick film metallization Conductive metallization applied to an insulating substrate.

thickness *1.* A physical dimension usually considered to represent the shortest of the three principal measurement axes. In flat products such as sheet or plate, for instance, it is the distance between top and bottom surfaces measured along an axis mutually perpendicular to them. 2. The distance from the external surface of a coating to the substrate, measured along a direction perpendicular to the coating-substrate interface. 3. The short transverse dimension of forged, rolled, drawn, or extruded stock.

thickness gage A device for measuring thickness of sheet material; it may involve physical gaging, but more often involves methods such as radiation absorption or ultrasonics.

thick plates Plates of steel or other material that are over two inches thick.

thin film A film of conductive or insulating material, usually deposited by sputtering or evaporation, that may be made in a pattern to form electronic components and conductors on a substrate or used as insulation between successive layers of components.

thin film metallization Conductive metallization applied to an insulating substrate by evaporating, sputtering, or plating metallic layers.

thin-film potentiometer A potentiometer in which the conductive element is a thin film of a cermet (metal mix), conductive plastic, or deposited metal; usually thin-film potentiometers are useful when a stepless output is desired.

thin-film strain gage A strain gage in which the gage is produced by depositing an insulating layer (usually a ceramic) onto the structural element, then depositing a metal gage element onto the insulation layer by sputtering or vacuum deposition through a mask that defines the strain gage configuration.

thin-layer chromatography (TLC) A form of chromatography in which a small amount of sample is placed at one end of thin sorbent layer deposited on a metal, glass, or plastic plate; after washing a solvent through the sorbent bed by capillary action, the individual components of the sample may be detected by visual means, ultraviolet analysis, radiochemistry, or other suitable technique.

thinner(s) An organic liquid added to a mixture such as paint to reduce its viscosity and make it more free-flowing. *See also* solvent.

thixotropic Materials that ordinarily appear as a gel, but will fluidize when agitated.

thixotropic substance A substance whose flow properties depend on both shear stress and agitation; at a given shear stress, flow increases with increasing time of agitation; when agitation stops, internal shear stress exhibits hysteresis.

thixotropy The property of a nonsag material that permits it to be moved (stirred or extruded) with less force than would be required for a Newtonian fluid. Non-Newtonian pseudoplastic materials (such as the nonsag sealants) stand like whipped cream without seeking their own level, but flow easily from a sealant gun under relatively low pressure.

thorium A radioactive metallic element used as an alloying element.

thread A thin single-strand or twisted filament of natural or synthetic fibers, plastics, metal, glass, or ceramic.

three-quarters hard A temper of nonferrous alloys and some ferrous alloys that corresponds approximately to a hardness and tensile strength midway between those of half hard and full hard tempers.

threshold An impact level to trigger a system.

threshold sensitivity The lowest value of a measured quantity that a given instrument or controller responds to effectively.

threshold voltage The threshold energy necessary to remove an electron from the bound position to the conduction band in solid state devices.

throwing power *1*. The relationship between the current density at a point on a surface and its distance from the counter electrode. The greater the ratio of the surface resistivity shown by the electrode reaction to the volume resistivity of the electrolyte, the better is the throwing power of the process. *2*. The relative ability of an electroplating solution or electrophoretic paint to cover irregularly shaped parts with a uniform coating. Contrast with covering power.

thulium A rare earth element.

thyristor A bistable semiconductor device comprising three or more junctions that can be switched from the off state to the on state or vice versa, such switching occurring within at least one quadrant of the principal voltage current characteristic.

tiger strips Continuous bright lines in the rolling direction.

tight *1*. Inadequate clearance, or barest minimum clearance between moving parts. *2*. In a pressurized or vacuum system, freedom from leaks. *3*. A class of fit having slight negative allowance, which requires light to moderate levels of force to assemble mating parts together.

TIG welding *See* gas tungsten-arc welding and tungsten inert-gas welding.

tile A preformed refractory, usually applied to shapes other than standard brick.

tile baffle A baffle formed of preformed refractory shapes.

tilt In electron microscopy, the angle of the specimen compared to the axis of the electron beam.

time A fundamental measurement whose value indicates the magnitude of an interval between successive events.

time base error In instrumentation tape recording and playback, the data error that results from a difference between tape recording speed and tape playback speed.

time base system A device to enable the determination of the time interval elapsed between any two recorded events.

time constant Generally, the time required for an instrument to indicate a given percentage of the final reading result from an input signal.

time-inverse A time-current relationship whereby the protective device opening time decreases as the current increases.

time lag Elapsed time between two events determined by measuring pulses initiated by the events.

time origin identification device A device for identifying the instant chosen as the time origin, usually the beginning of the impact.

time quenching Interrupted quenching in which the duration of holding in the quenching medium is controlled.

timer *1*. A device that automatically starts or stops a machine function, or series of functions, depending on either time of the day or elapsed time from an arbitrary starting point. *2*. An instrument that measures elapsed time from some arbitrary starting point. *3.* A device that fires the ignition spark in an internal combustion engine at a preset point in the engine cycle. *4.* A device that opens or closes a set of contacts, and automatically returns them to their original position after a preset time interval has elapsed. Also known as an interval timer. *5.* A device that provides the system with the ability to read elapsed time in split-second increments and to inform the system when a specified period of time has passed.

time, reaction The element of uptime needed to initiate a mission, measured from the time command is received.

time response The variation of an output variable of an element or a system, produced by a specified variation of one of the input variables.

time, testing The time required to determine whether designated characteristics of a material are within specified values.

tin A metallic element used in many commercial alloying systems.

tinning *1.* Coating with a thin layer of molten solder or tin to prevent corrosion or prepare a connection for soldering. *2.* Covering or preserving a metal surface with tin. *3.* A protective surface layer of tin or solder.

TIR Total indicator reading, or total indicator variation. The difference between the maximum and minimum indicator readings during a checking/inspection cycle.

titanium A metallic element used in applications requiring high-strength, low-density, and corrosion-resistant metal.

titration Analyzing the composition of a sample by adding fixed volumes of a fixed concentration reagent.

titration curve A plot with pH as the ordinate and units of reagent added per unit of sample as the abscissa.

Tl Chemical symbol for thallium.

TLC *See* thin-layer chromatography.

Tm Chemical symbol for thulium.

toe *1.* The junction between the face of a weld and the adjacent base metal. *2.* The portion of the base of a dam, earthwork, or retaining wall opposite to the retained material.

toe crack A crack in the weldment that runs into the base metal from the toe of a weld.

token *1.* The symbol of authority passed between nodes in a token protocol. Possession of this symbol identifies the node currently in control of the medium. *2.* Symbol representing the right to use the network medium.

token passing protocol A protocol in which a node that has communicated passes control of the bus, including the right to communicate to another node, at the end of the message via a token.

tol *See* tolerance.

tolerance (tol) *1.* Permissible variation in the dimension of material or a part. *2.* Permissible deviation from a specified value; may be expressed in measurement units or percent. *3.* The total amount by which a quantity is allowed to vary; thus, the tolerance is the algebraic difference between the maximum and minimum limits.

tolerance limits The extreme upper and lower boundaries of a specified range; it is computed from the nominal value and its tolerance.

tomography Technique of making radiographs of plane sections of a body or an object; its purpose is to show detail in a predetermined plane of the body, while blurring the images of structures in other planes.

ton *1.* A weight measurement equal to 2,000 lb (avoirdupois)—short ton; 2,240 lb (avoirdupois)—long ton; or 1,000 kg—metric ton. *2.* A unit volume of sea freight equal to 40 cubic feet. *3.* A unit of measurement for refrigerating capacity equal to 200 Btu/min, or about 3517 W; derived from the capacity equal to the rate of heat extraction needed to produce a short ton of ice having a latent heat of fusion of 144 Btu/lb from water at the same temperature in 24 hr.

tool steel Any of various steel compositions containing sufficient carbon and alloying elements to permit hardening to a level suitable for use in cutting tools, dies, molds, shear blades, metalforming rolls, and other tooling applications.

TOP Technical and Office Protocols.

top coat A material applied as a thin coating over the surface of applied sealant to protect it from the possible deleterious effects of fuel.

topology *1.* The configuration of the interconnected elements of a system. *2.* The logical interconnection between devices. Local area networks typically use either a broadcast topology (bus), in which all stations receive all messages, or a sequential topology (ring), in which each station receives messages from the station before them and transmits (repeats) messages to the station after them. Wide area networks typically use a mesh topology in which each station is connected to one or more other stations and acts as a bridge to pass messages through the network.

tor *See* torr.

torch A device used to control and direct a gas flame, such as in welding, brazing, flame cutting, or surface heat treatment.

torque A rotational effort produced by a force times a moment arm. A moment.

torque distribution *1.* Fixed—Output torque is fixed by the design of the device. *2.* Intermediate—Output torque distribution is not determined by the device, but by the input torque and tractive capability. *3.* Variable—Output torque distribution is variable by the design of the device.

torque-type viscometer An instrument that can measure the viscosity of Newtonian fluids, non-Newtonian fluids, and suspensions by determining the torque needed to rotate a vertical paddle or cylinder submerged in the fluid.

torr Also spelled tor. A unit of pressure equal to the pressure exerted by a column of mercury 1 mm high at 0°C.

torsiometer An instrument consisting of angular scales mounted around a rotating shaft to determine the amount of twist in the loaded shaft, and thereby determine the power transmitted. Also known as a torsionmeter.

torsion A twisting action resulting in shear stresses and strains.

torsional modulus The ratio of the torsion stress to the strain in the material over the range for which this value is constant.

torsional pendulum A device for performing dynamic mechanical analysis, in which the sample is deformed torsionally and allowed to oscillate in free vibration.

torsional rigidity The resistance of a fiber to twisting.

torsional stress The shear stress on a transverse cross section caused by a twisting action.

torsion balance An instrument for measuring minute magnetic, electrostatic, or gravitational forces by means of the rotational deflection of a horizontal bar suspended on a torsion wire whose other end is fixed.

torsion bar A type of spring that flexes by twisting about its axis rather than by bending or compressing.

torsion hygrometer An instrument for measuring humidity in which a substance sensitive to humidity is twisted or spiraled under tension in such a manner that changes in length of the sensitive element will rotate a pointer in direct relation to atmospheric humidity.

torsionmeter *See* torsiometer.

total absorption spectrometer An instrument that measures the total amount of x-rays absorbed by a sample and compares it to the amount absorbed by a reference sample; the sample may be solid, liquid, or gas.

total accuracy *See* accuracy, total.

total adjusted error The maximum output deviation from the ideal expected values. Expressed as LSBs or percent of full-scale range at a fixed reference voltage.

total carbon The measure of free and combined carbon in a ferrous alloy.

total case depth *1.* The distance (measured perpendicularly) from the surface of the hardened or unhardened case to a point where differences in chemical or physical properties of the case and core no longer can be distinguished. *2.* In some industries, the depth at which a stated hardness minimum is maintained.

total elongation Percent elongation measured after fracture in a tension test.

total emissivity The ratio of the total amount of thermal radiation emitted by a non-blackbody to the total amount emitted by a blackbody at the same temperature.

total failure The termination of the ability of an item to perform its required function.

total hydrocarbons The total of hydrocarbon compounds of all classes and molecular weights.

total range The portion of a system of units that is between an instrument's upper and lower scale limits, and therefore defines the values of the measured quantity that can be indicated or recorded.

total solids concentration The weight of dissolved and suspended impurities in a unit weight of boiler water, usually expressed in ppm.

total static deflection Of a loaded suspension system, the overall deflection under the static load from the position at which all elastic elements are free of load.

total temperature The temperature indicated by an error-free instrument having a fixed position in the gas stream; the sum of the static temperature and the temperature rise due to the conversion of kinetic energy to heat, as the compression occurs at the sensing element.

total transmission latency The time required to transfer a message from the transmitting node, measured from the moment it is prepared to send the message, until it is correctly received by the targeted receiver. It may include a retry strategy delay if the initial exchange is not successful.

total void Sum of plan voids and component voids.

toughness The ability of a material to deform plastically and absorb energy before fracturing.

tow An untwisted bundle of continuous filaments.

Townsend discharge A type of direct-current discharge between two electrodes immersed in a gas and requiring electron emission from the cathode.

T-peel strength The average load per unit width of bond line required to produce progressive separation of two bonded, flexible adherends under specific conditions.

TPI *See* thermoplastic polyimides.

TPUR *See* thermoplastic polyurethanes.

TPR *See* thermoplastic rubber.

trace *1.* A graphical output from a recorder, usually in the form of an ink line on paper. *2.* An interpretive diagnostic technique that provides an analysis of each executed instruction and writes it on an output device as each instruction is executed. *See also* sequence checking routine.

traceability The characteristic whereby requirements at one level of a design may be related to requirements at another level.

trace element Any element occurring in the specified material that was not intentionally added as an alloying addition and that may have a deleterious effect on mechanical properties.

tracer gas A gas used in connection with a leak-detecting instrument to find minute openings in a sealed vacuum system.

track *1.* A pair of parallel metal rails for use by a train, tram, or similar wheeled vehicle. *2.* A crawler mechanism for earth-moving equipment or military vehicles. *3.* A band of data on recording tape or the spiral groove in a phonograph record. *4.* An overhead rail for repositioning hoisting gear. *5.* To follow the movement of an object, for instance, by continually repositioning a telescope or radar set so its line of slight is always on the object. *6.* In data processing, a specific area on any storage medium that can be read by drive heads.

tracking error In lateral mechanical recording equipment, the angle between the vibration axis of the pickup and a plane that is both perpendicular to the record surface and tangent to the unmodulated recording groove at the point where the needle rides in the groove.

tracking system Any device that continually repositions a mechanism or instrument to follow the movement of a target object.

tractive force The force available from the driving wheels at the driving wheel/ground interface.

transducer *1.* Microphone, pressure transducer, or other device along with integral conditioning used for pressure transduction. *2.* The first device in a data channel, used to convert a physical quantity to be measured into a second quantity (such as an electrical voltage) which can be processed by the remainder of the channel.

transfer *1.* The conveyance of control from one mode to another by means of instructions or signals. *2.* The conveyance of data from one place to another. *3.* An instruction for transfer. *4.* To copy, exchange, read, record, store, transmit, transport, or write data. *5.* An instruction that provides the ability to break the normal sequential flow of control. Synonymous with jump.

transfer chamber In plastics molding, an intermediate chamber or vessel for softening a thermosetting resin with heat and pressure before admitting it to the mold for final curing.

transfer function A mathematical, graphical, or tabular statement of the influence that a system or element has on a signal or action compared at input and output terminals.

transfer lag *See* capacity lag.

transfer layer The portion of a communication protocol specifying sequential arrangement and meaning of bits or bit fields within a message. Eventually, the transfer layer may also cover the arbitration scheme, error detection, and handling and strategies for fault confinement.

transfer rate Information bits per unit of time during transmission, equivalent to bit rate.

transferred electron devices (TED) Electronic equipment utilizing diodes exhibiting negative conductance and susceptance.

transfer time The time interval between the instant the transfer of data to or from storage commences and the instant it is completed.

transfer vector A transfer table used to communicate between two or more programs. The table is fixed in relationship with the program for which it is the transfer vector. The transfer vector provides communication linkage between that program and any remaining subprograms.

transform To change the form of data according to specific rules.

transformation-induced plasticity *See* TRIP.

transformation ranges (transformation temperature ranges) The ranges of temperature within which austenite forms during heating and transforms during cooling. The two ranges are distinct, sometimes overlapping but never coinciding. The limiting temperatures of the ranges depend on the composition of the alloy and on the rate of change of temperature, particularly during cooling.

transformation temperature The temperature at which a change in phase occurs. The term is sometimes used to denote the limiting temperature of a transformation range. The following symbols are used for iron and steels: (1) Ac_{cm}—In hypereutectoid steel, the temperature at which the solution of cementite in austenite is completed during heating. (2) Ac_1—The temperature at which austenite begins to form during heating. (3) Ac_3—The temperature at which transformation of ferrite to austenite is completed during heating. (4) Ac_4—The temperature at which austenite transforms to delta ferrite during heating. (5) Ae_1, Ae_3, Ae_{cm}, Ae_4—The temperatures of phase changes at equilibrium. (6) Ar_{cm}—In hypereutectoid steel, the temperature at which precipitation of cementite starts during cooling. (7) Ar_1—The temperature at which transformation of austenite to ferrite or to ferrite plus cementite is completed during cooling. (8) Ar_3—The temperature at which austenite begins to transform to ferrite during cooling. (9) Ar_4—The temperature at which delta ferrite transforms to austenite during cooling. (10) M_s—The temperature at which transformation of austenite to martensite starts during cooling. (11) M_f—The temperature, during cooling, at which transformation of austenite to martensite is substantially completed. *Note*: All of these changes except the formation of martensite occur at lower temperatures during cooling than during heating, and depend on the rate of change of temperature.

transformed beta A local or continuous structure comprised of decomposition products arising either by martensitic or by nucleation and growth processes during cooling from either above the beta transus or some temperature high in the alpha-beta phase field.

transformer A device consisting of a winding with tap or taps, or two or more coupled windings with or without a magnetic core for introducing mutual coupling between electric circuits.

transgranular corrosion A slow mode of failure that requires the combined action of stress and aggressive environment, where the path of failure runs through the grains producing branched cracking.

transient *1.* A phenomenon caused in a system by a sudden change in conditions, which persists for a relatively short time after the change. *2.* A temporary increase or decrease of the voltage or current. Transients may take the form of spikes or surges.

transient analyzer An electronic device used to capture a record of a transient event for later analysis.

transient digitizer A device that records a transient analog waveform and converts the information it has collected into digital form.

transient emission test The test procedure prescribed by the U.S. EPA for emission testing of heavy-duty diesel engines. This test is performed in an engine test cell and requires the use of a dynamometer with motoring capability.

transistor A three-terminal solid-state semiconductor device that can be used as an amplifier, switch, detector, or wherever a three-terminal device with gain or switching action is required.

transition The switching from one state (for example, positive voltage) to another (negative) in a serial transmission.

transitional flow Flow between laminar and turbulent flow; generally between a pipe Reynolds number 2000 and 7000.

transition fatigue life The life at which elastic and plastic components of the total strain are equal; the life at which the plastic and elastic strain-life lines cross.

transition frequency *See* crossover frequency.

transition pressure The pressure at which phase transition occurs.

transition temperature Related to impact data characteristics, an observed temperature (usually identified by a test temperature-impact data curve) within the temperature range in which metal fracture characteristics change rapidly from primarily crystalline (cleavage) to primarily fibrous (shear). Favorable or unfavorable service temperatures can thereby be identified.

transit time The time it takes for a particle, such as an electron or atom, to move from one point to another in a system or enclosure.

translate To convert from one language to another language.

translator *1.* A program whose input is a sequence of statements in some language and whose output is an equivalent sequence of statements in another language. *2.* A translating device.

transmission electron microscopy (TEM) Analyzing the surface of a material using an electron beam mechanism.

transmission factor The ratio of the transmitted light to the incident light.

transmission loss The reduction in the magnitude of some characteristic of a signal, between two stated points in a transmission system.

transmittance (T) *1.* The fraction of light transmitted from a source, through a smoke-obscured path, that reaches the observer instrument receiver, expressed in percent (%). *2.* The ratio of the radiant flux transmitted by a medium or a body to the incident flux. *3.* The ratio of transmitted electromagnetic energy to incident electromagnetic energy impinging on a body that is wholly or partly transparent to the particular wavelength(s) involved.

transmitter A device that converts information or data signals to electrical or optical signals so that these signals can be sent over a communication medium (signal bus).

transmutation A nuclear reaction that changes a nuclide into a nuclide of a different element.

transparent *1.* The quality of a substance that permits light, some other form of electromagnetic radiation, or particulate radiation to pass through it. *2.* In data processing, describes a programming routine that allows other programs to operate identically regardless of whether the transparent instructions are installed or not installed.

transparent glass Glass having no apparent diffusing properties. Varieties of such glass are referred to as flint, crown, crystal, and clear.

transpiration The passage of gas or liquid through a porous solid (usually under conditions of molecular flow).

transponder Combined receiver and transmitter whose function is to transmit signals automatically when triggered by an interrogator.

trans stereoisomer A stereoisomer in which atoms are arranged on opposite sides of an atom chain.

transverse *1.* Perpendicular to longitudinal. *2.* In a uni-directional composite or prepreg tape, in the material plane and perpendicular to the fibers.

transverse loads Loads that are not perpendicular to the facings.

transversely isotropic Describes a material exhibiting a special case of orthotropy in which properties are identical in two orthotropic dimensions, but not the third.

transverse oscillation Oscillation in which the direction of motion of the particles is perpendicular to the direction of advance of the oscillatory motion, in contrast with longitudinal oscillation, in which the direction of motion is the same as that of advance.

transverse response *See* transverse sensitivity.

transverse sensitivity Fraction of accelerometer sensitivity that will apply to acceleration orthogonal to its principle axis. Also referred to as cross-axis sensitivity.

transverse strain Strain in a plane that is perpendicular to the loading axis.

transverse vibration *See* transverse oscillation.

transverse waves Waves in which the direction of displacement at each point of the medium is parallel to the wave front.

trap *1.* Conditional jump to a known location, automatically activated by hardware or software, with the location from which the jump occurred recorded. Often a temporary measure taken to determine the source of a computer bug. *2.* A vertical S-, U-, or J-bend in a soil pipe that always contains water to prevent sewer odors from backing up into the building. *3.* A device on the intake, or high-vacuum side, of a diffusion pump to reduce backflow of oil or mercury vapors from the pumping medium into the evacuated chamber. *4.* A receptacle for the collection of undesirable material.

trapped-air process A method of forming closed blow-molded plastics objects, in which sliding machine elements pinch off the top of the object after blowing to form a sealed, inflated product.

traverse lines Lines closely spaced across the full width of a sheet and running in the direction of rolling.

tread The portion of a tire that comes into contact with the road.

tread arc width The distance measured along the tread contour of an unloaded tire between one edge of the tread and the other. For tires with rounded tread edges, the point of measurement is the point in space that is at the intersection of the tread radium extended until it meets the prolongation of the upper sidewall contour.

tread chord width The distance measured parallel to the spin axis of an unloaded tire between one edge of the tread and the other. For tires with rounded tread edges, the point of measurement is the point in space that is at the intersection of the tread radium extended until it meets the prolongation of the upper side wall contour.

tread contact length The perpendicular distance between the tangent to edges of the leading and following points of road contact and parallel to the wheel plane.

tread contact width The distance between the extreme edges of road contact at a specified load and pressure, measured parallel to the y-axis at zero slip angle and zero inclination angle.

tread contour The cross-sectional shape of tread surface of an inflated unloaded tire neglecting the tread pattern depressions.

tread depth The distance between the base of a tire tread groove and a line tangent to the surface of the two adjacent tread ribs or rows.

tread pattern The molded configuration on the face of the tread. It is generally composed of ribs, rows, grooves, lugs, and the like.

treatment Single level assigned to a single factor during an experimental run.

treatment combination One level for each and every factor, during an experimental run.

trend analysis A technique in which deviation of recorded data and signature characteristics with respect to time are used to diagnose and prognosticate a malfunction or failure.

trending A technique that presents deviation of recorded data from a baseline or signature characteristic with respect to a time scale.

tribology Science of friction, wear, and lubrication.

triboluminescence The emission of light caused by application of mechanical energy to a solid.

trichlorethylene A volatile liquid used as a cleaning and degreasing agent.

trimetal A type of bearing construction in which a heavy-duty bearing material (lining) is bonded to a steel back and then a thin layer of softer bearing material is applied to the ID of the high-strength bearing material. Normally, this surface layer is obtained by electroplating and is referred to as the overlay or overlay plate. It is thin enough that the high strength of the intermediate layer determines the ultimate bearing strength from a fatigue standpoint. This type of bearing is normally used in heavy duty applications.

trimming *1.* Removing irregular edges from a stamped or deep-drawn part. *2.* Removing gates, risers, and fins from a casting. *3.* Removing parting-line flash from a forging. *4.* Adding or removing small amounts of R, L, or C from electronic circuits to cause minor changes in the circuit performance or to bring into specification.

triazine resins *See* cyanate resins.

triode An electron tube containing three electrodes: an anode, a cathode, and a control electrode, or grid.

TRIP (steel) Transformation-induced plasticity. A phenomenon occurring chiefly in certain highly alloyed steels that have been heat-treated to produce metastable austenite or metastable austenite plus martensite, whereby, on subsequent deformation, part of the austenite undergoes strain-induced transformation to martensite. Such steels are highly plastic after heat treatment, but exhibit a very high rate of strain hardening and thus have high tensile and yield strengths after plastic deformation at temperatures between 70 and 930°F (20 and 500°C). Cooling to −320°F (−195°C) may or may not be required to complete the transformation to martensite. Tempering usually is done following transformation.

triple point *1.* A temperature at which all three phases of a pure substance—solid, liquid and gas—are in mutual equilibrium. *2.* In tool steel metallography, the intersection of three grain boundaries. Evidence of fusion at this point would be an indication of overheating.

tritium An isotope of hydrogen having atomic weight of 3 (one proton and two neutrons in the nucleus).

true bond width On a bonded device, the maximum width of the beam lead in the bonded area directly over the conductor film metallization.

true complement *See* complement.

true mass flow A measurement that is a direct measurement of mass and independent of the properties and the state of the fluid.

true ratio A characteristic of an instrument transformer equal to root-mean-square primary current (or voltage) divided by root-mean-square secondary current (or voltage) determined under specified conditions.

true strain The natural logarithm of the ratio of gauge length at the moment of observation to the original gauge length for a body subjected to an axial force.

true stress The stress along the axis calculated on the actual cross section at the time of observation instead of the original cross-sectional area.

true value The actual value of the parameter being measured.

truncation error(s) The error resulting from the use of only a finite number of terms of an infinite series, or from the approximation of operations in the infinitesimal calculus for operations in the calculus of finite differences.

trunk *See* bus.

truth table A table that describes a logic function by listing all possible combinations of input values and indicating, for each combination, the true output values.

T-T-T diagram A diagram that shows the time- and temperature-dependent character of phase transformations.

tube A hollow product that is long in relation to its cross section and has uniform wall thickness except as affected by corner radii.

tubercle A localized scab of corrosion products covering an area of corrosive attack.

tuberculation The formation of localized corrosion products scattered over the surface in the form of knoblike mounds.

tube seat The part of a tube hole with which a tube makes contact.

tumbling *1.* A process for smoothing and polishing small parts by placing them in a barrel with wooden pegs, sawdust, and abrasives, or with metal slugs, and rotating the barrel about its axis until the desired surface smoothness and luster is obtained. *2.* Loss of control in a two-frame free gyroscope due to a slowing of the wheel.

tundish A pouring basin for molten metal.

tungsten A metallic element used to make filament wires and as an alloying element in tool steels.

tungsten carbide A very hard ceramic used in cutting tools.

tungsten inert-gas welding A non-preferred term for gas tungsten-arc welding. Also known as TIG welding. *See* gas tungsten-arc welding.

turbidity The optical obstruction to the passing of a ray of light through a body of water, caused by finely divided suspended matter.

turbulence A state of fluid flow in which the instantaneous velocities exhibit irregular and apparently random fluctuations so that in practice only statistical properties can be recognized and subjected to analysis.

turbulent boundary layer The layer in which the Reynolds stresses are much larger than the viscous stresses. When the Reynolds number is sufficiently high, there is a turbulent layer adjacent to the laminar boundary layer.

turbulent flow Fluid motion in which random motions of parts of the fluid are superimposed upon a simple pattern of flow. All or nearly all fluid flow displays some degree of turbulence. The oppose of laminar flow. *See* laminar flow. *See also* viscous flow.

turnover frequency *See* crossover frequency.

turns per inch (tpi) The amount of twist used to manufacture a reinforcement yarn or thread.

Twaddle scale A specific gravity scale that attempts to simplify measurement of liquid densities heavier than water, such as industrial liquors; the range of density from 1.000 to 2.000 is divided into 200 equal parts, so that one degree Twaddle equals a difference in specific gravity of 0.005; on this scale, 40° Twaddle indicates a specific gravity of 1.200.

twilled binding A weave in which each shute wire passes successively over two and under two warp wires and each warp wire passes successively over two and under two shute wires.

twin Two portions of a crystal having a definite crystallographic relationship; one may be regarded as the parent, the other as the twin. The orientation of the twin is either a mirror image of the orientation of the parent about a "twinning plane" or an orientation that can be derived by rotating the twin portion about a "twinning axis."

twist The number of spiral turns around an axis per length of a textile reinforcement strand.

two-film technique A procedure in which two films of different relative speeds are used simultaneously to radiograph both the thick and the thin sections of an item.

two's complement *1.* A method of representing negative numbers in binary; formed by taking the radix complement of a positive number. *2.* A form of binary arithmetic used in most computers to perform both addition and subtraction with the same circuitry, where the representation of the numbers determines the operation to be performed.

two-sided sampling plan Any statistical quality control method in which acceptability of a production lot is determined against both upper and lower limits.

Tyndall effect Effect whereby particles suspended in a fluid can be seen readily if illuminated by strong light and viewed from the side, even though they could not be seen when viewed from the front in the same light beam; this effect is the basis for nephelometry, which involves measurement of the intensity of side-reflected light.

typical basis The ordinary, or average property values for a material.

U

U Chemical symbol for uranium.

UF *See* urea formaldehyde.

U finish A designation indicating that the material is to be used for an unexposed part for which surface finish is unimportant.

UHF Ultrahigh frequency.

UHMWPE *See* ultrahigh molecular weight polyethylene.

UL *See* Underwriters Laboratories.

ultimate analysis *See* analysis, ultimate.

ultimate cycle method *See* Ziegler-Nichols method.

ultimate elongation The elongation at rupture.

ultimate strength *1. See* tensile strength. *2. See* strength.

ultimate tensile strength (S_u) The ultimate or highest stress sustained by a specimen in a tension test. The engineering stress at maximum load.

ultrahigh frequency (UHF) The frequency band between 300 and 3000 MHz.

ultrahigh molecular weight polyethylene (UHMWPE) Polyethylene resins with a molecular weight in the 1.5 to 3.0 million range.

ultrasonic atomizer A type of atomizer that produces uniform droplets at low feed rates by flowing liquid over a surface that is vibrating at ultrasonic frequency.

ultrasonic bonding A method of joining two solid materials by subjecting a joint under moderate clamping pressure to vibratory shearing action at ultrasonic frequencies until a permanent bond is achieved; it may be used on both soft metals and thermoplastics.

ultrasonic cleaning Removing soil from a surface by the combined action of ultrasonic vibrations and a chemical solvent, usually with the part immersed.

ultrasonic coagulation A process that uses ultrasonic energy to bond small particles together, forming an aggregated mass.

ultrasonic densimeter Density measuring instruments utilizing ultrasonic devices (sensors).

ultrasonic density sensor A device for determining density from the attenuation of ultrasound beams passing through a liquid or semisolid.

ultrasonic detector Any of several devices for detecting ultrasound waves and measuring one or more wave attributes.

ultrasonic frequency Any frequency for compression waves resembling sound where the frequency is above the audible range—that is, above about 15 kHz.

ultrasonic machining A machining method in which an abrasive slurry is driven against a workpiece by a tool vibrating axially at high frequency to cut an exact shape in the workpiece surface.

ultrasonic material dispersion Using ultrasound waves to break up one component of a mixture and disperse it in another to create a suspension or emulsion.

ultrasonics Technology associated with the production and utilization of sound having a frequency higher than about 15 kHz.

ultrasonic stroboscope A device for producing pulsed light by using ultrasound to modulate a light beam.

ultrasonic testing A nondestructive testing method in which high-frequency sound waves are projected into a solid to detect and locate flaws, to measure thickness, or to detect structural differences. There are two types, namely: (1) contact, during which the transducer contacts the test piece, and (2) immersion, during which the test piece is submerged in water through which the sound waves are transmitted.

ultrasonic thickness gage Any of several devices that use either resonance or pulse-echo techniques to determine the thickness of metal parts.

ultrasonic welding Same as ultrasonic bonding.

ultrasonoscope An instrument for displaying an echosonogram on an oscilloscope, and sometimes for providing auxiliary output to a chart recorder.

ultraviolet The portion of the light spectrum with wavelengths shorter than the visible (less than 3900 angstrom).

ultraviolet degradation Degradation caused by long-time exposure of a material to sunlight or other ultraviolet rays.

ultraviolet-erasable read-only memory (UVROM) A type of computer memory that can be erased or changed only by exposure to ultraviolet light.

ultraviolet light *See* ultraviolet radiation.

ultraviolet radiation The radiation beyond the violet end of the visible spectrum with wavelengths less than 400 nanometers. It is divided, for convenience, into: UVA—transmitted by glass, 400–320 nanometers; UVB—sunburning region of sunlight, 320–280 nanometers; UVC—transmitted by quartz, 280 nanometers.

ultraviolet spectrophotometry Determination of the concentration of various compounds in a water solution or gas stream based on characteristic absorption of ultraviolet rays; ultraviolet absorption patterns are not as distinctive "fingerprints" as their infrared counterparts, but in many cases the former are more selective and sensitive for use in process control applications.

unaccounted-for loss The portion of a boiler heat balance that represents the difference between 100 percent and the sum of the heat absorbed by the unit and all of the classified losses expressed as percent.

unbonded flash Flash that does not properly adhere to the mating material to which it is intended to be bonded.

unbonded strain gage A type of wire strain gage sometimes used in transducer applications, where strain is determined from elastic tension developed across the gage between mechanical end connections.

uncertainty The half-range of an interval within which the true value is expected to lie.

unconditional branch *See* unconditional transfer.

unconditional jump *See* unconditional transfer.

unconditional transfer An instruction that switches the sequence of control to some specified location. Synonymous with unconditional branch and unconditional jump. Loosely, jump.

uncoupled modes Modes of vibration that can exist in systems concurrently with and independently of other modes.

underaging Aging at any combination of time and temperature insufficient to produce maximum strength or hardness. Note: This treatment is used to improve workability in some precipitation hardening copper alloys.

underbead crack A crack in the heat-affected zone of a weldment that does not extend to the base metal surface.

under-cure A degree of cure less than desired.

undercut *1.* An unfilled groove in the base metal along the toe of a weld. *2.* A groove or recess along the transition zone from one cross section to another, such as from a hub to a fillet, that leaves a portion of one cross section undersized.

underfill A condition in which the face of a weld is lower than the position of an adjacent base metal surface.

underfilm corrosion Corrosion that occurs under organic films in the form of randomly distributed threadlike filaments or spots.

underflow Pertaining to the condition that arises when a machine computation yields a nonzero result that is smaller than the smallest nonzero quantity that the intended unit of storage is capable of storing.

understressing Repeatedly stressing a part at a level below the fatigue limit or below the maximum service stress to improve fatigue properties.

Underwriters Laboratories (UL) An independent testing and certifying organization.

undetected failure A failure that is not identifiable until a second and detectable failure has occurred.

uniaxial *1.* Having a single axis. *2.* In crystallography, having one optic axis or direction within the crystal, along which a ray of light can proceed without being bifurcated.

uniaxial compacting Compacting a metal powder along one axis.

uniaxial load Stressing in one direction along the centerline of a component.

uniaxial strain Strain along a single axis. *See* axial strain.

uniaxial stress *1.* Load on a specimen divided by the area through which it acts. *2.* A stress factor in which two of the three principal stresses are zero.

unidirectional *1.* Filament orientation in the lengthwise direction only. *2.* Moving in only one direction, as rectified electric current.

unidirectional concentric stranding A stranding in which each successive layer has a different lay length, thereby retaining a circular form without migration of strands from one layer to another.

unidirectional laminate A fiber-reinforced laminate in which the fibers are oriented in the same direction.

unidirectional lay A variation of concentric lay in which all of the helical layers of strands comprising the concentric conductor have the same direction of lay. The construction includes normal unidirectional lay, in which each successive layer has a greater lay length than the preceding layer, and unidirectional equal lay (unilay), generally limited to 19 strands, in which all helical layers have the same length of lay.

unidirectional pulse A wave pulse in which intended deviations from the normally constant values occur in only one direction.

uniform corrosion Chemical reaction or dissolution of a metal characterized by uniform receding of the surface.

uniform elongation (uniform strain, E_u) *1.* In a tension test, the percent elongation at the onset of necking, usually taken as the strain to maximum load. *2.* Applies to materials in which the cross section decreases uniformly under load.

uniform quality An expression meaning that all specified mechanical and metallurgical properties and conditions of the material must not exhibit variations exceeding the commonly recognized industry standards.

uniform strain *See* uniform elongation.

unilateral tolerance A method of dimensioning in which either the upper or lower limit of the allowable range is given as the stated size or location, and the permissible variation is given as a positive or negative tolerance from that size, but not both.

unilay Synonymous with unidirectional lay.

unilay strand A conductor constructed with a central core surrounded by more than one layer of helically laid wires, with all layers having a common length and direction of lay.

unimeric Describes a single molecule that is not monomeric.

unit die In die casting, a die block that contains several cavity inserts for making different kinds of castings.

unit sensitivity The specific amount that a measured quantity must rise or fall to cause a pointer or other indicating element to move one scale division on a specific instrument.

universal instrument *See* altazimuth.

universal load cell *See* bidirectional load cell.

universal mill A rolling mill in which rolls with a vertical axis roll the edges of the metal stock between some of the passes through the horizontal rolls.

unsaturated compound Any compound having more than one bond between two adjacent atoms, usually carbon atoms, and capable of adding other atoms at that point to reduce it to a single bond.

unsaturation A state in which the atomic bonds of the chain or ring of an organic compound are not completely satisfied (not saturated); unsaturation usually results in a double bond (as for olefins) or a triple bond (as for the acetylenes).

unsprung weight The portion of a vehicle's gross weight that is comprised of the wheels, axles, and various other components not supported by its springs.

update *1.* To put into a master file changes required by current information or transactions. *2.* To modify an instruction so that the address numbers it contains are increased by a stated amount each time the instruction is performed.

upsetting *See* cold heading.

uranium A radioactive metallic element used as a nuclear fuel and a source for plutonium.

urea formaldehyde (UF) A widely used thermosetting material that is hard, rigid, and scratch resistant.

urethane plastics Plastics derived from organic isocynates with resins that contain hydroxyl groups that are abrasive and impact resistant.

USASCII U.S. Standard Code for Information Exchange. The standard code, using a coded character set consisting of 7-bit coded characters (8 bits including parity check), used for information exchange among data processing systems, communications systems, and associated equipment. The USASCII set consists of control characters and graphic characters. Often referred to as simply ASCII.

U.S. Standard Code for Information Exchange *See* USASCII.

UT Symbol for the ultrasonic method of nondestructive testing.

utility *1.* Any general-purpose computer program included in an operating system to perform common functions. *2.* Any of the systems in a process plant or manufacturing facility not directly involved in production; may include any or all of the following: steam, water, refrigeration, heating, compressed air, electric power, instrumentation, waste treatment, and effluent systems.

utility program *See* utility routine.

utility routine A standard routine used to assist in the operation of a computer; for example a conversion routine, a sorting routine, a printout routine, or a tracing routine. Synonymous with utility program.

UVA *See* ultraviolet radiation.

UVB *See* ultraviolet radiation.

UVC *See* ultraviolet radiation.

UV erasable PROM (EPROM) Memory whose contents can be erased by a period of intense exposure to UV radiation. *See also* EPROM.

UVROM *See* ultraviolet-erasable read-only memory.

V

v *See* volt.

V *See* velocity.

vacuum *1.* A given space filled with gas at pressures below atmospheric pressure. *2.* A low-pressure gaseous environment having an absolute pressure lower than ambient atmospheric pressure.

vacuum arc melting A melting technique in which a previously cast electrode is remelted under vacuum through electrical resistance heating across an air gap between the electrode and the ingot. The ingot is cast into a cylindrical, water-cooled copper mold.

vacuum degassing A process of refining liquid steel in which the liquid is exposed to a vacuum as part of a special refining technique for the purpose of removing impurities or for decarburizing the steel.

vacuum deposition A process for coating a substrate with a thin film of metal by condensing it on the surface of the substrate in an evacuated chamber. *See also* vacuum plating.

vacuum filtration A process for separating solids from a suspension or slurry by admitting the mixture to a filter at atmospheric pressure (or higher) and drawing a vacuum on the outlet side to assist the liquid in passing through the filter element.

vacuum forming A method of forming sheet plastics by clamping the sheet to a stationary frame, then heating it and drawing it into a mold by pulling a vacuum in the space between the sheet and mold.

vacuum fusion A laboratory technique for determining dissolved gas content of metals by melting them in vacuum and measuring the amount of hydrogen, oxygen, and sometimes nitrogen released during melting; the process can be used on most metals except reactive elements such as alkali and alkaline-earth metals.

vacuum hot pressing (VHP) A method of processing powder metals at low atmospheric pressure.

vacuum induction melting A melting technique employing electrically induced current to heat and melt the charge under vacuum. Normally, the molten metal is also poured under vacuum into a cylindrical or rectangular cast iron mold and allowed to solidify statically.

vacuum melting Melting in a vacuum to prevent contamination from air, as well as to remove gases already dissolved in the metal; the solidification may also be carried out in a vacuum or at low pressure.

vacuum metallizing *See* vacuum deposition.

vacuum photodiode A vacuum tube in which light incident on a photoemissive surface (cathode) frees electrons, which are collected by the positively biased anode.

vacuum plating A process for producing a thin film of metal on a solid substrate by depositing a vaporized compound on the work surface, or by reacting a vapor with the surface, in an evacuated chamber. Also known as vapor deposition.

validated adhesive An adhesive that has satisfactorily passed its evaluation tests and has a current usable life.

validation The process of demonstrating, through testing in the real environment, or an environment as real as possible, that the system satisfies the user's requirements.

validity The correctness, especially the degree of closeness by which iterated results approach the correct result.

validity check A check based on known limits or on given information or computer results, for example, a calendar month will not be numbered greater than 12, and a week does not have more than 168 hours.

value The attribute of color perception by means of which an object is judged to appear light or dark to an object of the same hue and chroma.

vanadium A metallic element used as an alloying element in many tool steels and high-strength steels.

vapor(s) *1.* Gases whose temperatures are below their critical temperatures, so that they can be condensed to the liquid or solid state by increase of pressure alone. *2.* The gaseous product of evaporation.

vapor barrier A sheet or coating of low gas permeability that is applied to a structural wall to prevent condensation and absorption of moisture.

vapor degreasing A cleaning process that uses the hot vapors of a chlorinated solvent to remove soils, especially oil, grease, waxes, fingerprints, etc.

vaporimeter *1.* An apparatus in which the volatility of oils is estimated by heating them in a current of air. *2.* An instrument used to determine alcohol content by measuring the vapor pressure of the substance.

vaporization The change from liquid or solid phase to the vapor phase.

vaporization cooling A method of cooling hot electronic equipment by spraying it with a volatile, nonflammable liquid of high dielectric strength; the liquid absorbs heat from the electronic equipment, vaporizes, and carries the heat to enclosure walls or to a radiator or heat exchanger. Also known as evaporative cooling.

vapor phase epitaxy A crystal growth process in which an element or a compound is deposited in a thin layer on a slice of substrate single-crystal material by the vapor phase technique.

vapor pressure *1.* The pressure exerted by the molecules of a given vapor. For a pure confined vapor, it is the pressure of the vapor on the walls of its containing vessel; and for a vapor mixed with other vapors or gases, it is the contribution of the vapor to the total pressure (i.e., its partial pressure). *2.* The pressure (for a given temperature) at which a liquid is in equilibrium with its vapor. As a liquid is heated, its vapor pressure will increase until it equals the pressure above the liquid; at this point, the liquid will begin to vaporize.

vapor pressure thermometer A temperature transducer in which the pressure of vapor in a closed system of gas and liquid is a function of temperature.

vaportight So enclosed that vapor will not enter the enclosure.

var A unit of measure for reactive power; it is calculated by taking the product of voltage, current, and the sine of the phase angle.

VAR Abbreviation for vacuum arc remelting.

variable In testing, the characteristic under examination that can have many values.

variable-length record format A file format in which records are not necessarily the same length.

variable, measured *See* measured variable.

variable word-length Having the property that a machine word may have a variable number of characters. It may be applied either to a single entry whose information content may be changed from time to time or to a group of functionally similar entries whose corresponding components are of different lengths.

varnish A transparent coating material consisting of a resinous substance dissolved in an organic liquid vehicle.

vector(s) *1.* Quantities such as force, velocity, or acceleration, which have both magnitude and direction at each point in space, as opposed to scalar, which has magnitude only. *2.* A one dimensional matrix. *See* matrix.

vector quantity A property or characteristic that is completely defined only when both magnitude and direction are given.

vegetable oil Any oil from plant origin including the fixed oils of plants, which are also known as glycerides, specifically triacylglycerols.

veil An ultra-thin mat similar to a surface mat, often composed of organic fibers as well as glass fibers.

velocimeter An instrument for measuring the speed of sound in gases, liquids, or solids.

velocity (V) Rate of motion. Rate of motion in a straight line is called linear speed, whereas change of direction per unit time is called angular speed.

velocity head The pressure, measured in height of fluid column, needed to create a fluid velocity. Numerically, velocity head is the square of the velocity divided by twice the acceleration of gravity.

velocity meter A flowmeter that measures rate of flow of a fluid by determining the rotational speed of a vaned rotor inserted into the flowing stream; the vanes may or may not occupy the entire cross section of the flowpath.

velocity pressure The measure of the kinetic energy of a fluid.

vendor An organization that offers to perform work, sell material, or provide services to a purchaser in accordance with an agreement or specification, and agrees to be responsible for the performance of the materials or parts in accordance with the agreement or specification.

Venn diagram A graphical representation in which sets are represented by closed areas. The closed regions may bear all kinds of relations to one another, such as being partially overlapped, being completely separated from one another, or being contained totally one within another.

verification The evaluation of an implementation of requirements to determine that they have been met.

verify *1.* To determine whether a transcription of data or other operation has been accomplished accurately. *2.* To check the results of keypunching.

vermiculite A granular filler material used in composites.

Verneuil process Method of single-crystal growth in which powder is dropped through an oxy-hydrogen flame, falling molten on crystal seed.

vernier A short auxiliary scale that slides along a main instrument scale and permits accurate interpolation of fractional parts of the least division on the main scale.

vertical axis Also referred to as the normal axis. A line perpendicular to both the longitudinal axis and the lateral axis. Angular movement about the vertical axis is called yaw.

vertical boiler A fire-tube boiler consisting of a cylindrical shell, with tubes connected between the top head and the tube sheet forms the top of the internal furnace. The products of combustion pass from the furnace directly through the vertical tubes.

vertical firing An arrangement of a burner such that air and fuel are discharged into the furnace, in practically a vertical direction.

vertical motion simulators Vibration machines that produce mechanical oscillations parallel to the vertical axis.

VHP *See* vacuum hot pressing.

vibration Motion due to a continuous change in the magnitude of a given force which reverses its direction with time. Motion of an oscillating body during one complete cycle; two oscillations.

vibration damping Any method of converting mechanical vibrational energy into heat.

vibration machine A device for determining the effects of mechanical vibrations on the structural integrity or function of a component or system—especially electronic equipment. Also known as a shake table.

vibration meter *1.* A device for measuring vibrational displacement, velocity, and acceleration; it consists of a suitable pickup, electronic amplification circuits, and an output meter. *2. See* vibrometer.

vibration tests The conditioning of an item (that is, inflators, modules) to vibration of specified frequencies, amplitudes, and durations while being maintained at a given temperature.

vibrometer A device for measuring the amplitude of a mechanical vibration. Also known as a vibration meter.

Vicant softening point The temperature at which a needle will penetrate a plastic specimen 1 millimeter.

Vickers hardness *See* diamond-pyramid hardness.

Vickers hardness number A number derived from Vickers hardness test results.

Vickers hardness test A light-load hardness test using a diamond indenter and varying loads.

VIM VAR Vacuum induction melting followed by vacuum arc remelting.

virgin metal Pure metal obtained directly from ore.

virtual leak A gradual release of gas by desorption from the interior walls of a vacuum system in a manner that cannot be accurately predicted; its effect on system operation resembles that of an irregularly variable physical leak.

virtual processor Software that allows an individual user to consider a computer's resources to be entirely dedicated to him. A computer can simulate several virtual processors simultaneously.

viscoelastic damping The absorption of oscillatory motions by materials that are viscous while exhibiting certain elastic properties.

viscoelastic flow *See* viscoelasticity.

viscoelasticity A property involving a combination of elastic and viscous behavior. A viscoelastic material is considered to combine the features of a perfectly elastic solid and a perfect fluid.

viscometer An instrument that measures the viscosity of a fluid.

viscosity The ratio between the applied shear stress and rate of shear. It is sometimes called the coefficient of viscosity. It is thus a measure of the resistance of flow of the liquid. The SI unit of viscosity is the pascal-second; for practical use a sub-multiple (millipascal-second) is more convenient. The centipoise is 1 mPa. and is customarily used.

viscosity index A measure of a the change of viscosity of a fluid with temperature.

viscous damping *1.* The dissipation of energy that occurs when a particle in a vibrating system is resisted by a force that has a magnitude proportional to the magnitude of the velocity of the particle and direction opposite to the direction of the particle. *2.* A method of converting mechanical vibration energy into heat by means of a piston attached to the vibrating object which moves against the resistance of a fluid—usually a liquid or air—confined in a cylinder or bellows attached to a stationary support.

viscous flow *1.* The flow of a fluid through a duct under conditions in which the mean free path is very small in comparison with the smallest dimensions of a transverse section of the duct. This flow may be either laminar or turbulent. *See* laminar flow and turbulent flow. *2. See* leakage.

visible radiation *See* light.

visible spectrum *1.* The range of wavelengths of visible radiation. *2.* A display or graph of the intensity of visible radiation emitted or absorbed by a material as a function of wavelength or some related parameter.

vitreous enamel A coating applied to metal by covering the surface with powdered alkaliborosilicate glass frit and fusing it onto the surface by firing at a temperature of 800 to 1600°F (425 to 875°C). Also known as porcelain enamel.

vitreous silica A glass used to produce high-strength silica fibers.

vitreous slag Glassy slag.

vitrification Formation of a glassy or noncrystalline material.

vitrified wheel A grinding wheel with a glassy or porcelainic bond.

void content Volume percentage of voids, usually less than 1% in a properly cured composite.

voids *1.* A term generally applied to paints to describe holidays, holes, and skips in the film. Also used to describe shrinkage in castings or welds. *2.* Gas that has been trapped into a material after solidification.

void seal A seal used to fill holes, joggles, channels, and often other voids caused by the build-up of structure in a fuel tank. The void seal provides continuity of sealing where fillet seals are interrupted by such structure gaps.

vol *See* volume.

volatile *1.* With regard to a liquid, having appreciable vapor pressure at room or slightly elevated temperature. *2.* With regard to a computer, having memory devices that do not retain information if the power is interrupted.

volatile content The percent of volatiles that are present in a plastic or an impregnated reinforcement.

volatile loss Weight loss by evaporation.

volatile matter Products given off by a material as gas or vapor, determined by definite prescribed methods.

volatile memory Memory whose contents are lost when the power is switched off.

volatile storage *1.* A storage device in which stored data are lost when the applied power is removed, for example, an acoustic delay line. *2.* A storage area for information subject to dynamic change.

volatility The tendency of a substance to evaporate readily at ordinary temperatures and pressures.

volatilization The act of volatilizing or the condition of being volatilized. *See* vaporization.

volatilize To make volatile; to cause to pass off in vapor.

volt *1.* The unit of voltage or potential difference in SI units. The volt is the voltage between two points of a conducting wire carrying a constant current of one ampere, when the power dissipated between these points is one watt. *2.* A unit of electromotive force which, when steadily applied to a conductor whose resistance is one ohm, will produce a current of one ampere.

voltage In electricity, electromotive force or difference in electric potential expressed in volts. *See* electric potential.

voltage drop The amount of voltage loss from original input in a conductor of given size and length.

volt-ohm-milliammeter A test instrument having different ranges for measuring voltage, resistance, and current flow (in the milliampere range) in electrical or electronic circuits. Also known as a circuit analyzer, multimeter, or multiple-purpose meter.

volume (vol) *1.* The magnitude of a complex audio-frequency current measured in standard volume units on a graduated scale. *2.* The three-dimensional space occupied by an object. *3.* A measure of capacity for a tank or other container in standard units. *4.* A mass storage media that can be treated as file-structured data storage.

volume resistivity The resistance measured across opposite faces of a 1-meter cube of material. The units of measure are ohm-meters (Ωm).

volume swell Increase in physical size caused by the swelling action of a liquid, generally expressed as a percent of the original volume.

volumetric analysis Quantitative analysis of solutions of known volume but unknown strength by adding reagents until a color change or precipitation is reached.

volumetric flow rate (q) The volume of fluid moving through a pipe or channel within a given period of time.

vortex *1.* The swirling motion of a liquid in a vessel at the entrance to a discharge nozzle. *2.* The point in a cyclonic gas path where the outer spiral converges to form an inner spiral and where the two spirals change general direction by 180°.

vortex column A whirling mass of water forming a vacuum at its center, into which anything caught in the motion is drawn. *See* vortices.

vortex disturbances *See* vortices.

vortex tubes An imaginary tubular surface formed by drawing vortex lines through all points of a closed curve.

vortices In fluids, circulations drawing their energy from flows of much larger scale and brought about by pressure irregularities.

vulcanization An irreversible process during which a rubber compound, through a change in its chemical structure, for example, crosslinking, becomes less plastic and more resistant to swelling by organic liquids, and the elastic properties are confined, improved, or extended over a greater range of temperature.

vulcanizing Producing a hard, durable, flexible rubber product by steam curing a plasticized mixture of natural rubber, synthetic elastomers, and certain chemicals.

W

w (lb/hr) *See* flow rate.

W *1. See* watt. *2.* Chemical symbol for tungsten.

wafer *1.* A thin disc of a solid substance. *2.* A thin part or component, such as a filter element.

Wallner lines A pattern of intersecting parallel lines in brittle fracture surfaces.

wall thickness *1.* The dimensional difference between the inside and outside diameter of a cylindrical component. *2.* A term expressing the thickness of a layer of applied insulation.

WAN Wide area network.

wandering sequence A welding technique in which increments of a weld bead are deposited along the seam randomly in both increment length and location.

warm gas Gas at 1000°F (540°C) to 2500°F (1400°C) and typically available from the decomposition of a liquid monopropellant or from burning liquid propellants at nonstoichiometric conditions.

warm setting adhesives Term used synonymously with intermediate-temperature-curing adhesives.

warm working Plastic deformation of a material at temperatures above room temperature, but below recrystallization temperature.

warp *1.* Yarn that is running lengthwise in a woven material. *2.* A strand of yarns of long length running parallel. *3.* A change in dimension of a laminate.

warp clock A composite fabrication and engineering expression used as reference for aligning the warp yarns or tows in the desired direction.

warp direction The direction of the warp yarns or tows in a fabric or tape.

warp surface The surface of a fabric that has a majority of warp fibers woven above the fill fibers.

wash *1.* A stream of air or other fluid sent back along the axis of a propeller or jet engine. *2.* A surface defect in castings caused by heat from the metal rising in the mold, which induces expansion and shear of interface sand in the cope cavity. *3.* A coating applied to the face of a mold prior to casting. Also known as a mold wash. *4.* To remove soil from parts, especially using a detergent or soap solution 5. To remove cuttings or debris from a hole during drilling by introducing a liquid stream into the borehole and flushing it out.

wash primer A thin inhibiting paint, usually chromate pigmented with a polyvinyl butyrate finder.

water Dihydrogen oxide (molecular formula H_2O). The word is used ambiguously to refer to the chemical compound in general and to its liquid phase; when the former is meant, the term water substance is often used.

water absorption ratio Ratio of the weight of water absorbed to the weight of the original dry material.

water break test A test in which water is applied to the prepared surface and should remain in a continuous film over the whole area for at least 30 seconds.

water calorimeter A device for measuring radio-frequency power by determining the rise in temperature of a known volume of water in which the radio-frequency power is absorbed.

water column A vertical tubular member connected at its top to the steam and at its bottom to the water space of a boiler. The water gage, gage cocks, high- and low-level alarms, and fuel cutoff may be connected to the water column.

water gas Gaseous fuel consisting primarily of carbon monoxide and hydrogen made by the interaction of steam and incandescent carbon.

waterproof Impervious to water. Compare with water resistant.

waterproof grease A viscous lubricant that does not dissolve in water and that resists being washed out of bearings or other moving parts.

waterproofing agent A substance used to treat textiles, paper, wood, and other porous or absorbent materials to make them shed water rather than allow it to penetrate.

water resistant Slow to absorb water or to allow water to penetrate, often expressed as a maximum allowable immersion time. Compare with waterproof.

water vapor *1.* Water (H_2O) in gaseous form. Also called aqueous vapor. *2.* A synonym for steam, usually used to denote steam of low absolute pressure.

watt (W) *1.* A unit of the electric power required to do work at the rate of one joule per second. It is the power expended when one ampere of direct current flows through a resistance of one ohm. *2.* The SI unit of power, equal to one joule per second. *3.* Metric unit of power. The rate of doing work or the power expended, equal to 10^7 ergs/second, 3.4192 Btu/hour, or 44.27 foot-pounds/minute.

watt-hour meter An integrating meter that automatically registers the integral of active power in a circuit with respect to time, usually providing a readout in kW-h.

wattmeter Instrument for measuring the magnitude of the active power in an electric circuit.

wave Variation of a physical attribute of a solid, liquid, or gaseous medium in such a manner that some of its parameters vary with time at any position in the medium, while at any instant of time the parameters vary with position.

wave analyzer An electronic instrument for measuring magnitude and frequency of the various sinusoidal components of a complex electrical signal.

waveforms The graphical representations of waves, showing variation of amplitude with time.

wave front *1.* Of a wave propagating in a bulk medium, any continuous surface where the wave has the same phase at any given instant in time. *2.* Of a wave propagating along a continuous surface, any continuous line where the wave has the same phase at any given instant in time.

waveguide lasers Pump sources for deuterium oxide lasers.

wavelength In any periodic wave, the distance from any point on the wave to a point having the same phases on the next succeeding cycle; the wavelength, λ, equals the phase velocity, v, divided by the frequency, f.

wavelength dispersive spectroscopy (WDS) An x-ray type analysis that uses a crystal spectrometer to determine characteristics of x-ray wavelengths.

wave number The number of waves per length.

wave soldering A soldering technique used extensively to bond electronic components to printed circuit boards; soldering is precisely controlled by moving the assemblies across a flowing wave of solder in a molten soldering bath.

waviness A wave-like deviation from a perfectly flat surface.

Wb *See* weber.

w.c. *See* water column.

wear The mechanical removal of surface material by adhesion or abrasion.

wear resistance The ability to resist wear or abrasion.

weathering The surface deterioration of a material during outdoor exposure, such as checking, cracking, crazing, or chalking.

weatherometer A test apparatus used to estimate the resistance of materials and finishes to deterioration when exposed to climatic conditions; it subjects test surfaces to accelerated weathering conditions such as concentrated ultraviolet light, humidity, water spray, and salt fog.

weatherproof Capable of being exposed to an outdoors environment without substantial degradation for an extended period of time.

weather resistance The relative ability of a material or coating to withstand the effects of wind, rain, snow, and sun on its color, luster, and integrity.

web *1.* The vertical plate connecting upper and lower flanges of a rail or girder. *2.* The central portion of the tool body in a twist drill or reamer. *3.* A thin section of a casting or forging connecting two regions of substantially greater cross section.

webbing A narrow fabric woven with continuous filling yarns and finished salvages.

weber (Wb) Metric unit for magnetic flux.

wedge tensile strength The maximum tensile stress that a fastener is capable of sustaining when loaded eccentrically.

weeping Very slow leaks that show as dampness on a surface.

weft Yarn that is running perpendicular to the length in a woven material.

Weibel instability An instability of collisionless plasmas characterized by the unstable growth of transverse electromagnetic waves and large magnetic field fluctuation brought about by an anisotropic distribution of electronic velocities.

Weibull modulus The variation in strengths of brittle material, such as glass and ceramics.

weight (wt) (as in place weight) The multiplier value associated with a digit because of its position in a set of digits. The second place from the right has a place weight of 10 in decimal and 16 in hexadecimal, so a 2 in this location is worth 2×10 in decimal and 2×16 in hexadecimal. *See also* significance.

weighting Artificial adjustment of a measurement to account for factors peculiar to conditions prevailing at the time the measurement was taken.

weight percent Percent composition by weight rather than atomic percent.

weir An open-channel flow measurement device analogous to the orifice plate-flow constriction.

weir pond *See* stilling basin.

weldability The ability of a material to be welded within the parameters of what the weld was to accomplish.

welded strain gage A type of foil strain gage especially designed to be attached to a metal substrate by spot welding; used almost exclusively in stress analysis.

welding Producing a coherent bond between two similar or dissimilar metals by heating the joint, with or without pressure, and with or without filler metal, to a temperature at or above their melting point.

welding force *See* electrode force.

welding rod Welding or brazing filler material usually in the form of wire or rod.

welding stress Residual stress resulting from heating the base metal during welding.

weldment A structure or assembly whose parts are joined together by welding. More narrowly, the weld joint itself.

weld metal The metal in the fusion zone of a welded joint.

wet-and-dry bulb thermometer *See* psychrometer.

wet assay Determining the amount of recoverable mineral in an ore or metallurgical residue, or the amount of specific elements in an alloy, using flotation, dissolution, and other wet-chemistry techniques.

wet basis The more common basis for expressing moisture content in industrial measurement, in which moisture is determined as the quantity present per unit weight or volume of wet material; by contrast, the textile industry uses dry basis or regain moisture content as the measurement standard.

wet-bulb temperature The lowest temperature that a water-wetted body will attain when exposed to an air current. This is the temperature of adiabatic saturation.

wet-bulb thermometer A thermometer whose bulb is covered with a piece of fabric such as muslin or cambric that is saturated with water; it is most often used as an element in a psychrometer.

wet classifier A device for separating solids in a liquid-solid mixture into fractions by making use of the difference in settling rates between small and large particles.

wet dielectric test A voltage dielectric test in which the specimen to be tested is submerged in a liquid, and a voltage potential placed between the conductor and the liquid. Also called a tank test.

wet lay-up Applying liquid resin to a reinforced laminate set-up.

wet leg The liquid-filled low-side impulse line in a differential pressure level measuring system.

wetness A term used to designate the percentage of water in steam. Also used to describe the presence of a water film on heating surface interiors.

wet-out A woven material in which all the voids are saturated with resin.

wet-out rate The time required for a resin to fill the interstices of a reinforcement material and wet the surface of the reinforcement fibers, usually determined by optical or light-transmission means.

wet spinning The production of synthetic and man-made filaments by extruding the chemical solution through spinnerets into a chemical bath where they coagulate.

wet steam Steam containing moisture.

wet strength The strength of an adhesive joint determined immediately after removal from a liquid in which it has been immersed under specified conditions of time, temperature, and pressure.

wetting A formation of a continuous film of a liquid on a surface.

wetting agent A surface agent that reduces cohesion within a liquid.

wet winding With regard to filament winding, the process of winding glass on a mandrel in which the strand is impregnated with resin just before contact with the mandrel.

Wheatstone bridge A device that measures the electrical resistance of a conductor. The bridge consists of two known resistances and a third that is adjustable so that the resistance can be changed. These three conductors are connected in a circuit with the unknown resistor. The circuit also consists of a battery and a galvanometer. The adjustable resistor is then varied until the galvanometer shows no flow of current in the circuit. The value of the unknown can then be computed via use of an applicable formula.

whisker A short, single-crystal fiber used as a reinforcement or filler.

white cast iron Cast iron in which the carbon content is in the form of cementite. A fresh fracture of this material has a white appearance.

white glass Highly diffusing glass having a nearly white, milky, or gray appearance. The diffusing properties are an inherent, internal characteristic of the glass.

white light A mixture of colors of visible light that appears white to the eye. A mixture of the three primary colors is sufficient to produce white light.

white metal Low-melting point metals such as lead, tin, or zinc.

white radiation *See* Bremsstrahlung.

wicking *1.* Distribution of a liquid by capillary action into the pores of a porous solid or between the fibers of a material such as cloth. *2.* Flow of solder under the insulation on wire, especially stranded wire.

wide-area network In data processing, the interconnection of computers that may be miles apart, in contrast with computers interconnected within one building.

wideband radiation thermometer A low-cost pyrometer that responds to a wide spectrum of the total radiation emitted by a target object.

Widmanstatten structure A type of microstructure. *See* basketweave for a description.

Wien bridge A type of a-c bridge circuit that uses one leg containing a variable resistance and a variable capacitance to achieve balance with a leg containing a resistance and capacitance in parallel; although a Wien bridge can be used for measurements of this type, its more usual application is in determining frequency of an unknown a-c excitation signal.

Wightman theory *See* quantum theory.

window *1.* A defect in thermoplastic sheet or film, similar to a fisheye, but generally larger. *2.* An aperture for passage of magnetic or particulate radiation. *3.* An energy range or frequency range that is relatively transparent to the passage of waves. *4.* A span of time when specific events can be detected, or when specific events can be initiated to produce a desired result.

windowing A process equivalent to weighting a time domain acceleration record prior to transformation to the frequency domain.

wind tunnel(s) Tubelike structures or passages, sometimes continuous, together with their adjuncts, in which high-speed movements of air or other gases are produced, as by fans, and within which objects such as engines or aircraft, airfoils, rockets (or models of these objects), are placed to investigate the airflow about them and the aerodynamic forces acting upon them.

winning Recovering a metal from an ore or chemical compound.

wiped joint A type of soldered joint in which molten filler metal is applied, then distributed between the faying surfaces by sliding mechanical motion.

wire *1.* A thin, continuous length of metal with a circular cross section. *2.* A continuous length of an electrical conductive material.

wire cloth Screening made of wires crimped or woven together.

wired program computer A computer in which the instructions that specify the operations to be performed are specified by the placement and interconnection of wires.

Wobbe index The ratio of the heat of combustion of a gas to its specific gravity. For light hydrocarbon gases the Wobbe index is almost a linear function of the specific gravity of the gas.

Wollaston wire Extremely fine platinum wire used in electroscopes, microfuses, and hot wire instruments.

Wood's glass A type of glass that is relatively opaque to visible light, but relatively transparent to ultraviolet rays.

wool Fiber from the fleece of sheep or lamb, or hair of the angora or cashmere goat (or the so-called specialty fibers from the hair of the camel, alpaca, llama, and vicuna) that has never been reclaimed from any woven or felted wool product.

work hardening Increase in resistance to further deformation with continuing distortion. *See* strain hardening.

working fluid(s) Fluids (gas or liquid) used as the medium for the transfer of energy from one part of a system to another.

working life The period of time during which a liquid resin or adhesive, after mixing with catalyst, solvent, or other compounding ingredients, remains usable.

working space *See* working storage.

working standard Standardized material used during routine testing to ensure that testing equipment is properly calibrated for the material that requires testing.

working storage A portion of the internal storage reserved for the data upon which operations are being performed. Synonymous with working space. Contrast with program storage.

working temperature The temperature of the fluid immediately upstream of a primary device.

working zone The portion of the enclosed volume of a piece of thermal processing equipment occupied by parts or raw material during the soaking portion of a thermal treatment.

work life The time during which a liquid resin or adhesive remains usable after mixing a solvent or catalyst.

work softening The phenomenon whereby the yield strength of a metal drops when it has been strained or cold worked at low temperature and subsequently strained at an elevated temperature to cause the dislocations to become unstable.

World Federation The joining together of three international regions: (a) The Americas (Canadian MAP Interest Group and U.S. MAP/TOP Users Group) and Western Pacific (Australian MAP Interest Group), (b) Asia (Japan MAP Users Group), and (c) Europe (European MAP Users Group).

wormy alpha *See* stringy alpha.

woven fabric composite A major form of advanced composite in which the fiber constituent consists of woven fabric.

woven roving A glass fiber fabric used for reinforcement.

wrap forming *See* stretch forming.

wringing fit A type of interference fit having zero to slightly negative allowance.

wrinkle A surface defect in a laminate that appears as a crease or fold in the reinforcement closest to the surface.

wrinkling *1.* A paint-sealer incompatibility that causes the surface of a topcoat to cure at a different rate, resulting in the formation of ridges. These ridges can vary in absolute size, but generally are uniform in size on any one particular panel. *2.* Small buckles that occur in drawing sheet metal as it passes over the drawing ring radius.

wrought alloy A metallic material that has been plastically deformed, hot or cold, after casting to produce its final shape or an intermediate semifinished product.

wt *See* weight.

X

x-axis In polymer laminates, an axis on the plane of the laminate used as a 0° reference for the angle of the laminates.

xenon A noble gas used as a filler gas in light bulbs.

xenon chloride lasers Rare gas-halide lasers using XeCl as the active material.

xenon fluoride lasers Lasers using XeF as the active material.

XPS *See* x-ray photoelectron spectroscopy.

x-ray diffraction analyzer Any of several devices for detecting the positions of monochromatic x-rays diffracted from characteristic scattering planes of a crystalline material; used primarily in detecting and characterizing phases in crystalline solids.

x-ray diffractometer An instrument used in x-ray crystallography to measure the diffracted angle and intensity of x-radiation reflected from a powdered, polycrystalline, or single-crystal specimen.

x-ray emission analyzer An apparatus for determining the elements present in an unknown sample (usually a solid) by bombarding it with electrons and using x-ray diffraction techniques to determine the wavelengths of characteristic x-rays emitted from the sample; wavelength is used to identify the specific atomic species responsible for the emis-

sion, and relative intensity at each strong emission line can be used to quantitatively or semiquantitatively determine composition.

x-ray fluorescence analyzer An apparatus for analyzing the composition of materials (solid, liquid, or gas) by exciting them with strong x-rays and determining the wavelengths and intensities of secondary x-ray emissions.

x-ray goniometer An instrument for measuring the angle between incident and refracted beams of radiation in x-ray analysis.

x-ray imagery Reproduction of an object by means of focusing penetrating electromagnetic radiation (wavelengths ranging from 10^{-5} to 10^3 angstroms) coming from the object or reflected by the object. Analogous to infrared imagery, radar imagery, and microwave imagery using the IR, radar, and microwave frequencies.

x-ray microscope An apparatus for producing greatly enlarged images by projection using x-rays from a special ultra-fine-focus x-ray tube, which acts essentially as a point source of radiation.

x-ray monochromator A device for producing an x-ray beam having a narrow range of wavelengths; it usually consists of a single crystal of a selected substance mounted in a holder which can be adjusted to give proper orientation.

x-ray NDI Nondestructive inspection using x-rays.

x-ray photoelectron spectroscopy (XPS) A method of measuring the energy of electrons emitted from the surface of a material exposed to x-rays.

x-rays Short-wavelength electromagnetic radiation, having a wavelength shorter than about 15 nanometers, usually produced by bombarding a metal target with a stream of high-energy electrons.

x-ray spectrograph A photographic device to record x-ray emissions.

x-ray spectrometer An instrument using spectrographic analysis to characterize elements that are present.

x-ray spectrometry A device that measures the wavelengths of x-rays.

x-ray thickness gage A device used to continually measure the thickness of moving cold-rolled sheet or strip dur-

ing the rolling process; it consists of an x-ray source on one side of the strip and a detector on the other; thickness is proportional to the loss in intensity as the x-ray beam passes through the moving material.

x-ray tubes Vacuum tubes designed to produce x-rays by accelerating electrons to a high velocity by means of an electrostatic field, then suddenly stopping them by collision with a target.

XY plotter A device used in conjunction with a computer to plot coordinate points in the form of a graph.

XY recorder A recorder for automatically drawing a graph of the relationship between two experimental variables; the position of a pen or stylus at any given instant is determined by signals from two different transducers that drive the pen-positioning mechanism in two directions at right angles to each other.

Y

Y *1.* The chemical symbol for yttrium. *2. See also* expansion factor.

Yang-Mills theory Mathematical theory for describing interactions among elementary particles, based on the idea of gauge invariance under a non-Abelian group.

yarn A group of twisted filaments or fibers that form a continuous length suitable for weaving.

y-axis In polymer laminates, an axis on the plane of the laminate that is perpendicular to the x-axis.

Yb Chemical symbol for ytterbium.

yellowing With regard to paint, a particular form of straining in which a yellowish stain appears, usually associated with light color topcoats.

yield *1.* Deformation resulting from a single application of load in a relatively short period of time. *2.* The quantity of a substance produced in a chemical reaction or other process, e.g., metal production, from a specific amount of incoming material, usually expressed as a percentage.

yield point A special case of yield strength applicable when yielding occurs discontinuously. An upper and sometimes a lower value can be determined. The upper yield point is defined as the first stress in the material at which an increase in strain occurs without an increase in stress. The lower yield point is defined

as the lowest point between the first stress and the onset of continuous plastic behavior, at which an increase in strain occurs without an increase in stress.

yield point elongation Percent elongation at the end of nonhomogeneous yielding in a tension test.

yield strength *1.* The stress at which a material exhibits a specified limiting deviation from the proportionality of stress to strain. At the point of limiting deviation, the yield strength is expressed in units of stress and is referenced to a particular strain. The units of stress and strain should be defined as either engineering stress and strain or true stress and strain. *2.* The lowest stress at which a material undergoes plastic deformation. Below this stress the material is elastic; above it, viscous.

yield strength-offset The distance along the strain coordinate between the initial portion of the curve and a parallel line that intersects the stress-strain curve at a value of stress that is used as a measure of the yield strength.

yield stress The force per unit area at the onset of plastic deformation, as determined in a standard mechanical-property test such as a uniaxial tension test.

yield/tensile ratio The yield strength divided by the tensile strength.

Young's modulus The ratio of the stress to the resulting strain within the elastic region. *See* modulus of elasticity.

ytterbium A rare earth element used as gamma-ray source in radiography.

yttrium A metallic element, the oxides of which are used in the manufacture of ceramics.

Z

Z *See* compressibility factor.

ZAF corrections An x-ray program that corrects for atomic number, absorption, and fluorescence effects in a matrix material examination.

zap In data processing, a slang word meaning to erase or wipe out data.

z-axis In polymer laminates, the reference axis common to the xy plane of the laminate.

Z-buffering Technique used in hidden-surface algorithms in which the z-depth value of each pixel being examined is compared against the value stored in the buffer to see which pixel is closer to the viewer. The pixel closest to the viewer is then displayed. High-depth z-buffering is accomplished by using additional bits in the z-buffer (24 bits).

Z direction tensile *See* cohesive strength.

Zeeman effect The splitting of degenerate electron energy levels into a condition of different energies in an external magnetic field.

Zener effect A reverse current breakdown due to the presence of a high electrical field at the junction of a semiconductor or insulator.

zeolites Any of a group of hydrated silicates of aluminum with alkali metals, used for the molecular sieve properties.

zero a device To erase all of the data stored on a volume and reinitialize the format of the volume.

zero adjuster A mechanism for repositioning the pointer on an instrument so that the instrument reading is zero when the value of the measured quantity is zero.

zero bias A positive or negative adjustment to instrument zero to cause the measurement to read as desired.

zero code error A measure of the difference between the ideal (0.5 LSB) and the actual differential analog input level required to produce the first positive LSB code to transition (00…00 to 00…01).

zero elevation Biasing the zero output signal to raise the zero to a higher starting point. Usually used in liquid level measurement for starting measurement above the vessel connection point.

zero gas A calibrating gas used routinely to check instrument zero.

zero suppression *1.* The elimination of non-significant zeros in a numeral. *2.* Biasing the zero output signal to produce the desired measurement. Used in level measurement to counteract the zero elevation caused by a wet-leg.

zero/zero out The procedure of adjusting the measuring instrument to the proper output value for a zero-measurement signal.

Ziegler-Nichols method A method of determining optimum controller settings when tuning a process-control loop. Also called the ultimate cycle method. It is based on finding the proportional gain that causes instability in a closed loop.

zincblende Zinc sulfide, ZnS; a cubic crystal.

zinc chlorides Reaction products of hydrochloric acid and zinc; white crystals soluble in water and alcohol, with a melting point of 290°C.

zinc plating An electroplating coating of zinc on a steel surface which provides corrosion protection in a manner similar to galvanizing.

zinc rich coating A single-component zinc-rich coating which can be applied by brush, spray, or dip, and dries to a gray matte finish. ZRC is accepted by Underwriters Laboratories, Inc., as the equivalent to hot-dip galvanizing.

Z_o *See* characteristic impedance.

ZRC *See* zinc rich coating.

Zyglo test method A technique for liquid-penetrant testing to detect surface flaws in a metal using a special penetrant that fluoresces when viewed under ultraviolet radiation.